F Labrosse

The Navigation of the Pacific Ocean, China Seas, etc.

F Labrosse

The Navigation of the Pacific Ocean, China Seas, etc.

ISBN/EAN: 9783337319731

Printed in Europe, USA, Canada, Australia, Japan

Cover: Foto ©berggeist007 / pixelio.de

More available books at **www.hansebooks.com**

No. 58.

UNITED STATES HYDROGRAPHIC OFFICE,
BUREAU OF NAVIGATION.

THE NAVIGATION

OF THE

PACIFIC OCEAN, CHINA SEAS, ETC.

TRANSLATED AT

THE UNITED STATES HYDROGRAPHIC OFFICE,

FROM THE FRENCH OF

MONS. F. LABROSSE,

BY

J. W. MILLER,

LIEUTENANT U. S. NAVY.

WASHINGTON:
GOVERNMENT PRINTING OFFICE.
1875.

ADVERTISEMENT.

The Navigation of the Pacific Ocean has been translated at this office from the French of M. F. Labrosse, as a valuable addition to our knowledge of the winds and currents of, and routes through, this ocean.

R. H. W.

U. S. Hydrographic Office,
April 20, 1874.

NOTE.

The bearings are *true*. The distances are expressed in *nautical miles*. The orthography of the geographical names is in accordance with the latest English and American standards.

TABLE OF CONTENTS.

PART I.

CALMS, WINDS, TYPHOONS, USE OF THE BAROMETER, CURRENTS, ICEBERGS.

CHAPTER I.

Calms, winds, typhoons, cyclones, barometer.

CHAPTER II.

Currents, icebergs.

PART II.

OBSERVATIONS ON THE PRINCIPAL ROUTES ACROSS THE PACIFIC OCEAN.

CHAPTER I.

Routes from south to north on the western coast of America.

CHAPTER II.

Routes from north to south on the western coast of America.

CHAPTER III.

Routes from the western coast of America across the Pacific.

CHAPTER IV.

Routes from Europe to Australia, New Caledonia, and Tahiti, and return routes.

CHAPTER V.

Routes from the ports of Australia or Asia to the east.

CHAPTER VI.

Routes from the ports of Oceania.

CHAPTER VII.

Routes from Europe to China, and return routes.

CHAPTER VIII.

Routes to the northward in the China sea.

CHAPTER IX.

Routes to the southward in the China sea.

CHAPTER X.

Routes from the Australian ports to Asia and China.

CHAPTER XI.

Routes from China and Asia to Australia.

PART I.

CALMS, WINDS, TYPHOONS, USE OF THE BAROMETER, CURRENTS, ICEBERGS.

CHAPTER I

CALMS, WINDS, TYPHOONS, CYCLONES, BAROMETER.

§ 1. CALM-BELTS, EQUATORIAL AND TROPICAL CALMS.— In the author's *Instructions for the Navigation of the Atlantic*, tables were given indicating, for every season, the percentage of calms experienced in each square of 5 degrees.

The following tables are prepared in a similar manner:

As may be observed, information is wanting for the central portion of the North Pacific, comprised between the meridians 165° W. and 150° E. Directions for the other parts, including the most frequented routes, are, on the contrary, as full as could be desired.

Table showing the chances of calms.

Latitudes.	Longitudes east.																Longitudes west.																					
	From 100° to 105°.	From 105° to 110°.	From 110° to 115°.	From 115° to 120°.	From 120° to 125°.	From 125° to 130°.	From 130° to 135°.	From 135° to 140°.	From 140° to 145°.	From 145° to 150°.	From 150° to 155°.	From 155° to 160°.	From 160° to 165°.	From 165° to 170°.	From 170° to 175°.	From 175° to 180°.	From 180° to 175°.	From 175° to 170°.	From 170° to 165°.	From 165° to 160°.	From 160° to 155°.	From 155° to 150°.	From 150° to 145°.	From 145° to 140°.	From 140° to 135°.	From 135° to 130°.	From 130° to 125°.	From 125° to 120°.	From 120° to 115°.	From 115° to 110°.	From 110° to 105°.	From 105° to 100°.	From 100° to 95°.	From 95° to 90°.	From 90° to 85°.	From 85° to 80°.	From 80° to 75°.	From 75° to 70°.
JANUARY, FEBRUARY, MARCH.																																						
From 55° to 60° north																																						
50° to 55° north																					0																	
45° to 50° north							4	4	...	...	...	...	...	...	...	...	...	...	...	0	1	1	...	...	...	...	4											
40° to 45° north		C	H	I	N	A	4	4	4	...	...	...	...	...	...	...	...	...	...	...	...	...	...	...	...	...	1				A	M	E	R	I	C	A	
35° to 40° north							7	7	10	10	...	...	...	...	...	...	...	...	...	2	0	0	5	5	2	2	1	1										
30° to 35° north					0	3	7	7	10	10	...	...	...	...	...	...	...	...	...	2	0	0	5	5	2	2	1	1	12									
25° to 30° north					3	3	2	2	5	5	...	...	...	...	...	...	...	...	...	0	3	3	1	1	2	2	1	1	2	2								
20° to 25° north			0	0	3	3	2	2	5	5	...	...	...	...	...	...	...	...	...	9	3	3	1	1	2	2	1	1	2	2	2							
15° to 20° north		0	1	1	0	0	1	1	1	1	...	...	...	...	...	...	...	...	...	0	3	3	0	0	0	0	1	1	1	1	6	6	8	8				
10° to 15° north	0	0	1	1	0	0	1	1	1	1	...	...	...	...	...	...	...	...	...	0	3	3	0	0	0	0	1	1	1	1	6	6	8	8	0			
5° to 10° north		0	0	...	1	1	1	1	...	...	...	...	...	...	...	...	...	...	...	0	2	2	0	0	3	3	4	4	4	4	4	4	3	3	5	5	5	
0° to 5° north	1	2	...	...	1	1	1	1	...	...	...	...	...	...	...	...	...	...	...	0	2	2	0	0	3	3	4	4	4	4	4	4	3	3	5	5	3	
0° to 5° south															4	10	3	6	1	2	1	0	1	0	1	2	3	5	0	1	1	0	3	0	5	4		
5° to 10° south															1	14	13	6	6	6	5	0	6	2	1	0	0	2	0	0	0	0	0	0	1	3		
10° to 15° south															0	1	2	6	14	0	4	4	4	4	3	1	6	0	0	0	0	0	0	0	0	1	5	
15° to 20° south			A	U	S	T	R	A	L	I	A				0	0	10	2	8	6	1	7	10	8	...	7	12	2	0	0	0	0	0	0	2	0	0	10
20° to 25° south														0	3	...	4	1	5	3	2	5	6	6	...	5	5	7	0	5	0	0	0	0	3	5	4	10
25° to 30° south												0	2	0	0	4	1	1	6	6	5	4	15	11	...	4	8	3	0	4	2	0	11	0	9	7	3	2
30° to 35° south											2	3	10	3	3	2	1	6	12	5	2	1	2	6	12	0	6	10	6	10	12	9	5	6	5	5	4	8
35° to 40° south							6	7	2	4	2	4	2	2	4	6	2	10	9	2	1	6	6	4	6	1	3	2	0	6	3	4	4	7	4	5	3	7
40° to 45° south							2	2	5	5	1	0	0	0	3	2	7	2	1	0	0	1	0	6	6	5	0	0	6	2	0	0	0	2	5	5	3	4
45° to 50° south							1	0	1	2	1	0	1	2	3	2	0	2	2	1	0	3	1	0	2	1	1	4	2	2	4	2	0	1	3	4	2	

Table showing the chances of calms—Continued.

Latitudes.	Longitudes east.																Longitudes west.																					
	From 100° to 105°.	From 105° to 110°.	From 110° to 115°.	From 115° to 120°.	From 120° to 125°.	From 125° to 130°.	From 130° to 135°.	From 135° to 140°.	From 140° to 145°.	From 145° to 150°.	From 150° to 155°.	From 155° to 160°.	From 160° to 165°.	From 165° to 170°.	From 170° to 175°.	From 175° to 180°.	From 180° to 175°.	From 175° to 170°.	From 170° to 165°.	From 165° to 160°.	From 160° to 155°.	From 155° to 150°.	From 150° to 145°.	From 145° to 140°.	From 140° to 135°.	From 135° to 130°.	From 130° to 125°.	From 125° to 120°.	From 120° to 115°.	From 115° to 110°.	From 110° to 105°.	From 105° to 100°.	From 100° to 95°.	From 95° to 90°.	From 90° to 85°.	From 85° to 80°.	From 80° to 75°.	From 75° to 70°.
APRIL, MAY, JUNE.																																						
From 55° to 60° north										0	...	...	...	...	...	...	...	...	...	...	...	...	5	5	5													
50° to 55° north										0	...	...	...	...	...	...	...	...	...	3	2	2	5	5	1	1												
45° to 50° north								2	5	5	...	...	...	...	...	...	...	...	...	3	2	2	2	2	2	2	1											
40° to 45° north	C	H	I	N	A		12	12	5	5	...	...	...	...	...	...	...	...	...	2	2	2	2	2	2	2	1			A	M	E	R	I	C	A		
35° to 40° north						9	8	8	1	1	...	...	...	...	...	...	...	...	...	5	3	3	3	3	4	4	2											
30° to 35° north						4	8	8	1	1	...	...	...	...	...	...	...	...	...	5	3	3	3	3	4	4	2	2										
25° to 30° north					0	0	3	3	6	6	...	...	...	...	...	...	...	...	...	4	1	1	2	2	3	3	0	0	3	3								
20° to 25° north			1	6	0	0	3	3	6	6	...	...	...	...	...	...	...	...	...	0	1	1	2	2	3	3	0	0	3	3	2							
15° to 20° north		0	2	2	2	2	3	3	3	3	...	...	...	...	...	...	...	...	...	0	2	2	0	0	0	0	0	0	6	6	80	80						
10° to 15° north		2	2	2	2	2	3	3	3	3	...	...	...	...	...	...	...	...	...	0	2	2	0	0	0	0	0	0	6	6	80	80	15					
5° to 10° north	3	3	3	3	1	1	0	...	0	...	...	...	...	...	...	...	...	...	...	4	2	2	2	2	2	2	3	3	4	4	2	2	4	4	2	2	7	
0° to 5° north	5	5	3	3	1	1	...	...	...	...	...	...	...	...	...	...	...	...	...	0	2	2	2	2	2	2	3	3	4	4	2	2	4	4	2	2	2	
0° to 5° south													0	4	17	1	2	0	0	5	2	3	0	0	0	0	0	3	3	2	0	0	0	0	1	1		
5° to 10° south															0	0	0	4	3	12	4	4	0	0	3	0	0	2	0	0	0	0	0	0	0	1		
10° to 15° south													0	...	0	0	0	6	4	0	9	10	2	4	0	0	4	1	0	0	0	0	0	3	1	2	3	
15° to 20° south		A	U	S	T	R	A	L	I	A	0	0	0	...	0	4	5	3	5	10	0	5	6	6	0	...	3	0	15	0	0	0	0	0	4	2	1	9
20° to 25° south											0	5	6	0	7	1	6	5	9	5	2	3	3	2	0	0	1	6	0	0	0	0	2	7	6	0	2	6
25° to 30° south											0	3	2	7	2	1	1	1	3	3	8	10	2	0	0	0	6	6	0	0	9	0	0	0	5	15	2	3
30° to 35° south											0	6	0	0	3	3	2	5	2	5	3	0	0	0	0	12	0	0	0	14	1	12	0	4	4	3	5	7
35° to 40° south							2	4	3	1	3	7	2	0	4	13	3	0	0	0	0	4	...	0	0	0	0	0	0	0	15	3	2	4	0	5	3	7
40° to 45° south							1	0	1	0	0	2	0	0	9	4	10	0	4	6	2	0	0	5	6	0	4	0	0	2	0	7	0	4	3	3	2	
45° to 50° south							0	0	2	0	0	12	1	0	8	1	0	0	0	0	1	0	...	9	6	4	0	0	1	0	0	3	0	0	1	3	4	

Table showing the chances of calms—Continued.

Latitudes.	Longitudes east.																Longitudes west.																					
	From 100° to 105°.	From 105° to 110°.	From 110° to 115°.	From 115° to 120°.	From 120° to 125°.	From 125° to 130°.	From 130° to 135°.	From 135° to 140°.	From 140° to 145°.	From 145° to 150°.	From 150° to 155°.	From 155° to 160°.	From 160° to 165°.	From 165° to 170°.	From 170° to 175°.	From 175° to 180°.	From 180° to 175°.	From 175° to 170°.	From 170° to 165°.	From 165° to 160°.	From 160° to 155°.	From 155° to 150°.	From 150° to 145°.	From 145° to 140°.	From 140° to 135°.	From 135° to 130°.	From 130° to 125°.	From 125° to 120°.	From 120° to 115°.	From 115° to 110°.	From 110° to 105°.	From 105° to 100°.	From 100° to 95°.	From 95° to 90°.	From 90° to 85°.	From 85° to 80°.	From 80° to 75°.	From 75° to 70°.
JULY, AUGUST, SEPTEMBER.																																						
From 55° to 60° north......									0	0	...	...	...	...	...	...	...	...	...	12	3	3	4	4	3	3												
50° to 55° north......									0	0	...	...	...	...	...	...	...	...	...	2	3	3	4	4	3	3												
45° to 50° north......								3	4	4	...	...	...	...	...	...	...	...	...	2	2	2	2	2	3	3	3	3										
40° to 45° north......	C	H	I	N	A		10	17	4	4	...	...	...	...	...	...	...	...	...	0	2	2	2	2	3	3	3	3		A	M	E	R	I	C	A		
35° to 40° north......							6	6	2	2	...	...	...	...	...	...	...	...	...	2	7	7	3	3	4	4	5	5										
30° to 35° north......							6	6	2	2	...	...	...	...	...	...	...	...	...	4	7	7	3	3	4	4	5	5	16									
25° to 30° north......					2	2	0	0	3	3	...	...	...	...	...	...	...	...	...	1	2	2	2	2	1	1	4	4	6	6	11							
20° to 25° north......			3	0	2	2	...	...	3	3	...	...	...	...	...	...	...	...	...	0	2	2	2	2	1	1	4	4	6	6	6							
15° to 20° north......		12	1	1	2	2	1	1	4	4	...	...	...	...	...	...	...	...	...	0	3	3	2	2	7	7	8	8	18	18	6	0						
10° to 15° north......		4	1	1	2	2	1	1	4	4	...	...	...	...	...	...	...	...	...	...	3	3	2	2	7	7	8	8	18	18	0	23	...	0				
5° to 10° north......	0	2	0	0	3	3	8	...	...	...	...	...	...	...	...	...	...	...	...	...	2	2	2	2	1	1	0	0	0	0	0	0	5	5	1	1	18	
0° to 5° north......	0	3	0	0	3	3	4	...	...	...	...	...	...	...	...	...	...	...	...	1	2	2	2	2	1	1	0	0	0	0	...	0	5	5	1	1	3	
0° to 5° south......							4	...	...	...	14	12	18	16	7	5	4	1	0	0	0	0	0	0	0	0	0	0	0	0	0	0	0	0	0	1		
5° to 10° south......							0	0	...	...	13	...	10	32	0	0	8	1	0	6	0	0	0	0	0	0	0	0	0	0	0	0	0	0	1	1	0	
10° to 15° south......							0	0	0	0	...	...	0	...	0	0	0	1	0	0	3	4	0	0	0	1	0	3	0	0	0	0	0	0	0	0	2	
15° to 20° south......												0	0	0	0	1	6	2	5	0	2	4	5	0	0	0	0	6	...	9	0	0	0	0	0	0	0	6
20° to 25° south......											0	0	10	2	4	4	0	1	4	2	4	2	1	3	0	0	0	0	40	18	4	15	3	1	8	2	2	10
25° to 30° south......		A	U	S	T	R	A	L	I	A	2	0	1	2	3	0	0	4	2	1	1	0	2	0	0	6	0	0	...	0	9	0	10	2	10	12	6	6
30° to 35° south......											4	5	12	1	2	8	2	4	...	4	1	5	0	4	0	0	25	3	0	1	0	2	5	5	2	4	4	4
35° to 40° south......							0	2	7	11	3	0	3	0	0	8	0	6	...	0	0	1	2	2	0	15	0	0	0	1	0	0	4	1	2	4	3	5
40° to 45° south......							1	1	2	2	0	18	0	0	9	5	0	...	...	0	0	0	0	0	0	0	3	0	0	0	...	0	0	8	4	0	3	3
45° to 50° south......							0	0	0	0	...	...	0	...	3	0	0	...	0	0	0	0	0	0	0	0	0	0	0	0	0	0	0	0	1	2	1	

Table showing the chances of calms—Continued.

Longitudes east.

Latitudes.	From 100° to 105°.	From 105° to 110°.	From 110° to 115°.	From 115° to 120°.	From 120° to 125°.	From 125° to 130°.	From 130° to 135°.	From 135° to 140°.	From 140° to 145°.	From 145° to 150°.	From 150° to 155°.	From 155° to 160°.	From 160° to 165°.	From 165° to 170°.	From 170° to 175°.	From 175° to 180°.
OCT., NOV., DEC.																
From 55° to 60° north......																
50° to 55° north......																
45° to 50° north......																
40° to 45° north......										7	...	...	...	...	...	...
35° to 40° north......			C	H	I	N	A				...	...	...	...	...	...
30° to 35° north......									0	0	...	...	...	...	...	...
25° to 30° north......					30	...	...	...	0	0	...	...	...	...	...	...
20° to 25° north......			1	1	2	2	4	4	0	0	...	...	...	...	...	...
15° to 20° north......			1	1	2	2	4	4	0	0	...	...	...	...	...	...
10° to 15° north......		2	2	2	2	2	1	1	1	1	...	...	...	...	...	...
5° to 10° north......	12	4	3	3	2	2	3	3	0	0	...	...	...	...	...	...
0° to 5° north......	2	5	3	3	2	2	3	3	0	0	...	...	...	...	...	...
0° to 5° south......											0	0	0	11	4	5
5° to 10° south......							30	...	...	...	28	12	9	0	...	12
10° to 15° south......												8	...	2	0	0
15° to 20° south......												0	0	...	0	0
20° to 25° south......		A	U	S	T	R	A	L	I	A		4	0	0	0	0
25° to 30° south......												3	11	0	0	4
30° to 35° south......											6	1	2	10	3	4
35° to 40° south......							1	3	5	1	2	5	7	10	10	8
40° to 45° south......							1	1	1	3	2	0	0	0	11	5
45° to 50° south......							0	2	2	2	7	0	0	0	1	1

Longitudes west.

Latitudes.	From 180° to 175°.	From 175° to 170°.	From 170° to 165°.	From 165° to 160°.	From 160° to 155°.	From 155° to 150°.	From 150° to 145°.	From 145° to 140°.	From 140° to 135°.	From 135° to 130°.	From 130° to 125°.	From 125° to 120°.	From 120° to 115°.	From 115° to 110°.	From 110° to 105°.	From 105° to 100°.	From 100° to 95°.	From 95° to 90°.	From 90° to 85°.	From 85° to 80°.	From 80° to 75°.	From 75° to 70°.
OCT., NOV., DEC.																						
From 55° to 60° north......																						
50° to 55° north......																						
45° to 50° north......				0	...	...	3	3	0	0	0	0										
40° to 45° north......	...	...	...	0	3	4	3	3	0	0	0	0		A	M	E	R	I	C	A		
35° to 40° north......	...	...	...	0	2	2	2	2	2	2	5	5										
30° to 35° north......	...	...	...	2	2	2	2	2	2	2	5	5	4									
25° to 30° north......	...	...	...	6	3	3	4	4	4	4	3	3	3	3								
20° to 25° north......	...	...	...	0	3	3	4	4	4	4	3	3	3	3	7							
15° to 20° north......	...	...	...	1	5	5	0	0	1	1	1	1	4	4	9	9	12					
10° to 15° north......	...	...	...	0	5	5	0	0	1	1	1	1	4	4	9	9	5	3				
5° to 10° north......	...	...	...	0	4	4	4	4	2	2	3	3	2	2	2	2	2	2	1	1	8	
0° to 5° north......	...	...	...	0	4	4	4	4	2	2	3	3	2	2	2	2	2	2	1	1	12	
0° to 5° south......	11	6	3	2	0	2	0	0	0	0	0	0	0	0	0	0	0	0	0	0		
5° to 10° south......	8	36	2	3	4	6	1	2	0	0	0	0	0	0	0	0	0	0	0	0	0	
10° to 15° south......	1	5	1	0	5	9	1	4	1	0	0	0	0	0	0	0	0	0	0	0	2	
15° to 20° south......	8	2	4	2	2	3	2	6	0	0	0	6	6	0	0	0	0	2	0	0	1	10
20° to 25° south......	1	2	6	3	3	5	6	0	0	15	0	0	8	12	0	20	0	12	8	3	4	1
25° to 30° south......	3	4	3	2	9	4	3	0	...	...	0	2	...	12	2	2	0	14	11	1	1	3
30° to 35° south......	2	3	3	3	3	2	8	0	6	4	0	3	5	6	6	2	11	6	8	5	4	6
35° to 40° south......	0	3	3	4	5	3	8	0	0	6	0	2	6	2	1	1	0	5	2	4	3	4
40° to 45° south......	5	4	1	5	2	4	3	4	3	3	3	0	3	0	0	1	1	6	5	3	3	6
45° to 50° south......	0	3	0	2	3	2	2	4	0	1	0	10	1	0	1	0	1	2	5	2	2	

An inspection of the preceding tables proves the existence of a clearly defined region of calms, lying between the two sets of trades in the eastern part of the Pacific. These calms are termed equatorial. It is also shown that the zone in which the navigator is most exposed to detention from calms, takes the form of a wedge, the base resting on the coasts of Guatemala and Mexico, between 5° and 25° N., and the apex extending to the westward for a distance varying according to the season.

January, February, and March. Thus, in January, February, and March, calms are common on the western coast of America from the equator to 20° N. Vessels making passage to the northward, east of 110° W., will find a calm-belt about 20° wide, when they will be liable to from 4 to 6 per cent. of calms.

Between 110° and 130° W. the belt is only 10° wide; here there are only 4 per cent. calm chances. Finally, west of 130° W. the calm-belt may be said to cease, vessels usually passing from one set of the trades to the other without being appreciably detained.

April, May, and June. In April, May, and June, the calm-belt extends from the 120th meridian to the coast of America, causing navigation on the Mexican coast, from the gulf of Tehuantepec to cape Corrientes, to be almost impossible for sailing-vessels. The calms in this locality often last for several successive weeks. Well defined equatorial calms are not encountered west of 120° W., or at the farthest 130° W.

July, August, and September. The equatorial calms are of greater duration during these months and prevail north of 10° N. They extend from the coast of Mexico to 140° W. East of 130° W. they extend as far north as the 30th parallel; between 130° and 140° W. the calm-belt is only 10° broad, while west of 140° W. calms are no longer common.

October, November, and December. In October, November, and December, a calm-belt extends from the Mexican coast to 120° W., and from 10° to 20° or 25° N. Farther to the westward several calm-regions exist, but they have none of the attributes of genuine *calm-belts*. Equatorial calms also exist in the West Pacific, especially between the equator and 10° S.; in the Central Pacific, however, though the numerous groups of islands interrupt the trades, calms rarely prevail to any great extent.

Tropical calms. Tropical calms are those which prevail on the polar borders of the trades. The calms of the tropic of Cancer are only well defined in the eastern part of the Pacific, and during the months comprised between April and September. They are common between the parallels 30° and 40° N. East of Japan, however, and between the same parallels, the calms are of greater duration, except during October, November, and December.

The calms of the southern hemisphere, near the tropic of Capricorn, are especially prevalent from October to April. During this season they occupy a belt extending over nearly the whole breadth of the Pacific, and reaching from 25° to 40° S.

During the rest of the year they are not comprised in so well defined a region. But it is important to note that in the eastern portion tropical calms exist at all seasons in a more marked manner, and have a greater width in latitude, appearing to increase on approaching the coast of America.

Later will be given, in the descriptions of the various routes, more detailed information on the chances of detention by calms.

§ 2. Northeast trade-winds.—The trade-winds of the Pacific Ocean blow, in the northern hemisphere, from a general northeasterly direction. They are usually stronger than the NE. trades of the Atlantic Ocean.

We have already had occasion to observe, in the *Instructions on the Atlantic*, how difficult it is to solve the question of the limits of the trades. We again repeat that it is not best to rely too much on the mean latitudes indicated in the *Instructions*, because there will sometimes exist a difference of 600, and even 900, miles between these mean limits and those which may be found in attempting to enter or leave the trades.

We think it best, however, to give below the parallels between which the regular NE. trades are most often encountered:

In January, February, and March, the trade-winds blow from NE., between 6° and 25° N. January, February, and March.

In April, May, and June, they blow between 7° 30′ and 29° N. April, May, and June.

In July, August, and September, between 14° 30′ and 28° N. July, August, and September.

October, November, and December. In October, November, and December, between 9° and 25° N.

The trades do not begin to be well defined within 300 miles of the western coast of America. On coming closer to the coast variable winds are found, according to the season of the year. (Vide §§ 23 and 24.)

Though the western limits of the NE. trades can hardly be definitely fixed, it is generally conceded that these winds extend, all the year round, as far as the Caroline and Mariana Islands. Monsoons prevail westward of these groups. (Vide § 4.)

Finally, the NE. trades of the Pacific attain their greatest force while the sun is in the southern hemisphere, or, in other words, from October to March. This remark applies to all trade-winds, for they are always stronger when the sun is in the opposite hemisphere; when the sun is in their own hemisphere, they are, on the contrary, always weak, baffling, and often changed to monsoons blowing from an opposite direction.

§ 3. SOUTHEAST TRADE-WINDS.—The southeast trades are especially prevalent in the eastern part of the Pacific, between the following parallels:

January, February, and March. In January, February, and March, between 4° and 31° S.

April, May, and June. In April, May, and June, between 2° 30′ N. and 27° S.

July, August, and September. In July, August, and September, between 5° 30′ N. and 25° S.

October, November, and December. In October, November, and December, between 3° N. and 26° S.

These limits are, however, merely approximative, especially for the SE. trades, which are much less regular than the northeast.

Settled SE. winds will be found from 250 or 300 miles off the coast of America to 108° or 118° W., while west of these meridians the wind shifts to E. and ESE. Beyond 138° W. the trades become exceedingly variable, undergoing such changes, especially from October to April, that some authors consider them to possess all the characteristics of genuine monsoons.

This question will be reconsidered in the second part of § 4.

North of the equator. § 4. PREVAILING WINDS OF THE WEST PACIFIC.—Information concerning the prevalent winds north of the

Line is far from being complete; consequently too much confidence must not be placed in the following remarks:

NE. winds prevail in the West Pacific during this season. Though called the NE. monsoon, they are in reality only the steady trade-wind. Near the Caroline islands the NE. monsoon does not set in steadily until January, while north of this group and among the Mariana islands it sometimes begins in November and lasts till May or June; in short, the duration of the monsoon varies considerably. The NE. winds are generally accompanied with good weather. October to April.

All authorities agree in stating that steady SW. winds prevail at this season; near the Philippine islands, from May to September, and in the Mariana group, from July to August, they are especially common. This is the rainy season in these localities. SW. winds are frequent in the neighborhood of the Bonin islands from April to July, but too much reliance must not be placed on the chance of meeting winds from this direction, in the western portion of the North Pacific; and even in the neighborhood of the Mariana and Caroline groups, due allowance should be made for variations in the SW. monsoon. April to October.

Hurricanes are common among the Marianas, and to eastward of the group, especially in June, July, August, December, and January. They follow the laws governing cyclones in the northern hemisphere. (Vide § 26.) Hurricanes.

In § 3 we stated that the SE. trades are unsettled to westward of 138 °W.; this is especially the case from October to April. South of the Equator.

The winds are very variable during this season, throughout the region included between 138° W. and about 170° E., and from the Equator to 25° S. October to April.

Westerly winds are here nearly as frequent as easterly winds; that is, the squares on the wind-charts, where the westerly winds predominate, are almost as numerous as those containing winds from an opposite quarter. It follows, therefore, that at this season passages from west to east are made under sail with less difficulty than during the rest of the year.

The NW. monsoon prevails, at this season, to the westward of about 170° E.; and from the Equator or 1° N., to 15° S., or even 19° S. This monsoon begins in October, with winds shifting from N. to NW., and from W. to SW., Monsoons.

accompanied by storms, calms, and rain. The west winds begin to be regular after the month of November and last until March. The NW. wind, or rather NW. monsoon, prevails, during January and February, along the coasts of New Guinea and the adjacent islands, and extends to 170° E., but after March the direction of the wind again varies considerably.

Storms. It can be stated in general terms, that the bad season of the whole extent of the South Pacific, west of 138° W., lasts from October to April; the weather is then rainy and stormy and the wind exceedingly variable. Storms are common at this season between 10° and 25° S., and commence as simple squalls, the wind not changing in direction. Near the Tonga group they appear to follow the law of cyclones for the southern hemisphere, (vide § 26.) The worst gales are from November to April, but fortunately they are of rare occurrence.

The good season. The good season of the South Pacific lasts from April to October. The trades are then found to westward of 138° W. They are, however, variable for different localities, shifting from SSE. to E., and even to NE. Voyages from east to west are then easily made. Even at this season, however, the winds are scarcely ever steady and strong from SE., except in the extreme west of the Pacific, near the New Hebrides, Solomon, and New Guinea, during the months of June, July, and August. In these quarters the SE. trades may be justly termed the SE. *monsoon*, in contradistinction to the NW. *monsoon*, which begins to blow there in November. The change occurs during September and October. In September the SE. winds blow gently. In October squalls and variable westerly winds set in.

§ 5. ZONES OF GENERAL WESTERLY WINDS.—The *Instructions on the North Atlantic* (§ 3) and on the *South Atlantic* (§ 4) contain descriptions of the prevailing westerly winds, which are applicable in all respects to the Pacific. We shall merely give a summary review of those descriptions, adding a few further details.

General westerly winds are encountered at some distance beyond the polar limits of both sets of trades, the trade-wind limit itself following the changes in the sun's declination, (vide §§ 2 and 3.) Two principal antagonistic currents in the atmosphere—called the *polar* and *tropical* winds—

exist in the region of variables. A rotary movement in the atmosphere is produced by the meeting of these currents, causing cyclones or revolving storms.

Navigators at the present time are, however, a little too apt to fall into the error of imagining that every storm they encounter is a cyclone. This remark is especially applicable to the Pacific, where real circular storms, with the exception of the typhoons of the China seas, are very rare.

Warnings which precede polar and tropical winds.

In either hemisphere the wind from the direction of the adjacent pole is cold, dry, and squally; while the wind from the equator, or *tropical wind*, is warm, damp, and rainy. Thus, when the wind is blowing from SW., in the *northern hemisphere*, and the sky shows signs of clearing to the NW. or N., while at the same time the thermometer falls, and the deck, sails, and rigging dry rapidly, it follows that the wind will soon come out from NW. and N. These signs are verified by a rise in the barometer.

The same rule applies for the *southern hemisphere*, when the wind is from NW., and the sky is inclined to clear to the SW. or S., the wind then of course shifting to the southward.

On the other hand, the N., NE., and E. winds of the *northern hemisphere* will change to SE., S., and SW., when the sky becomes overcast to the southward, the weather warmer, and the deck, rigging, &c., covered with dampness; while at the same time the glass falls.

The corresponding changes for the *southern hemisphere* are from S. to E., and then to the northward and westward, with the northern horizon overcast.

In spite of their apparent irregularity, the so-called *variable winds* follow a general law in all their changes of direction. In the northern hemisphere they all rotate with the hands of a watch, thus W. ○ E.; in the southern, their rotation is in a contrary direction, thus, W. ○ E. When, as sometimes occurs, the winds act contrary to this law of rotation, they are said to *back;* and in such an event it is prudent to take precautions against bad weather, and to watch the barometer and the horizon.

A fact of equal importance, and one well known to sailors, is, that when the westerly winds are well from the southward in the northern hemisphere, or well to the northward in the southern, they do not attain much strength until they have blown for some time. On the contrary, the shifts of wind toward the adjacent pole are generally quick, and the subsequent weather bad. On this account it is advisable for vessels encountering bad weather from the SW., in the *northern hemisphere*, to lie to on the starboard tack, in order to guard against sudden changes of wind from SW. to NW. On the port tack they would be liable to be taken aback, and their spars endangered. In the *southern hemisphere*, on the other hand, with bad weather from the NW., it is advisable to lie to on the port tack in order to keep the sails full when the wind shifts violently from NW. to SW.

Observations of Captain Prouhet, on the usual course of storms in the southern hemisphere, (Ann. Hydr., vol. 31:)

"Besides numerous gales, we experienced continued bad weather, while in the high latitudes of the Pacific.

"The SE. gales began and ended at nearly the same compass point; the wind blowing furiously, but scarcely ever changing more than from 4 to 6 points; SSE. and E. were the extreme limits. At times the wind sprang up, increased, and died away from the same point; at others, it shifted to E. before falling, when it quickly fell calm.

"The shifts of wind did not take place as they should, from the northward to the westward; the gales, on the contrary, being preceded by a retrograde motion of the wind. If the wind, after shifting from W. or NW. toward N., got as far round as SE., and the horizon became overcast, a southeaster was almost sure to follow. The wind very rarely shifted in a contrary direction as far as E., for when it reached the SE. point it died away altogether, and afterward sprang up from NE., blowing a gale. In passing to the northward it increased slowly, blowing violently when it reached the NW. point; after having blown from this direction for a little while, it inclined toward W. and SW., and sometimes suddenly shifted to the latter point. Here the squall was at its worst, though it fortunately lasted for only a short time. If it continued to rotate toward S. the weather quickly cleared, and the wind abated; though a gentle breeze from S. and SE. could be expected for a day or two. But, now and then, the wind did not even follow this law, and backed again toward the westward, after the first burst from the SW. In this case the wind returned with fresh vigor, after a considerable interval, during which numerous gusts showed that this was only a lull in the violence of the storm. After this the wind always came out very strong from SW.

"The squalls, which afterward blew fiercely from the W. and SW., began generally from W. and NW., with gusts, fog, or rain; they grew violent from SW., and continued to increase, swinging slowly around to the southward, ending at the SE. in furious blasts. Throughout they were more a succession of violent shocks than a steady gale. At times there were intervals during which the weather moderated and seemed on the point of clearing; but these lulls were followed by renewed squalls from SW., which blew with redoubled violence. Each new blast drew nearer to SE. than the preceding ones, until having finally reached this point the wind died away without returning to the southward; in fact, it more often fell after reaching E. We encountered these "bursters" in quite high latitudes for variable winds; between the latitudes 50° and 45° we were driven before one for five days."

§ 6. PREVAILING WINDS ON THE COAST OF AUSTRALIA.— On the NW. coast of Australia, between Melville island and cape Northwest, the winds are irregular from March to December. But in summer, that is to say, from December to March, the land-breezes are steady from E. to NE., and the sea-breezes from SSW. to SSE. The usual winds prevail along this coast. From October to April, the west monsoon, with variable gusts from SW. and NW.; from April to October, the east monsoon, with variable winds from E. and SE., when the dry season commences. These laws hold as far as the tropics; but on going farther south, from cape Cuvier to cape Leeuwin, the prevalent winds are from NW. to SW. The northerly winds blow only for short periods, but are excessively warm; those from NW. are common and often violent in the vicinity of cape Leeuwin. Next to the NW. winds the SW. are the strongest. In short, this part of the coast of Australia is exposed to very heavy winds and seas; the vapor in the atmosphere is condensed on approaching the coast and causes showers, squalls, and bad weather, particularly during the rainy season, from May to October. Land-breezes from E. and NE. are common during the summer months, while in with the land sea-breezes prevail from W. and SW. The latter are particularly to be noticed in December, January, and February.

The west coast.

The south coast. North of a line drawn from the Recherche islands to capes Northumberland and Bridgewater, regular land and sea breezes, varying from SE. to ENE., may be expected from the 15th January to the 15th April. During this season stiff blows are *sometimes* met with, which, in case of their shifting to SW., become violent. During the remainder of the year the prevalent direction of the wind to northward of the line given above is west.

Westerly winds prevail in the offing at all seasons, especially from April to November, when stiff gales, with a high sea and heavy swell setting to the eastward, are found. The gales usually begin from NW., and after an interval shift suddenly to SW.; frequently they work back to NW., following the general law given in the preceding paragraph.

During the summer, particularly in February, there will sometimes be an easterly wind favorable to vessels making passage from cape Otway to cape Leeuwin. Still, it would not be advisable to count upon such a wind.

The east coast. Dry weather and west winds prevail during the winter months, from May to September, from Bass strait to Sandy cape; rain and warm NW. and N. winds are, however, quite frequent. Well off the coast the weather is bad, and the prevalent wind from NE. to S.

During the summer months the wind on this part of the E. coast is usually from SE. and the weather fine. The land-breezes near the shore increase in steadiness near the tropic; but while running to the southward, past cape Howe, navigators should be on the lookout for southerly and southwesterly squalls and strong winds from N. to ENE., followed by rainy weather. The warm NW. summer winds of the east coast generally terminate in a sudden shift to SE. and SSW. A description of the prevalent winds of the coast from the tropic of Capricorn to Torres strait will be found in § 4. From the end of April to September the wind is SE., while during the remainder of the year the NW. monsoon prevails, with its accompanying bad weather, rains, and variable winds.

The following quotation is taken from the Ann. Hydr., vol. 31:

Southerly bursters or brickfielders. "Strangers sailing along the east coast of Australia should take every precaution against the tornadoes called 'southerly bursters,' which are common during the summer months. They generally set in after the ordinary NE. winds and blow a gale for from 2 to 12 hours. If the weather be clear, the wind NE., and black thunder-clouds and forked lightning are noticed to the SW., the barometer falling, it is necessary to shorten sail, for the wind, after blowing very stiffly from NE., will die away calm, and then after a few moments' calm come out from S. with great strength. After having blown from 2 to 12 hours it will pass from S. and SE. to E., after which the usual summer wind will set in with a rising barometer. Sometimes the wind blows violently from the S. for two or three days, afterward veering from E. to NE.

"Dangerous squalls may also be expected from NW. These winds are warm, accompanied by thick clouds charged with electricity, the barometer being low; they shift abruptly from W. to SSW. and S. with lightning and occasional rain, and blow violently for a short time. Forked lightnings on this coast invariably indicate more wind, or a sudden change of wind from the quarter in which it lightens, or at least unsettled weather.

"The winter squalls generally come from the west with a clear sky and blow furiously for two or three days, commencing at NNW. and passing to the westward with a low barometer. The barometer rises after the wind reaches SW., and the force of the storm gradually abates. Sometimes the wind jumps from NW. to SW., with a loss of strength.

"The easterly squalls are the most dangerous for strangers; they begin from SE. with the barometer at $30^{in}.00$ and overcast weather and heavy swell from the eastward; they pass from E. to ENE., then return to E. and ESE., blowing with great violence, accompanied by rain and a heavy cross-sea. They last from 24 to 48 hours. Sometimes they begin with gloomy and overcast weather, light baffling winds, an easterly swell, and forked lightning from all points of the compass. In this case vessels should put to sea to await good weather."

§ 7. PREVAILING WINDS IN BASS STRAIT.—W. and SW. winds prevail in this strait the whole year round. The winds are generally strong and the squalls frequent. The latter begin at NNW. with the barometer falling from $29^{in}.92$ to $29^{in}.68$, and even lower, they become very strong in shifting to W. and SW. When they back from SW. and W. toward NW. the weather generally becomes worse. January, February, and March are the only months during which a vessel may pass through the strait with a favorable easterly wind. When these winds blow the weather is ordinarily fine.

In conclusion, the system of winds is here the same as that of the south coast of Australia, described in the preceding paragraph. But at the eastern entrance of the strait, and E. of Van Diemen's Land, the winds are similar to those of the E. coast of Australia, also mentioned § 6. It is necessary to be extremely cautious against squalls from SW. to SE., especially those from the latter point. NE. winds are also common, but never attain any great strength.

§ 8. PREVAILING WINDS ON THE COAST OF NEW ZEALAND.—New Zealand includes three large islands: *the northern*, or Ika-Na-Mawi; *the middle*, or Tavaï-Pounamou; and *the southern*, or Stewart. The northern and middle islands are separated by Cook's strait; the middle and southern by Foveaux strait.

North island. On the east coast the weather generally moderates in summer. At this season, from cape North to cape East, the sea-breezes blow regularly from NE. during the day; and the land-breezes from W. during the night. On this part of the coast the winter winds prevail from NW. to SW.; they also blow strongly from NE. and NW., shifting at the end of 24 hours toward WSW. and SW. with clearing weather. The SE. winds are very cold, being frequent and violent near cape East. Between cape East and Cook strait squalls are common in winter from SE. and S.; the climate, however, is good on this part of the coast.

On the west coast of North island the winds are steady from NW. to SW. during the year. In winter the NW. wind is naturally rainy. The most violent squalls are in the spring and autumn. The great rains take place in July, August, and September; fogs are frequent in October and November, particularly in the morning.

Cook strait. In this strait it is almost always windy either from the NW. or SE. On approaching the strait the wind is nearly always from one or the other of these directions. It frequently blows a gale from NW. or SE. In crossing, the wind often shifts from cape to cape, especially on leaving the strait. There is then danger of being taken aback, and it would be advisable for steamers under sail to light their fires. April, August, November, and December are months of comparatively good weather. SE. winds preceded by a fall in the barometer are most frequent in winter, (May, June, and July;) at this season it is not advisable to pass through the strait from west to east. Violent squalls from NW., with a very high barometer, occur oftenest in spring and summer, from September to March.

The best months on the eastern coast are December, January, February, and March; winter is the season for squalls and rain, especially in June, July, and August. Middle island.

In summer, from December to April, NE. winds are common during the day; land-breezes do not blow steadily at night. In winter, NW. squalls blow sharply and with great force; they are preceded by an unusual clearness of the atmosphere, and by a falling barometer; like the sirocco of the Mediterranean, they cause an increase of ten degrees in temperature. Squalls from SSE. and SSW. are preceded by a falling barometer, which rises as soon as they begin. They make overcast weather, and in winter bring rain; in spring and autumn they are accompanied by strong gusts, hail, and sleet. On the west coast of Middle island the prevailing winds are from NW. to SW. The NW., bringing rain and cloudy weather, hauls often to the N. on nearing land. Rains are frequent on this coast. The SW. winds prevailing in summer are fine and clear. Vessels coming up the coast from Foveaux strait have to encounter a southerly current, and steady winds from NW. and NNW. until they have doubled cape West.

Strong winds from NW. to SW. blow here almost incessantly, particularly from the former point, accompanied by cloudy, rainy weather. It often happens that the wind blows from SW. on the eastern coast of Stewart island, and from NNW. on the western coast at the same time. The NW. wind is preceded by a falling barometer, and at the end of several days usually shifts to SW., with a rising barometer and clearing weather. The worst storms in this strait occur in July. Foveaux strait

Occasionally heavy SE. winds enable vessels to pass through the strait from east to west. These may be foretold by a rise in the barometer and a long strip of clouds in the SE., and a light haze over the mountains and horizon. These winds are frequently violent and bring a drizzling rain and overcast weather.

§ 9. PREVAILING WINDS IN NEW CALEDONIA.—The *wet season* lasts four months, from the end of December to April; rain is then frequent, particularly when the wind blows from ENE. to WSW. and shifts by the north point of the compass. The other eight months of the year are the *dry season*.

In the southern part of the island the winds during the winter are irregular and variable, often blowing very strongly; particularly in January and February,* when storms are to be feared. In the northern part the trades blow from SE. and ESE. nearly the whole year, except in September and October, when one is particularly exposed to violent westerly storms, preceded by lowering, foggy weather, and a dead calm.

To the southward of the island the ESE. trade-winds prevail, especially during the eight months of the dry season. From NW. and W. winds are frequent in the neighborhood of Noumea. They blow freshly for several hours, accompanied by rain, and preceded by a barometer at between $29^{in}\cdot68$ and $29^{in}\cdot80$; the wind afterward shifts to SW. and S., with clearing weather. When there are NW. winds in the vicinity of Noumea, it frequently blows from SE. in the eastern part of the island and in Havannah channel. Inversely, the wind is often from NW. in the SE. part of the island, while it is blowing from SE. near Noumea.

In conclusion, the trade-winds are especially prevalent on the E. coast; in June and December they occasionally blow violently.

If at this time the barometer fall, look out for heavy squalls and driving rain. A continued fall of the barometer indicates a change of wind to N. and NW., whence it will blow with great strength for several hours. Afterward, as we have already said in speaking of Noumea, it will pass to SW. with clearing weather.

§ 10. PREVAILING WINDS IN THE SOCIETY ISLANDS.—The Society or Tahiti group are situated in that part of the West Pacific where, as we have already stated in paragraphs 3 and 4, the SE. trades do not blow with regularity.

* Vide note on cyclones, § 26.

In the eastern part, the winds ordinarily prevail from SE. to E. and NE. during the whole year; they are now and then interrupted by breezes from W. and NW. These winds are frequently rainy, and blow strongest during April, May, and June. Rains are to be expected also in November, December, and January. When the SE. trades blow strongly, it also rains at times.

The trades blow in January, February, and March, but they vary greatly in direction. There are seven to ten per cent. chances of calms during these three months, (vide tables § 1.) In April, May, and June the chances of calms are from five to six per cent.; westerly winds are frequently felt in April; in May they blow steadily from W. and NW.; in June the trades are prevalent, though still frequently interrupted. In July, August, and September there are only four to five per cent. of calms; and the trades, though very variable, steadily increase. In September, October, and November, they are steadier than at any other time. For the last quarter of the year the table in § 1 indicates only from two to three per cent. of calm-chances.

§ 11. Prevailing winds in the Marquesas islands.—In the Marquesas group the trades, from April to October, blow from ESE., varying at times a few points to the southward or eastward. During the other six months of the year they often haul to ENE. and even to NNE.

When the wind passes the north point and comes out from NNW. the weather is squally.

It often rains in torrents during May, June, and July, and is windy and squally from S. and SSW. In January it rains hard, and violent NW. gales are then common.

Most of the bays, particularly the harbor Tai-o-hae, are completely sheltered from the prevailing winds. It is extremely difficult for sailing-vessels to enter or quit these bays, and they are often obliged to be towed by their boats, or at least to have them in readiness in case the vessel is taken aback by baffling breezes. Vessels with auxiliary steam-power will, however, experience no difficulty in coming to anchor or leaving the harbors, as the shore is high and steep-to.

§ 12. Prevailing winds in the Sandwich islands.—During the entire year the trades prevail from NE. The rainy season lasts from January to May; during which period occasional NW. and SW. gales occur. In the Hawaiian group the chances of calms never amount to more than from 2 to 3 per cent.

Land and sea breezes blow on the western coast of Hawaii. The damp or rainy season lasts from May to September, with occasional strong winds from SW. In December, January, and February the weather is dry, with prevailing northerly winds.

§ 13. PREVAILING WINDS IN THE JAVA SEA.—The Java sea is bounded on the north by the islands of Borneo and Celebes; on the south by the islands of Java, Bali, Lombok, Sumbawa, Sapi, and Flores. It communicates with the China sea by the straits of Banca, Gaspar, and Carimata, with the Celebes sea by the strait of Macassar, and with the Indian ocean by the straits of Sunda, Bali, Lombok, Allas, Sapi, and Flores.

In the Java sea two monsoons prevail; the SE., though frequently interrupted by calms, begins in April, becomes strong in May, and ends in October. During this month of transition calms occur. The NW. monsoon commences during the first fortnight of November, blows with full force in December, and lasts till the end of March. In April the winds are again light, accompanied by showers and squalls.

§ 14. PREVAILING WINDS IN THE BANDA, TIMOR, AND MOLUCCA SEAS.—The Banda sea is bounded on the north by the islands of Bouro, Amboina, and Ceram; on the south by the islands of Flores, Pantar, Ombay, Serwatty, and Timor-laut. The Molucca sea, situated to the north of Bouro and Ceram, is bounded by Waygiou, Gillolo, and Celebes; it communicates with the Pacific by Pitt, Gillolo, and Dampier straits. The Timor sea stretches from Timor, Serwatty, and Timor-laut, on the north, to Mellville island and
The Banda sea. the coast of Australia on the south. Here the monsoon generally blows from WNW. from October to March, and from ESE. from April to September. The latter is called the SE. monsoon, (notwithstanding the fact that it blows from nearer east than south.) It begins during April, and toward the end of May has set in steadily at Amboina, Ceram, and Banda islands, when it corresponds to the rainy season; but it is well to remark that, even at this season,

the weather is always good at Bouro island. Between Bouro and Ceram the SE. monsoon is sometimes very fresh.

The NW. monsoon generally brings squalls, overcast weather, rain, and easterly currents.

In the Molucca sea the NW. monsoon sets in during the first two weeks of November, and blows strongly from December to March. Then the transition period begins, with variable breezes, calms, squalls, and rain. In April we have the SE. monsoon, which prevails from May to the end of September. In October the transition period is repeated. The Molucca sea.

In the passage between Celebes and Gillolo the general direction of one monsoon is NNW., and of the other SSE. At times these directions are the same in Gillolo strait; but the monsoons do not blow at all regularly in this strait, and still less so in Dampier and Pitt straits; vessels can consequently often pass through against the monsoon.

In the Timor sea the westerly winds are prevalent from October to April, with squalls from the Banda sea. This monsoon extends to 15° or 16° S. and hauls toward the SW. The SE. monsoon sets in during the last of May, and lasts, in the neighborhood of Timor, until the 15th of November. Timor sea.

§ 15. PREVAILING WINDS IN THE SULU AND CELEBES SEAS.—The Celebes sea is bounded on the south by the northern coast of Celebes, on the west by the NE. coast of Borneo, on the north by the Sulu and Mindanao islands, and on the east by the Sanguir, Saddle, etc., chain of islands.

The Sulu sea is bounded on the south by the Sulu islands and the north coast of Borneo, on the west by Palawan, and on the east by the Philippines.

In the Celebes and Sulu seas the easterly monsoon is variable in October, though blowing strongly from NE. and E. from November to April. During the last days of May the westerly monsoon commences; it is very variable, both in strength and direction, with squalls and rain. In June it is well established, in July rainy and stormy, ending in September.

§ 16. PREVAILING WINDS IN THE ARAFURA SEA.—The Arafura sea lies between the northern coast of Australia and the southern coast of New Guinea; it is bounded on the west by the islands of Arru and Timor-laut, and on the east by Torres strait.

In the Arafura sea the westerly monsoon brings the rainy season, when thunder storms are violent and frequent. Captains having a proper regard for the health of their crews or passengers will avoid the locality at this season. Toward the end of December the NW. or W. winds are strong and regular; in February and March they commence to be variable. During the latter month the weather is often gloomy and overcast, with a variable wind from SW. After the beginning of April the winds are from SE. and E.; they become settled after eight or ten days of storm and rain. From May to August the monsoon blows strongly from ESE. to SE., and lasts irregularly until the month of October. In November the winds are uncertain and the calms frequent, until in December the westerly monsoon sets in.

§ 17. PREVAILING WINDS IN THE CHINA SEA.—The SW. monsoon prevails in the China sea from the 15th April to the 15th October. It is variable in force and direction, especially in May; on the south coast of China it is sometimes interrupted by a series of winds from S. and SSE., and by winds from E. and SE., especially in the northern part of the China sea.

Near Formosa and Formosa channel it is not uncommon to find winds from N. to E. during the months of July, August, and September, but it is not well to rely upon this chance.

The SW. wind is first felt in the neighborhood of the gulfs of Siam and Tonquin, toward the middle of April, but the true monsoon does not reach its full force until June, July, and August, when it brings dull and rainy weather. It should also be borne in mind that typhoons are frequent during these three months.

Violent squalls preceded by heavy clouds coming from the gulf of Siam are dangerous, as far as Pulo Sapata, from May to September.

Strong squalls, rotating from NW. to WSW. and S., and accompanied by heavy rain, are also frequent at this season off the gulf of Tonquin.

Vessels coasting from the gulf of Siam to cape Padaran between May and September meet with light land-breezes at night; slight calms in the morning and a steady monsoon from SW. during the day. This is especially the case beyond cape Padaran, and as far as the gulf of Tonquin. Instead of the SW. monsoon, a genuine sea-breeze often blows from SE. during the day, alternating with light land-breezes at night.

The SW. monsoon, which first becomes settled in the southern part of the China sea, lasts longer in this region than in the northern part. Thus during the month of September and a part of October strong breezes from SW., bringing overcast, rainy weather, are often found at the entrance of Balabac strait. Southerly winds frequently prevail between Singapore strait and Pulo-Sapata, from the 10th to the 15th of October, while NE. and E. winds are common in the northern part of the China sea.

The NE. monsoon begins in the north of the China sea toward the end of September or the beginning of October; but to the southward of Pulo-Sapata and off the western extremity of Palawan it seldom sets in before the month of November, on account of the greater persistence in these quarters of the variable southerly winds which end the SW. monsoon. Northeast monsoon.

Although the SW. winds blow in September, they have to contend against variable breezes from NE. to E. and SE. This state of things continues till the October full and change, at which time there is often a storm from SW., varying to W. and NW. Then the wind passes to NNE. and NE., sometimes bringing with it the NE. monsoon. Some years the weather during September and October is fine, and the NE. monsoon is not always preceded by a squall. During these months strong winds from ENE. and NE. sometimes blow for several successive days, on the coast of China.

In November the NE. monsoon is generally established; it attains its greatest force and regularity in December and January. During these two months heavy rains and seas are found, especially between Singapore, Pulo-Condore, and Pulo-Sapata. In October, November, and the commencement of December the weather is overcast and rainy; as the wind is variable it is possible at this season to sail either up or down the coast of Palawan. In February the NE. monsoon abates; in March it blows moderately; during these two months the weather is fine in the China sea.

Gulf of Tonquin. The NE. monsoon prevails from the latter part of September to the first of April in the gulf of Tonquin and along the coast to cape Padaran; ENE. winds, shifting from NNE. to SE., are then common on the southern coast of China.

Western coast. § 18. PREVAILING WINDS ON THE COAST OF LUZON.—On the western coast of Luzon the monsoon varies from N. to NNE. from November to April. The winds at this season sometimes shift to NW. and W., blowing violently, and accompanied by rain. During March and April the weather is usually fine, with land and sea breezes. The SW. monsoon begins in May, becomes settled in June, blows with full force in July, and lasts until the end of September. From June to September the weather is usually overcast and very rainy. The monsoon begins to abate after the first of October, and soon after gives way to that from the NE.

Eastern coast. The weather is fine on the eastern coast of Luzon during the SW. monsoon, from May to September, the rainy season occurring at different seasons on the eastern and western coasts of the island.

The monsoon blows quite as often from S. as it does from SW.; winds are steadier during June, July, and August; during these months, however, there is frequently a series of winds from SE., NE., and NW. In April and May it blows quite as often from NE. and E. as from SE., S., and SW. Finally, from the beginning of October to the end of March the wind is steady from NE., shifting only to NNE. and E.

Squalls near Luzon. Around the island of Luzon, and especially on the western coast, from June to October, and particularly during September and October, it is advisable to guard against strong squalls which begin from N. and NW., shifting to W., SW., and S. Though fortunately of short duration they blow with great strength, producing a very heavy sea and rain.

§ 19. PREVAILING WINDS IN THE SEA AND ISLANDS OF JAPAN.—The Japan sea is situated between the coast of Tartary and the Japan islands. The system of winds in this region is far from being accurately known. The winds are variable; those from the NE. are cold, with frequent fogs; those from the NW. and SW. bring fine weather. The dry season lasts from April to August, and the rainy season from September to March. Squalls appear to be frequent during the bad season, especially from September to December. They begin by winds from S., SE., or NE., which rotate to NW.

It is generally admitted that the monsoons blow *on the S., SE., and E. coast of Niphon* from NE., from September to April, and from SW., with bad weather, gusts, and storms, and water-spouts, from June to September.

The transition period is from the 15th of May to the 15th of June, when rains are abundant. Typhoons are prevalent from July to November.

But it is not advisable to rely too much on the monsoons of these latitudes. For the prevailing winds in January, February, and March off the eastern coast of Japan are from NNW., W., and WSW.; and in April, May, and June from S. and SW., being very variable, and often hauling as far as E., NE., and NW. as they leave the coast. In July, August, and September variable winds, mainly from a SE. direction, are to be expected; close to the islands they blow from NE. and N., while out at sea winds from SW., S., SE., and especially from E., are found. In October, November, and December, the winds off the coast prevail from NE. and N., and especially from NW.; they also sometimes blow from SW. and S.

§ 20. TYPHOONS OF THE CHINA SEAS.—Typhoons are circular storms, which, though of short duration, are often extremely violent. Two or three typhoons may be encountered during one year, while during others not one will occur.

They make their appearance generally in the northern part of the China sea, along the E. and S. coast of China; also on the coasts of Luzon; between this island and Formosa; in the Meiaco-Sima, Loo-choo, and Clursan islands, and on the coast of Japan, etc. They do not appear to have been found in the China sea southward of 12° S.

Out of forty-six typhoons mentioned by Piddington as occurring between 1780 and 1845, two occurred in June, five in July, five in August, eighteen in September, ten in October, and six in November.

The most dangerous months are therefore September, October, and November; it is best also to be cautious in June, July, and August; finally, from September to the end of May there is little danger of meeting these storms. According to Horsburgh, the most terrific typhoons happen in June, July, and during the September equinox, especially if the moon is new or at its perigee at the same time; the strongest kind do not occur in *May*, *November*, and *December*, except occasionally in the neighborhood of the Bashees and Formosa islands, where there are very violent squalls in November. The wind is strongest close to shore, and less outside and in the southern part of the China sea.

It is imprudent to trust to the chance of not meeting a typhoon during *May*, *November*, and *December;* unfortunately we can cite in support of this opinion the late example of the *Mongo*, a large screw mail-steamer, which was lost, with all hands, in November, during a typhoon, several days after leaving Saigon for Hong-Kong.

The fall of the barometer is the principal sign of the approach of a typhoon, (vide § 26.)

Great care is to be taken whenever any of the following signs are observed between the beginning of June and the first of December: the horizon clear in some points, the tops of the mountains and islands enveloped in heavy black clouds; on the NE. horizon a low, thick, black cloud of a coppery tinge, growing whiter toward the top. In this last case, when the cloud comes up rapidly, the typhoon bursts with rain, thunder, and lightning. The hurricane generally lasts for about twelve hours; afterward it falls calm for about an hour; the wind then shifts to SW., and blows as violently as before.

Finally, it is important to notice that typhoons often burst suddenly, and without any warning whatsoever.

Referring again to Horsburgh, we will quote the passage in which he states the manner in which the winds rotate, after which we shall enumerate the principal rules regulating these storms:

Horsburgh on typhoons.

"Many ships have been driven from the Grand Ladrone to the Mandarin's cap, and even to the Taya islands, near Hainan, during typhoons; for among the islands and near the coast these tempests generally commence between NW. and N., then veer suddenly to NE. and E., frequently blowing with inconceivable fury, and raising the sea in turbulent pyramids, which impinge violently against each other; the current at such times runs strong to the westward. From eastward, the wind veers to the SE. and southward, and then moderates. This rotary motion of the wind during typhoons is generally experienced contiguous to and within a moderate distance of the coast of China; but about 2° or 3° from the coast, a contrary motion often takes place. Here, as before, typhoons commence from the northward, but instead of veering to the NE. and eastward, as in the former case, the wind veers to NW. and W., blowing very severely; it afterward changes to SW. and S., where it gradually abates in violence."*

The diameter of the typhoons of the China seas varies from 60 to 200 miles, and their axial rotation is the same as that of the cyclones of the northern hemisphere, viz: *against* the hands of a watch, or ○. Consequently the bearing of the storm center is eight points to the right of the wind, your *face* being to it. Thus when the wind is E. the center is S.; and when the wind is WNW. the center is NNE. Observations on their diameter.

But the onward movement, or track of typhoons, the speed of which varies from 7 to 24 miles an hour, obeys neither the general law governing cyclones of the northern hemisphere, nor any fixed rule; and this uncertainty, added to the innumerable other dangers to which the navigator is liable in certain parts of the China sea, makes these typhoons especially dangerous. After a perusal of Piddington's researches, it must be admitted that the track of the typhoons lies between NNW. and SSW.; but it is best not to rely too confidently on this, as typhoons have been known to move toward the W., then toward the N., and afterward turn sharp around *toward the E.*, in the direction of the Bashee isles. On their track.

This being allowed, it is clear that no fixed rule can be given to avoid typhoons; but we shall confine ourselves to giving general advice by which the navigator can profit, modifying it more or less according to circumstances.

Therefore, if the wind is from NE., passing to ENE. and E., vessels are almost always in the *right* semicircle of the typhoon,† and, consequently, *if they lie-to it should be on the starboard tack.*

---

* This shifting of the wind simply shows the two different semicircles of the same typhoon.—*Translator.*

† The *right* or *left* semicircle is the one situated to the right or left of the track of the center, (vide § 26.)

On the contrary, if the winds are from NW., passing to WNW. and W., vessels are almost always in the *left* semicircle; and *if they lie-to it should be on the port tack.*

Cases in which vessels should run before the wind.

For example, in the first case, with violent winds from ENE. to E., and an hourly fall of $0^{in}.6$ in the barometer, there would be danger of passing in or near the typhoon's track. That is to say, in lying-to (on the *starboard* tack) there would be many chances of passing through the center, and being exposed to the greatest danger. Hence, *wind and sea permitting*, and if there be sea room, it is best to run to the N., or, if possible, to NNE., even if bound to the southern part of the China sea. In disregarding this advice, and continuing on a southerly course, vessels would be apt to fall into the storm-center.

In the second case, with violent winds from WNW. to W., for example, and an hourly fall of $0^{in}.6$ in the barometer, there would also be much danger of passing in or near the track. And a vessel lying-to (on the *port* tack) would be liable to be overtaken by the storm-center. Whence we conclude that, *wind and sea permitting*, and with no land or reefs in the neighborhod, it is best to run to the SE., even if the ship be bound to the northern part of the China sea. If in spite of these precautions the course should be shaped to the northward, the central and dangerous portion of the typhoon will be encountered.

Toward the beginning of a typhoon the wind nearly always blows from some point between ENE. and NE., or between WNW. and NW.

If, during any months between May and December, indications of an approaching typhoon are evident, the barometer and the direction of the wind should be attentively watched. As long as the barometer falls it shows that the center of the typhoon is approaching, and the position of the vessel becomes more and more critical until the barometer again rises.

If the direction of the wind does not perceptibly change, and if its strength increases while the mercury falls, the vessel is in the track of the cyclone. In this case, unless absolutely prevented, it is best to run before the wind at first; then bring it on the starboard quarter, and continue on the same compass course, no matter how the wind may happen to change. When the wind changes it will be in a direction contrary to that of the hands of a watch; for example, from NE. to N., or from N. to NW. The vessel will then be entering the left semicircle, or comparatively moderate part of the typhoon. If the vessel continue on this same course until the barometer rise, she will be gradually drawing away from the path of the storm. She will then shape her course with the wind on the starboard beam.

If the direction of the wind changes *with the hands of a watch*, it is safe to conclude that the vessel is in the right or dangerous semicircle. If forced to bring to it should be on the *starboard tack*. If the course can be shaped with the wind abeam, or close hauled on the starboard tack, there are more chances of avoiding the center. As soon as the barometer rises the vessel can run off with the wind a little free on the starboard tack. If the wind change in a *direction contrary to that of the hands of a watch*, the vessel is in the *left* or less dangerous semicircle. If it be decided to lie to, it should be on the *port tack*. But if land or reefs are not in the way it is certainly preferable to run off with the wind two or three points on the starboard quarter. After a little time the wind will change and moderate, but still continue as nearly as possible on the same course, bracing up sufficiently, until the barometer commences to rise. Then, by letting the vessel fall off, bringing the wind on the starboard quarter, you will be at a safe distance.

The above precautionary measures are the best which can be given. No more remains to be said, except that a typhoon is always to be feared, and that hardly any precaution can be considered as absolutely efficacious for avoiding their effects, (vide § 157.)

Ships overtaken by a typhoon near the E. coast of China can find shelter in the following ports: The island of Tamtu; Mirsbay; the isle of Ty-sami, (9 miles on a course E. 8° N. from Hong-haï island,) if the vessel does not draw more than 13 feet; the island of Namoa, abreast of Stewart's house; port Tung-shan; port Amoy; Quemoy island; port Makung, (in the SW. part of Ponghou, the largest of the Pescadores;) port Chinchew; Hungwa channel; strait of Haetan, (southern entrance;) port Pih-quan; port Bullock; port Kelung, (at the N. extremity of Formosa;) the inner and outer harbors of Ting-hae, (S. coast of Chusan;) port Chinkeamun, (SE. extremity of Chusan;) Harbors of refuge.

port Chin-keang, (W. coast of Chusan;) Chang-pih, or Fisher island, (off N. coast of Chusan;) port Ta-outse, (NW. of Kintang.)

§ 21. PREVAILING WINDS ON THE COAST OF CHILI.—From the island of Chiloe, where it rains nearly all the year round, to 35° S., the winds are very variable in strength, but prevail from the westward. During the months of December, January, February, and March they blow frequently from SSW. to WSW.

The winds prevail from SE. to SW., between 35° and 25° S., during the *dry season*, which lasts from the middle of September to the end of May. During this period of nearly nine months the wind between 35° and 30° S., generally blows from some point between S. and SW. Near the parallel 30° S. this general direction of the wind changes to S. or SE., but it is generally fixed between 30° and 25° S., at some point to the eastward of S. The weather, from September to May, is generally clear and little rain falls; sometimes, however, it rains south of 31° S. for two or three successive days, with strong northerly winds.

The strong southerly winds.

Along the coast, between 35° and 25° S., the dominant southerly winds are often so strong that vessels close-hauled are obliged to take two reefs in their topsails. When the wind sets in strong from this direction vessels are often prevented from gaining an anchorage—Valparaiso, for instance—notwithstanding the fact that they may have sent down top-gallant masts, and close-reefed the topsails.

During the *three months* of the *rainy season*, from the end of May to September, calms, variable breezes, and bad weather, are to be expected, as well as *squalls from the northward*, with *rain* and a *heavy sea*, on the coast, and far out at sea.

In *June* the winds blow as often from N. as from S., between 35° and 30° S.; between 30° and 25° S. they prevail from S. and SE. In *July* there are variable winds between 35° and 25° S.; they are nearly as frequent from the N. as S., though the majority, perhaps, come from SW. In *August*, from 35° to 30° S., the winds are variable, but prevail from S. and SE.; between 30° and 25° S. the general direction of the wind is from SE. and SW.

The northerly winds.

The northerly winds of the rainy season rarely amount to a squall. Years will occasionally elapse without a norther, while two or three will occur in a bad season. *At Valpa-*

raiso, from May to September, (especially in July and August,) it is well to look out for northers.

Overcast weather, a *swell* from the north, and a *falling barometer* are certain indications of the approach of these storms, during which a large number of vessels have been driven ashore. During the good season the southerly winds in this roadstead are fresh enough to make vessels drag their anchors. Northers are felt all along the coast, including the harbor of Copiapo, when at certain periods of the bad season they are sometimes tolerably strong from NW.

§ 22. PREVAILING WINDS ON THE COAST OF PERU.—From 25° S. to cape Blanco and Guayaquil the wind blows all the year round from SE., varying from SSW. to ESE. The SE. wind becomes gradually steadier as the coast is left. Within 100 miles of the shore the winds partake more of the character of land and sea breezes.

The wind rarely blows at sea from SSE. and SE. with sufficient strength to make vessels by the wind take a second reef in their topsails.

To the southward of the parallel of 16° S., and near cape Blanco, or Nazca point, it reaches its greatest strength.

The weather often looks bad, but the squalls which follow are almost always to be weathered with single-reefed topsails and courses. At all seasons the winds vary in force and direction within 300 or 400 miles of the coast; beyond this limit the trades are well established. But, from April to August in the zone comprised between 100 and 400 miles off the coast, the winds are particularly subject to changes. At this time, and particularly in July, between 20° and 25° S., N. and NW. winds and calms are frequent.

Winds along the coast.

The winds close to the coast are more variable; they are particularly so from Cobija to Callao; north of Callao they are steadier. A land-breeze, varying from SE. to ESE., begins ordinarily an hour or two after sunset and lasts till morning; during the night it is dewy, cool, and damp, but the breeze is never strong. A sea-breeze, varying from SE. to SSW., sets in between 10 and 11 a. m., and dies away during the afternoon; it sometimes blows quite stiffly; it is then best to be careful in entering the harbors, on account of the sudden violent squalls which sweep down from the high lands.

Occasionally dead calms, lasting for several hours, occur between the land and sea breezes. In April and August, light puffs of wind from N. and NW. sometimes happen; these, however, never last more than 5 or 6 hours.

The belt of winds we have just described varies in width, being from 100 to 120 miles broad on the parallel of Arica, and from 30 to 35 miles off Callao.

The coast of Peru is not subject to storms or tempests; thunder and lightning are unknown there; gentle rains are common from July to August; near the coast calms are frequent in the interval between the land and sea breezes. The percentage of chances of calms is shown in the tables, § 1.

From December to May, fogs are very thick and frequent, particularly on the southern part of the coast. They last sometimes from 24 to 36 hours.

§ 23. PREVAILING WINDS ON THE COAST OF COLOMBIA AND IN THE BAY OF PANAMA.—*Between cape Blanco and cape San Francisco*, situated about 1° N., SE. and S. winds prevail all the year round. They frequently shift from S. to SW., particularly during the months of February, March, July, August, October, and November.

Between cape San Francisco and point Guascama.

Between cape San Francisco and point Guascama, situated 24 miles to the southward of the island of Gorgona, the winds, though quite variable, generally blow from SW. and W. In January, February, and March alone northerly winds become rather frequent. Generally the SE. winds haul to the S. and SW. on going up the coast from cape Blanco and the bay of Guayaquil; and on passing cape San Francisco they shift still farther to the westward. This belt of winds extends seaward for 100 or 200 miles. Farther from land they gradually shift to S. and SE.

Between point Guascama and cape Corrientes.

Between point Guascama and cape Corrientes the breezes are very baffling; sometimes they come from SW.; in January, February, and March they are often from N. to NE. Calms are frequent, and rains persistent during the greater part of the year, especially during June, July, and August, when it rains in torrents. During these months, and even until November, violent squalls occur.

Between cape Corrientes and Panama.

Between cape Corrientes and Panama the prevalent winds are from NW., with squalls. Southwesterly winds and damp weather are frequent between June and December.

In the gulf of Panama the fine or *dry season* does not actually commence until December, and lasts till April. During this period the winds are regular and prevail from ENE.; near the coast, however, land and sea breezes exist. The land-breeze varies from N. to NNE., the sea-breeze coming from SSW. At this season the winds in the southern part of the gulf are often very fresh; and, abreast of the coast of Veragua, vessels are frequently obliged to take two reefs in their topsails. In April and May the weather is variable and squally; the NE. winds begin to be interrupted by calms and southerly or southwesterly breezes, which bring rain. The rainy season commences in June, with strong settled winds from S. to SSW., interrupted at times with those from NW. There is often much rain in June, July, and August, and sometimes in September. The month of October is damp, but in November the weather begins to improve, the wind setting in again from E. to N. Gulf of Panama.

We shall complete these instructions on the system of winds in the bay of Panama by quoting verbatim Maury's observations, (edition, 1859:)

"In the discussion of the winds as it is conducted for the pilot-charts, Panama and its approaches are included between the parallels of 5° and 10° N. Between these parallels, and east of 85° W., it appears, from the observations which have been discussed, that the prevailing winds in November, December, January, May, June, and July, are between NW. and SW. inclusive; that in December, January, February, and March they prevail about one-fifth of the time from the northward and eastward; that calms are least prevalent in the month of March, the prevailing wind for March being NW., and for June SW., though NW. winds are also frequent in June, and that for the other months the observations are too few to give any indication as to the prevailing winds.

"Between the same two parallels, but to the west of 85°, and as far as 95°, the prevailing winds are in December, January, and February NE.; in March and April they are variable, prevailing alternately from NE. and NW. From May to September they prevail from S. to SW., inclusive; in October from SE. to SW., inclusive. In November they are inclined to be variable, though from SE. by the way of S. to WSW. is the favorite quarter.

"It is, moreover, indicated that to the east of 80° the winds in December, January, and February, prevailing as they do from the northward and westward, are generally favorable for getting to the southward and westward, by steering SSW. or SW.; that in May calms are frequent, and the prevailing points of the wind are decidedly WSW., SW., and SE.; and in June W., WSW., SW., and NW.; but as the favorite point is W., and calms are not so frequent as in May, June appears to be a more propitious month than May for crossing the parallel of 5° N. by a southwardly course from Panama. Between 5° and 10° N. for the other months, I have not observations enough to the east of 80° to justify me in any remarks as to the winds.

"Neither have I observations enough for January, February, or March to the east of 80°, and between 0° and 5° N., to authorize deductions; but for all the other months of the year they are abundant. They show that, to the east of 80°, between the equator and 5° N., the winds are steady between SE. by the south to west, and that calms are most frequent in this part of the ocean during the months of December and April.' The points from which the winds most prevail are, in December, SW.; in April, SSW. and SW.; in May, June, and July, SW.; in August, SSW. and SW.; in September, SW.; in October and November, from SE. to WSW.

"Between 80° and 85° W., from the equator to 5° N., the prevailing direction of the wind all the year is between SE. and W. by the way of S.; though from March to August, inclusive, it is most inclined to be variable. In December, March, and April, calms are most frequent.

"Between 85° and 90°, the prevailing quarter for the wind all the year from the equator to 5° N., is between SE. and SW. It is most variable from January to June, inclusive. In March and June the NE. trades are frequently found here; calms are most prevalent in March.

"Continuing west between the same parallels, the region from 90° to 95° W. seems to be of all the most liable to calms the year round. From October to January, inclusive, they are not so frequent as in the other months, being less frequent in October. From SE. to SSW. is the ruling quadrant for the winds here all the year; though from January to June, inclusive, they go from NE. around by the way of E. to W.

"To the west of 95° they are steady between SE. and S., except from January to May, inclusive. In January, February, and March, they often get as far north as NE., and in April and May as far as ENE."

§ 24. PREVAILING WINDS ON THE COASTS OF GUATEMALA, MEXICO, AND CALIFORNIA.—From the gulf of Panama to cape Blanco (gulf of Nicoya, about 10° N.) the winds are variable, but prevailing between SW. and NE., passing around by the way of S. and E. in January, February, and March. The two principal directions are S. and SE. The winds are, as a rule, steady the rest of the year; they prevail from SW. to SE., particularly from the southward.

From cape Blanco to cape Corrientes.

From cape Blanco to cape Corrientes (about 20° 30′ N.) the prevailing winds in January, February, and March are, first, those from NE., varying from N. to NW., between cape Blanco and Acapulco; second, those from NW., varying to N. and NE., between Acapulco and cape Corrientes. The first three months of the year correspond to the *dry season;* at this time the winds often blow violently from NNE. to NE. In April, May, and June, calms are continuous, and the breezes light and variable. From cape Blanco to Acapulco the wind prevails from S. and E., and from Acapulco to cape Corrientes from NE. to N. and NW. July, August, and September are the bad season, which actually lasts from May to October. SE. and E. winds prevail as far as 15° N. Near Acapulco, and continuing as far as cape Corrientes, variable winds blow, principally from NW., but also often from SW., S., and particularly from SE., ESE., and E. At this season the winds from SSW. to SSE. bring heavy rains and tornadoes, followed by calms all along the coast. (Vide § 1.) Frequently the SW. winds are violent; sometimes the heavy squalls from this direction at Acapulco, San Blas, etc., render it dangerous for vessels to come to anchor or remain any length of time until December.

In October, November, and December prevailing winds are found from NE., N., and NW. to the northward of cape Blanco; these often haul to the westward between the gulf of Tehuantepec and Acapulco; afterward, between Acapulco and cape Corrientes the prevalent winds are found to be from NE., N., and especially NW. The fine season can be considered as commencing in December.

Between cape Corrientes and cape Mendocino.

Between cape Corrientes and cape Mendocino (about 40° 30′ N.) moderate weather may be counted upon. The winds generally prevail from NW. The southwesterly winds bring rain, especially in November. In January, February, and March the prevalent winds are from N. to NW. and W., with a few southerly breezes between capes Corrientes and San Lucas. From cape San Lúcas to about 30° N. they vary from NE. to N. and from NW. to SW., the greater part coming from NW. and N. Farther to the northward, from the parallel of 30° to cape Mendocino, variable breezes from all points of the compass may be encountered, especially from NW., N., and NE. Between Corrientes and San Lucas the winds in April, May, and June are from NW., varying to W. and SW.; from San Lucas to 30° N. from NW. varying to NNW. and N., while from 30° N. to cape Mendocino they still blow from NW., though varying to N. and NNE. In July, August, and September the winds between Corrientes and San Lucas are from NW., shifting very often to W. and SW.; from San Lucas to 30° N. they also prevail from NW., varying to N. and NE. Finally, from 30° N. to Mendocino, they come from NNW., frequently passing to N. and NNE. In October, November, and December the prevalent winds are from NW., varying to NNW. and N. between Corrientes and San Lucas, and also extending to 30° N.

Beyond this parallel they are dominant from NNW., N., and NE.; but during this season they often blow from some point between WNW. and SSW., in which case they are accompanied by rain. During the whole year thick fogs are frequent, to the northward of 30° N., extending to the 45th parallel. They constitute the principal danger to navigation, and to making a land-fall on the coast of California.

North of cape Mendocino.

North of cape Mendocino the winds are variable and present a certain analogy to those which prevail off the coasts of England and Ireland. They are, however, prevalent from NNW. from cape Mendocino to about 50° N., except in December, January, February, and March, when they blow oftenest from SE. to SW. During this winter season the weather is bad, with rain and strong winds which are especially to be feared when they *back* from WSW. toward the SW., S., and SE. Winds from WNW. and N. are dry; those from SW. and SE., foggy and damp. The strongest squalls seem to come from between SW. and SE. During all the year fogs are to be expected as far to the northward as 45° N.; they are less frequent in winter than during the rest of the year.

§ 25. USE OF THE BAROMETER.—In the *Instructions for the Atlantic Ocean* it has been stated that it is indispensable to know, in order to make use of the barometer, the *mean barometrical* height, or *mean level* of the mercury at the position of the observer. The data collected for the Pacific are not sufficient to give a table of barometrical heights which would be as reliable as that given for the Atlantic. But this want of definite and absolute information does not prevent us from taking advantage of barometrical observations in practical navigation.

Thus the mean barometric height in the part of the Pacific where the trades continually blow, varies ordinarily from 29in.84 to 29in.96. But it is worthy of remark that, especially in the western part of the inter-tropical region, the barometer rarely remains steady within the above limits.

Experience shows that decided falls precede or accompany the westerly winds, which sometimes take the place of the trades, during the summer months of each hemisphere. (Vide § 4.) Barometric heights are often observed between 29in.29 and 30in.08 in both the NE. and SE. trade regions; but bad weather is only to be feared when the fall is rapid.

At 30° N. the mean level appears to be about 29in.81; at 35° N. we find it 29in.69; at 40° N. only 29in.57; and from 40° to 45° N. it is very variable, ranging from 29in.45 to 30in.32.

In the southern hemisphere we obtain 30in.00 at 30° S.; 29in.88 at 35° S.; 29in.73 at 40° S.; 29in.65 at 45° S.; 29in.57 at 50° S.; 29in.53 at 55° S.; and 29in.45 at 57° S.

As stated in the *Instructions for the Atlantic* the mean barometric height at cape Horn is 29in.25.

We will now give the principal rules for obtaining the requisite information in regard to the working of the barometer in the Pacific.

We will not treat of the fall in the barometer which announces the approach of the cyclones and typhoons of the western portion of the inter-tropical zone and the China sea, as ideas on that subject will be given in a subsequent paragraph.

We will also omit the instructions on the use of the barometer near cape Horn, as they were given in the *Instructions for the Atlantic.*

In the region of general westerly winds the barometer ordinarily announces several hours in advance any important changes in the force or direction of the wind, (vide § 5.) Gales may nearly always be foretold twelve hours in advance.

In the northern hemisphere, when the wind is about to shift from E. to SE. and S., the barometer falls; if the wind passes the SW. point and comes out from W., NW., and N., the barometer rises. When the wind gets to NE. the glass ceases to rise, and begins to fall again with wind from E.

In the southern hemisphere the glass is affected in a similar manner when the wind is about to shift from E. to NE. and N., etc., or, in general terms:

The barometer is *high*, or *rises*, when the wind is from the adjacent pole; and *low*, or *falls*, when it blows from the opposite pole. When the barometric level remains stationary for five or six hours, changes in either the force or direction of the wind need not be apprehended.

If the barometric level *oscillates*, or, in other words, falls and rises alternately from .02 to .06 of an inch for the space of half a day, the weather will be uncertain.

If it rises gradually from .04 to .08 of an inch in five hours, less wind may be expected or colder and drier weather.

If it falls gradually from .04 to .08 of an inch in five hours, more wind or warmer and damper weather may be looked for.

As a general thing, if the barometric height remains stationary at .16 to .20 or even .39 to .47 of an inch above or below the mean level, the weather will be steady and moderate; from the northward in the first case and from the southward in the second.

The barometer rarely attains the height of .59 to .78 of

an inch or more, above or below the mean level, unless during or just before very bad and windy weather.

Nevertheless, the principle that extremes in the barometer denote wind is generally taken too literally.

The following rule is more certain :

It will blow a gale whenever the barometer rises or falls very suddenly; especially when the level reached is distant from the mean level.

It is best, then, in observing the barometer, to note not only if it is above or below the mean level, but particularly to observe the number of inches it has risen or fallen since evening, since morning, or during the three or four hours previous. Thus a moderate movement—say of .04 to .08 of an inch in four hours—indicates moderate winds; while a sudden rise or fall—say of .16 to .20 of an inch in five hours, or .5 to .8 of an inch in twenty-four hours—foretells a gale.

If the wind is from the direction of the elevated pole and the barometer stands very high, (at .6 to .8 of an inch, or more above the mean level,) a sudden fall, accompanied by a rise in the thermometer, shows that the wind will soon come out strong from the direction of the opposite pole.

If, on the contrary, the wind is from the direction of the equator, and the barometer stands very low, (at .6 to .8 of an inch below the mean level,) a sudden rise foretells a shift of wind toward the adjacent pole.

Cold weather and a sudden fall in the barometer below the mean level indicates snow.

Finally, according as the rise or fall of the barometer is more or less rapid the weather predicted is more or less close at hand, and will last a longer or shorter time.

OBSERVATIONS OF M. LE COMMANDANT PROUHET. *On the use of the barometer in the southern hemisphere (Ann. Hydr., vol.* 31:)

"It may be observed, as applicable to the whole zone, that the direction of the wind has more influence on the barometrical height than even the state of the weather. This influence is much stronger in the variable winds of the southern hemisphere than in those of the northern. It may be also observed that all meteorological phenomena are characterized more strongly in the southern hemisphere. On the parallels of 42° and 43° the barometer, which marked $29^{in}.88$, with squalls from SE., stood steady at $29^{in}.53$, with a gentle breeze from NE., varying to N.; and at $29^{in}.21$, with winds from W. and fine weather. Several hundredths of an inch fall in the barometer, when the wind is from SE., may indicate worse weather than a fall of twice the amount when the wind is from NE., or of three times the amount when it is from NW. This influence that the direction of the wind has over the barometer becomes more and more marked as the latitude increases. So that the standard established for our climates, and which is accurate enough as far south as 41° or 45° S., becomes altogether inexact in higher latitudes.

"It would appear to be almost necessary in these latitudes to graduate a scale for each of the four principal directions of the wind, viz., SE., NE., NW., and SW.; thus a level at $29^{in}.90$, with winds from SE., can be considered as very low, and in no sense gives the idea of variable weather, while $29^{in}.10$, with winds from NW., is a mean level, which answers very well to this kind of weather.

"On the parallels 56° and 57° S., a slow and uniform fall to $28^{in}.35$, with winds from NW. or W., causes less anxiety than a level of $29^{in}.50$ on the W. coast of France. If sudden oscillations occur, however, there is grave cause for alarm. Everywhere, I think, but principally in these latitudes, movements of this nature foretell bad weather, which is never caused by a slow and regular change. A change of .02 of an inch per hour, might here take place without exciting apprehension; a greater rapidity of movement is invariably succeeded by squalls, and if it attains a rate of .04 of an inch per hour, gusts of great violence are to be expected. There is, therefore, considerable difference in the laws of the barometer according to the latitude; half the above movement, in the neighborhood of 40°, indicated the approach of furious squalls,

"The line of demarcation which it is the practice to draw near the parallel 44° or 45° S., to denote the limit at which our barometers cease to be exact, would appear to be arbitrarily placed. In the South Atlantic, where the first observations on this subject were undoubtedly made, it appears to be quite accurate; but in the middle of the South Pacific, the change in the laws regulating the barometer is not appreciable until we reach 48° or 49° S. In mid-ocean, half way between the cape of Good Hope and the southern point of New Zealand, it is, however, very distinctly marked near 42° S. Now, whether this line of demarcation ought to be drawn parallel to the equator or not, it is still admissible to divide the zone into two parts, near the latitudes mentioned above, and this division will allow us to proceed in the following manner with our observations on the law regulating the rise and fall of the barometer.

"In the northern part of the zone of variable winds the barometrical indications are to be interpreted in the same manner as those of the corresponding zone north of the equator; provided, it be remembered that the southerly winds here correspond to the northerly ones of our climate. The easterly and westerly winds have the same respective influence in both hemispheres; the points of comparison in the barometric scale also remain nearly the same; but the variation of the wind has more effect upon the barometer in the southern hemisphere.

"In the southern part of the zone of variable winds the laws governing the movement of the barometer are the same as in the case last mentioned, depending upon the direction of the wind; a rise indicating pleasant weather without a change of wind, or that the wind tends toward SE.; a fall indicating that the wind tends toward NW., or, if it does not change, that the weather will grow worse; but the extent of rise or fall is not at all the same as it is in the other part of the zone, and we would be very apt to be misled if we relied upon a knowledge of the working of the barometer obtained in other localities."

§ 26. Cyclones of the Pacific ocean.—Independently of the *typhoons* of the China sea, a description of which is given § 20, cyclones are encountered in many other parts of the Pacific. We give below localities where cyclones are found, the laws they obey, the information to be deduced from watching the barometer, and the precautions necessary to take to weather these storms with the least possible damage.

Latitudes in which cyclones have been observed.

Cyclones have been reported in the Marianas and Caroline islands, and from these groups to the Sandwich islands, and the west coast of North America. They seem to travel in a curve, inclining generally to the northward; there are, however, but few examples. Thus, in the Radack islands, 10° N. and 170° E., SW. storms occur in September and October. Piddington appears to lean to the opinion that these have the characteristics of cyclones. A hurricane (not circular) was observed 13° N., 147° 40′ W. A squall, *rotating* from NNE. to E. and S., was felt in September, 15° N., 119° 40′ W.

Another genuine cyclone was observed in the beginning of October at about 27° N. and 135° W. On this latter occasion, the wind shifted successively from ESE. to SE., S., SW., and W. The track of the cyclone was first to NW., afterwards to N. and NE., following the general law of circular storms.

Cyclones in the southern hemisphere.

In the southern hemisphere it is certain that cyclones occur from Australia to the Paumotas, and even farther to the eastward, particularly between the equator and 25° S. We shall briefly quote the examples mentioned by Piddington.

At Viti-Levu (Fiji Islands) a cyclone passing to the southward was observed in February. At Apia (Samoa islands) a very violent cyclone occurred toward the end of December, and in the same group another cyclone was observed traveling to the southward and eastward. Between the Tonga and Samoa groups several vessels have been lost at different times, during these cyclones. At the Kingsmill islands, on the equator, there are sometimes violent tempests. At Vavu, (Tonga group,) in December, an American whaler was thrown ashore by a hurricane, but was floated off during a shift of the wind. At Rarotonga (Cook group) a circular storm has been observed; also one in December on the passage between Tahiti and Mangaia. Cyclones are unmistakably felt at New Caledonia and the Loyalty islands, and between the latter and New Hebrides. Cyclones are to be dreaded, especially from the 1st of December to the 15th of April, in the neighborhood of New Caledonia, and particularly in the channel between the mainland and the Loyalty islands.

The route followed by the circular storms, between New Caledonia and Australia, probably corresponds to the part of their curve, which, though first directed toward SW., is gradually inflected to S. and SE. Finally in New Zealand, and in all the space between Van Diemen's Land and cape Horn, gales, with all the appearances of circular storms, have been encountered. In conclusion, we repeat, that in the southern hemisphere, especially in the western part and between the tropics, it is advisable to look out for cyclones, particularly from November to April.

The cyclones of the Pacific seem to obey certain general laws, well known to sailors. General laws of cyclones.

In the northern hemisphere, the rotation of the wind is against the hands of a watch. In the southern hemisphere the winds rotate in a direction corresponding to that of the hands of a watch.

From these facts the following excellent rule has been deduced:

That in the *northern hemisphere the center* of the cyclone *bears* eight points to the *right of the wind, your face being to the point of the horizon whence the wind blows ;* or eight points *to the left* in the *southern hemisphere.*

The onward movement of a cyclone seems to be in a course dependent upon the bearing of the neighboring pole.

In the northern hemisphere, for instance, the cyclones seem to travel toward NW. from the equator to the outer limits of the tropics, then toward N., and finally toward NE., beyond the parallel 30° N.

In the southern hemisphere they seem to travel first toward SW. from the equator to the outer limits of the tropics; then toward S., and finally toward SE., beyond 30° S.

But it would be wrong to trust too implicitly to this track, which is theoretical to a certain extent; under many circumstances these revolving storms have deviated considerably from these *directions.*

With the instructions given in the preceding paragraph, no danger is to be apprehended, unless in the event of being surprised by a cyclone in the neighborhood of reefs or land without shelter.

Supposing the line followed by the storm-center to be traced upon a map, it invariably divides the cyclone into two semicircles: *the right,* or that to the right hand of an observer facing *with* the storm; and *the left,* or that to the left hand of an observer in this position.

In the northern hemisphere the *right* semicircle is the *dangerous,* and the left semicircle the moderate side.

Inversely, in the southern hemisphere the *right* semicircle is the moderate, and the *left* semicircle the *dangerous* side.

One example is sufficient to illustrate these observations.

Suppose a cyclone in the northern hemisphere, traveling to the northward; suppose at the same time two vessels on the northern edge of the storm. One of these vessels being to the right of the track, experiences winds from SE.; the sea and leeway set her toward the center; she is therefore on the dangerous side. The other vessel being to the left of the track, experiences winds from NE.; the sea and leeway set her away from the center; she is therefore on the moderate side. However, the conclusion must not be drawn that these terms, dangerous and moderate, imply that the weather is worse on the one side than on the other. This is not proved by facts.

The truth is that vessels are more often directly in front of a cyclone when they are overtaken by the storm; therefore as a general thing, if the vessel is to the right of the track in the *northern* hemisphere, or to the left in the *southern* hemisphere, she will have much more difficulty in escaping than if she were in the other semicircle. Her position is therefore more critical, whence the name of dangerous given to the side in question.

Aboard a vessel lying to or close-hauled, in either hemisphere, or in any latitude, when the shifts of wind occur *to the right of N.* you are in the *right-hand semicircle;* on the contrary, when the wind shifts *to the left of north*, you are in the *left-hand semicircle.*

Thus, either N. or S. of the Line, if the wind changes in the direction NE. SW., (that is to say to the right of the N.,) you are on the *right-hand side* of the cyclone. If, on the contrary, the wind veers in a contrary direction, NW. SE., (that is to say to the left of N.,) the vessel is certainly in the *left* semicircle; and we again repeat, that these directions hold true in either hemisphere and in all latitudes.

The law can be put under a different form, thus:

1°. In the *northern hemisphere*, a vessel being close-hauled or lying to, if the wind shifts to the right of N., or with the hands of a watch, she is in the right or *dangerous* side. When the wind shifts inversely the vessel is in the left or moderate side.

2°. In the *southern hemisphere*, a vessel being close-hauled or lying to, if the wind changes to the left of N., or in the direction contrary to that of the hands of a watch, she is in the left or *dangerous* side. When the wind shifts inversely the vessel is in the right or moderate side.

In both hemispheres and in any latitude, when a vessel *directly in the path of a cyclone* is obliged to lie to, she receives the *wind* constantly from *the same direction*, and the barometer falls, with a speed which increases in proportion as the center of the tempest approaches. The moment the vessel is reached by the center, the wind falls for a short time, sometimes for a period of one or two hours; the barometer then reaches its lowest level. A fall of $1^{in}.97$ to $2^{in}.16$ has been often observed, also one extraordinary fall of $2^{in}.79$ at the center of a cyclone. The sea is then wild and furious, and the short breathing space allowed by the storm is none the less mentioned as the most frightful position by those who have been fortunate enough to escape. Suddenly the wind comes out with equal fury from the opposite quarter; if the vessel be able to bear this new shock, she will soon be out of harm's reach, as the center leaves with the same speed as it came. This speed, which is that of the movement of translation, is very variable; in the regions between the tropics it is generally about six miles per hour. Some of the cyclones observed had an onward movement of only two or three miles per hour, while others moved at the rate of from ten to twelve miles. Beyond the tropics, the speed is greater, and often reaches from fifteen to twenty or twenty-five miles per hour. As a rule, vessels finding themselves in the path of a cyclone should calculate on an hourly speed of from ten to twelve miles.

The barometer always gives sufficient warning of the approach of a cyclone, and of the distance that the ship is from the center of the revolving storm. Use of the barometer.

In cyclones the barometric level becomes lower as you near the center. Therefore a vessel is forewarned of the approach of the center if the barometer falls, and of its departure if the barometer rises.

In the tropics, where the accidental variations are relatively small, (vide § 25,) it is best to be always on the lookout when the barometer is from .4 to .6 of an inch below the mean level, especially when the hourly fall is great. Numerous observations seem to show that, with the barometer at .8 of an inch below its mean level, it generally blows strong enough to take three reefs in the topsails. When the glass stands at an inch or more below the mean level, vessels are compelled to lie to or run before the wind.

At the storm-center the barometer often sinks to $1^{in}.6$ or 2^{in} and even to $2^{in}.4$ below the mean level. In the temperate zones beyond the trades the fall of the barometer should be greater by from .2 to .4 of an inch on the approach of a cyclone. The conclusion is to be drawn from the facts collected by Piddington, that every observer placed in the track of a cyclone will notice the barometer fall from .02 to .06 of an inch per hour, when the distance of the center is from 150 to 250 miles. When the hourly fall of the barometer is from .06 to .08 of an inch, the center is distant 150 or 100 miles; when the hourly fall is from $0^{in}.08$ to $0^{in}.12$, the center is from 100 to 80 miles off; finally, with an *hourly fall* of from $0^{in}.12$ to $0^{in}.15$, the center is at a distance of not more than 80 or 50 miles.

In certain cases the barometer has been known to fall .50 and even .75 of an inch in an hour.

In general, when the barometer does not fall more than .08 of an inch per hour, a vessel placed in the path of a cyclone may avoid the center by running off before the wind. But it becomes nearly impossible to escape it when the hourly fall is more than from .08 of an inch to .1 of an inch.

After the center has passed, the barometer rises as fast as it fell; that is to say, it first rises very rapidly, then slower and slower, as the center recedes.

On board a vessel lying to, in either the dangerous or moderate side, the barometer falls quickly as the storm-center approaches; then rises as the center departs. These movements, however, are not so sudden as in the preceding case, because the center of the cyclone does not pass over the vessel as before, but at a greater or less distance from it.

When a vessel, with the wind on the quarter, is in a cyclone, the barometer varies moderately. If it rises, the ship is leaving the center, and the course is a good one; if it falls more than .08 of an inch an hour, the ship is approaching the center, and will be unable to clear or outsail it. The course is, therefore, a bad one. Finally, if a vessel is running before the wind, (which ought not to occur, unless she were directly in the path of the cyclone,) she will revolve with the storm, without being able to get out. In this case the distance of the vessel from the center of the cyclone would not vary, and the barometer would remain nearly stationary. Several instances of this nature have been known to occur to ships which persisted in running before the wind in a cyclone.

We have given above reliable information on the regular changes of the wind, for vessels lying to in cyclone. All sailors know that if the wind hauls ahead, when a vessel is lying to, there is danger of being taken aback, and even under low sail of making stern-board, in which case there would be great danger. On the contrary, if the wind *draw aft*, a vessel gathers way, and afterward luffs to her course without difficulty. Maneuvers during cyclones.

In order to be sure that the wind will draw aft, while lying to in a cyclone, it is best to conform to the following rule, taken from Reid's work on storms: "*In both hemispheres*, lie to on the *starboard tack*, when you are in the *right semicircle* of the cyclone; or on the *port tack* in the *left semicircle*."

1°. When the signs, furnished by the steadiness of the wind and the fall of the barometer, show that you are *directly in the track of the storm-center*, run off at once before the wind, or with the wind a little on the quarter, unless neighboring land or reefs absolutely prevent.

If you are unable to run off before the wind, the only alternative left is to let the center pass over you.

As soon as you judge that you have escaped the track of the center, continue on the same compass course, with the wind on the quarter, no matter how it may happen to change. In the *northern* hemisphere the wind should be kept on the *starboard quarter;* in the *southern* on the *port.* In both cases the wind will gradually haul ahead, but continue on the same course, if possible, until the barometer rises, or at least ceases to fall. As soon as the barometer shows any signs of rising, (or before, if the wind and sea permit,) more sail may be set, and the vessel brought gradually to the *starboard* tack in the northern hemisphere, and to the *port* in the southern.

2°. Suppose the observer to be in the N. hemisphere, and admitting that the order in which the wind shifts proves that the ship is in the right semicircle of the cyclone, or dangerous side. Then, if you are forced to lie to, it should be on the starboard tack. But, if it be possible to shape the course with the wind on the starboard beam, or even a little closer, the chances are greater that you will avoid the track of the storm-center. When the barometer commences to rise, you can run off a little, with *the same tacks aboard.*

3°. Suppose the observer to be in the N. hemisphere, and admitting that the order in which the wind shifts proves that the ship is in the left or moderate semicircle of the cyclone. Then, if it be decided to lie to, it should be on the port tack. But, if land or reefs are not in the way, it is certainly preferable to run the ship off at once with the wind two or three points on the starboard quarter. After a little the wind will change and commence to haul ahead, but still continue on the same course, as nearly as possible, bracing up as the wind hauls, until the barometer begins to rise. Then, if the wind and sea permit, make sail, and run off again with the wind free on the starboard tack. If the observer be in the *southern* hemisphere, substitute *starboard* for *port* and *right* for *left* in the two preceding cases.

CHAPTER II.

CURRENTS—ICEBERGS.

§ 27. THE EQUATORIAL CURRENT.—According to the researches of Captain Duperrey, the waters of the Pacific ocean show in the tropical regions a tendency to drift toward the west with a variable rate; the mean rate of this movement being about 24 miles per day. This vast stream, about three thousand miles wide, is called *the equatorial current.* It appears to have been particularly observed between the parallels 26° S. and 24° N.

A counter-current has been proved to exist, setting to the eastward, at some distance north of the line, and especially in the western part of the Pacific. This counter-current, of which a description will be given, § 28, divides the great equatorial current into two branches, which set to leeward in both the NE. and SE. trade regions, and which are distinguished by the names of the *northern equatorial current,* and the *southern equatorial current.*

The northern equatorial current.

The *northern equatorial current* begins in the neighborhood of 126° W., and sets toward the W. and WSW. with a mean speed of about 1 knot per hour; it is especially observed between 10° and 24° N., to the eastward of the Sandwich islands, and between 10° and 19° N., to the westward of this group. South of 10° N. its speed is less, losing itself in the northern limit of the counter-current, which is ordinarily felt between 10° and 5° N. The temperature of the water increases from the 24th parallel, where it stands at 74°.5, to the equator, where it reaches about 81°.7 in the eastern portion of the Pacific; 83°.1 in the central part; 88°.5 in the western part, and in the neighborhood of New Guinea. The northern equatorial current is felt in the West Pacific as far as the Loochoo islands, (China.) It is bounded on the meridian of 142° E. by the parallels 26° and 12° N. Its southern limit passes between Guam island (Mariana group) and the islands of Oulouthi, situated about 360 miles to the SW. Beyond the meridian 142° E. the drift is toward WNW., and the bed of the current contracts as it approaches the island of Formosa. At some distance from this island the current inclines to the northward, forming a circuit analogous to that of the Gulf stream in the Atlantic. From this point the stream comes under the head of the *Japan current*, (vide § 37.)

Alternate currents, depending upon the prevailing monsoon, exist between the equator and the southern limit of the northern equatorial current, and from the meridian of 142° E. to the Waygiou, Gillolo, and Philippine islands. These are sometimes called the currents of the Caroline monsoons, because the maximum speed of this current, toward NE. and ENE., is observed a little to the westward of the Caroline islands during the SW. monsoon, (from June to October;) but during the NE. monsoon, (from October to May,) the current sets to SW. and WSW., and forms a prolongation of the northern equatorial current.

The southern equatorial current.

The southern equatorial current begins near 88° W., and flows toward W. and WNW., with a mean and constant speed of about one knot per hour; it reaches from the equator to the tropic of Capricorn, and even to 26° S., in the part comprised between the Paumotas and Tongas. But the drift of the stream is no longer regular beyond the meridians of the Samoa and Tonga islands; this may be accounted for by the variable winds which prevail from November to March in the western part of the South Pacific, (vide § 4.) Between 20° and 26° S. and west of 178° W. the waters of the southern equatorial current divide into two branches, called respectively the Rossel current (vide § 31) and the Australian current (vide § 29.) The maximum temperature of the southern equatorial current appears to be reached on each meridian between the parallels 8° and 15° S. This maximum is about 78°.8 between 105° and 120° W., 80°.6 between 120° and 135° W., and 84°.2 near 178° W.

§ 28. THE EQUATORIAL COUNTER-CURRENT.—The equatorial counter-current is an irregular stream, setting toward the east; it is about three hundred miles wide, and lies between the northern and southern equatorial currents. Though the limits of the counter-current are imperfectly defined, it appears to be comprised between the equator and the parallel 8° N.; the greatest width of its bed is never more than 5° of latitude. In the eastern part of the Pacific it génerally keeps between 5° and 8° N., while in the Central Pacific it is generally nearer the equator.

It sets toward the east with a speed of sometimes two knots and a half per hour; but its rate is ordinarily much less; the mean speed being about .6 of a knot.

The existence of an easterly counter-current in all that part of the Pacific which lies between the Carolines and the coast of America is very doubtful. But the motion of the sea appears, without doubt, to be constant in the western part of the ocean to the southward of the Carolines, and even as far as the Mulgrave islands. Therefore, sailing-vessels can make their "eastings" without much difficulty by keeping a little to the northward of 2° N., which appears to be the limit of the southern equatorial current. Naturally navigation is here rendered still easier between June and October, when the SW. monsoon may be expected. This wind sometimes extends even beyond the Caroline islands. The reader should also remember what was said in § 27 upon the subject of the alternate currents which predominate to the westward of the Carolines, and which certainly sustain the equatorial counter-current during the SW. monsoon.

§ 29. THE AUSTRALIAN CURRENTS, (EAST COAST.)—A distinction should be made between the great *ocean* current and the *coast* current.

The ocean current off the eastern coast of Australia is only the prolongation of one of the branches of the southern equatorial current, which divides a little to the southward of the Fiji islands and west of 178° W. As has been stated in § 27, one branch forms the Rossel current, (vide § 31,) the other the Australian current.

This latter current first sets toward the west and passes to the southward of New Caledonia, then it turns to the SW. and passes to the westward of Norfolk island. It continues to run to the SW., setting toward Howe island and to about 28° or 30° S. Here the current is found about 300 miles from the coast, and extends to about 480 miles that is, its breadth is about 180 miles. It stretches toward the south, particularly between 152° and 157° E., and inclines to the SE. after passing the parallel of the extreme southern limit of Tasmania. The strength of this current varies from 6 to 19 miles per day; its temperature, which decreases rapidly as the waters advance to the southward has been found to be (in the center of the current) 70° on the parallel of port Jackson; 60°.8 on the parallel of Bass strait; and from 53°.6 to 55°.4 on the parallel of Tasmania.

Currents on the eastern coast of Australia.

On the eastern coast of Australia and to the southward of 28° S., as far as Tasmania, there are two sets of opposite currents, one within 21 miles of the land, the other beyond that distance from the coast.

Thus, during the *southern summer*, from 15th August to 15th April, the current, within 21 miles of the coast, generally sets toward S. by W., with a mean speed of 6 miles per day. At the same time and beyond 21 miles, a current is found setting to N. by E., with a speed of about 18 miles per day.

On the contrary during the *southern winter*, from 15th April to 15th August, the direction of the current is toward N. by E., with a daily speed of 6 miles, within 21 miles of the coast; while, beyond 21 miles, it runs toward S. by W., with a speed of 18 miles. It should be added that the prevailing current, which exists beyond 21 miles from the coast, from 15th August to 15th April, and which sets toward N. by E., rarely extends beyond 60 miles. Farther from the coast, the ocean current, alluded to at the commencement of the present paragraph, is found.

§ 30. THE AUSTRALIAN CURRENTS, (SOUTH COAST, FROM CAPE LEEUWIN TO BASS STRAIT.)—On the southern coast of Australia a tolerably regular current exists, flowing to the eastward. This motion appears to be the natural effect of the permanent winds which blow from NW. to SW., as has been stated in § 6.

Off cape Leeuwin the stream coming from W. and SW. appears to be divided into two branches: one striking to the northward, along the western coast of Australia; the other to the eastward, with a variable speed, influenced no doubt by the force and direction of the prevalent winds. In the vicinity of cape Leeuwin, and as far as King George's sound, the rate is often more than a knot, say 28 or 29 miles per day, and even more than a knot and a half, say 36 miles per day, in the part of the ocean comprised between point D'Entrecasteaux and King George's sound. Farther to the eastward the speed diminishes, and it is especially variable and feeble in the large gulf lying between the Recherche islands on one side, and capes Northumberland and Bridgewater on the other. Nor does it seem astonishing that, in this large bight, when the wind no longer blows from the westward with any regularity, (vide § 6,) the current should not set to the eastward with the same strength as between the parallels 38° and 40° S. It would seem only natural that eddies and return westerly currents should be found in the bight and along the land, particularly from January to April, when the winds are here from SE. and ENE.

Beyond cape Bridgewater and in approaching Bass strait the easterly current again becomes manifest with a speed of about 1 knot per hour.

The body of the current passes to the southward of Tasmania, though a part crosses Bass strait. Therefore a ship is nearly always set to the eastward along the coast of Australia, between cape Otway and Wilson promontory. After passing 2 or 3 degrees beyond the longitude of the Furneaux islands easterly currents are found; their existence has also often been proved between Wilson promontory and cape Howe, near which they flow to the eastward with a speed of about 1 knot per hour.

The tides can scarcely be relied upon in *Bass strait.* In this passage the *flood* tide sets to the *westward;* as may be seen by comparing the establishments of the port for different points on the coast of Australia and Tasmania. Thus, at full and new moon the tide is high successively at cape Howe at 0^h and at Wilson promontory (Refuge creek) at 12^h. Again, at the same epochs, it is high water in the Furneaux islands, between 10^h and 11^h; at port Dalrymple, at $12^h\ 5^m$; and in Franklin roads, (King island,) at 1^h. The ebb tide naturally makes in an opposite direction, or to the eastward, in King strait, and upon the adjacent coast of Australia.

From these observations it will be seen that a vessel making passage through *Bass strait,* from west to east, has the flood tide longer than the ebb; this is unfavorable, as the flood tide sets to the westward. Inversely, ships making passage to the westward have the flood in their favor.

§ 31. THE ROSSEL CURRENT.—As has been stated in § 27, the waters of the southern equatorial current, comprised between 20° and 26° S., and to the southward of the Fiji islands, divide into two branches. One of these branches stretches to the SW., increasing the current off the eastern coast of Australia, (vide § 29;) the other sets toward NW., and is called the Rossel current.

The waters of this current pass between New Caledonia and the New Hebrides, and about 150 miles to the eastward of this group. They flow toward NW. and pass south of Vanikoro island and the Solomon group, when they change their direction to W. and WNW., and set toward Torres strait. The speed of the Rossel current varies generally from 4 to 18 miles per day; its mean rate is from 8 to 10 miles. The temperature of the water is about 78°.

§ 32. GENERAL CURRENTS IN THE "SEAS OF PASSAGE."—In the "seas of passage," that is, in the Java, Celebes, Banda, Timor, and Arafura seas, the currents generally set in the same direction as the monsoons. The water appears to be put in motion by the wind. The currents usually flow in a westerly direction while the NE. and SE. monsoons prevail; on the contrary, their direction is toward the east during the SW. and NW. monsoons.

The direction of the current varies, not only with the change in direction of the wind, which occurs at times during a monsoon; but also, in consequence of the impulse received in the neighborhood of the straits from the currents which run rapidly in all narrow channels.

The usual speed is rarely over 1½ knots per hour, except through the straits, where it is often much more. Thus, in the straits of Sunda, Bali, Lombok, Allas, Sapi, Flores, Alloo, Pantar, and Ombay, the currents are often very rapid; they depend upon the direction of the wind, and especially of the tide. Though they are uncertain and irregular, they ordinarily set toward the east in the strait of Sunda from January to April, and in the opposite direction during the rest of the year, with a speed of often 3½ knots per hour.

§ 33. THE GREAT ANTARCTIC DRIFT-CURRENT.—This name is given in the Pacific, as well as in the Atlantic and in the Indian oceans, to the great body of water moving toward the east, between 40° and 60° S., with a constancy analogous to that of the prevalent westerly winds.

This current is particularly noticed in the Pacific, between 45° and 55° S., from Tasmania, and the S. point of Stewart island, (New Zealand,) to about 118° or 108° W. At this longitude a portion branches off and forms the *Mentor cur-*

rent, (vide § 34,) which flows to the NE., toward St. Ambrose islands, near 78° W. and 26° S. The greater part of the main current continues to drift to the eastward, as far as 84° or 86° W., where the waters of this southern branch divide into two currents, between the parallels 42° and 47° S.; one bears to the northeast in the direction of Valdivia and Valparaiso, forming the *Chile current*, (vide § 40;) the other tends to ESE. and SE., in the direction of the gulf of Peñas and the strait of Magellan, and forms the cape Horn current, (vide § 42.)

The Antarctic drift-current has apparently a mean speed of about 20 miles per day; but the figures on this subject are not thoroughly reliable.

A much stronger current is sometimes noticed setting toward NE. or SE., generally after a series of strong westerly winds. Under other circumstances there is either no current at all, or one in an opposite direction, particularly during the southern summer, and after easterly winds. Generally the current is strongest and most favorable for vessels crossing the Pacific near 50° S.

§ 34. THE MENTOR CURRENT.—It has been seen above (§ 33) that a part of the Antarctic drift-current sets toward ENE., near 118° or 108° W., in the direction of the St. Ambrose islands. This branch, which is called the *Mentor current*, has a mean speed of 0.7 of a knot per hour. It is comprised in a zone of several degrees of latitude, lying to the northward and southward of a line drawn from 42° S. and 128° W., to 30° S. and 85° W. Beyond 85° W. the waters shape their course to the northward; first, toward NE., then toward N., after they reach the parallel of the St. Ambrose islands. Beyond this latitude the current bends rapidly to the NW., passing the parallel 20° S., between 83° and 88° W.; it then bears to the westward and becomes merged in the southern equatorial current.

The Mentor current, in the eastern part of the Pacific, unites the Antarctic drift-current to the southern equatorial current, in the same manner as these two currents are united by the Australian current (vide § 29) in the western Pacific.

In the northern hemisphere the equatorial current also describes a complete circuit by uniting the Japan current (vide § 37) to the California current, (vide § 39.)

§ 35. THE CURRENTS OF THE CHINA SEA.—The currents of the China sea appear to be caused by the wind; during the NE. monsoon they tend generally to the SW., and during the SW. monsoon to the NE.

This fundamental principle being admitted we shall mention as briefly as possible the *principal exceptions*, and the most important facts which have been observed and which demand attention.

Current during the NE. monsoon. During the NE. monsoon the direction of the current on the S. coast of China is WSW.; its speed is sometimes 2, 3, and even 4 knots, particularly in shoal water, and within 60 miles of the shore. Between Pulo-Canton and cape Padaran (Cochin-China) the current is southerly, and sometimes reaches, near land, a speed of from 2 to 2.5 knots. From cape Padaran to Pulo-Obi it flows to SW., but it should be added that on this coast the *ebb tide* sets to the NE., and the flood to the SW.; from Pulo-Obi to Pulo-Capas and Pulo-Brala there are SSW. and S. currents, running at a rate of 2 or 2.5 knots per hour. On the northern coast of the Malay peninsula the currents flow to SSE.; between Bintang and Borneo the general direction is southerly.

On the NW. coast of Borneo, and NE. of the Natunas, a counter-current often flows to NE. and N. This can be often utilized while beating up the China sea against the monsoon. On the west coasts of Palawan and Luzon the currents are variable, and ordinarily dependent on the prevailing winds; they often set to the northward, especially along the coast of Luzon. When these currents are strong as far as cape Bojeador, they then bend to NE. and ENE. toward the Babuyan islands, where they meet the SW. current coming from the north of the China sea; and also the westerly current coming from the northern equatorial current. Consequently, eddies and variable currents are here found flowing in different directions, and often obtaining considerable speed.

In Formosa channel the current sets to the southward.

Current during SW. monsoon. During the SW. monsoon the flow of the current varies on the E. coast of China, from NNE. to ENE.; its speed being sometimes 3 or even 4 knots. At the Pescadores isles, a current has been noted during the month of August, setting to the northward at the rate of 4 knots. On the southern coast of China the current flows to the eastward. The waters of the Canton river make out to sea in a direction between WSW. and WNW., forming a genuine current with a speed which often reaches 1 or 2 knots between Macao and St. John. Still it is best not to rely on this current, as it sometimes amounts to little or nothing or is replaced by an easterly one during the SW. monsoon. Between cape Padaran and Hainan the stream is very irregular and feeble, especially along the coast between Padaran and Pulo-Canton. Abreast of the gulf of Tonquin, and during NW. and W. gales, the current flows to SW. and S., and as it sets across the direction of the monsoon, a heavy and broken sea is produced. From Padaran to Pulo-Obi the current flows toward E., but the tide must be taken into account along this coast; the flood setting to SW. and the ebb to NE. Abreast of the gulf of Siam the currents are to the N. or NE.; on the northern coast of the Malay peninsula they flow to the N.; on the western coasts of Palawan and Luzon a moderate current sometimes flows to the N. Northerly currents sometimes make out from near cape Bojeador with great swiftness, and tend to the NE., toward the Calayan and Camiguin islands, (Babuyans;) one part of the stream follows the north coast of Luzon, flowing ESE.; then, meeting cape Engaño, bears to the northward. The waters are then affected by the northern equatorial current, which impels them to the NW. They have sometimes a speed of 5 knots per hour, which slackens quickly some distance beyond cape Engaño. The current which we have already noted as existing during the NE. monsoon, causes strong eddies among the Babuyans islands, but rarely reaches to the Bashees. In this last group of islands the current flows swiftly in a northerly direction; sometimes it tends toward the east when there are strong westerly breezes.

§ 36. THE CURRENTS OF THE JAPAN SEA.—It has been seen, § 19, that information is not yet complete concerning the system of winds prevalent in the Japan sea, which lies between the coast of Tartary and the islands of Japan. This remark is equally true of the currents. The only thing which can be positively asserted is that both winds and currents in this sea are variable, and that it is necessary for navigators to be extremely careful while passing from Corea channel to Tsugar strait.

Vol. 11 of the North Pacific Pilot contains the following remarks by Lieut. Silas Bent, U. S. Navy:

"I am inclined to believe that a current from the Arctic ocean exists, running counter to the Kuro-Siwo, and which passes to the westward through the strait of Tsugar, down through the Japan sea, between Corea and the Japanese islands, and forms the hyperborean current on the east coast of China, which is known to flow to the southward, through the Formosa channel into the China sea. For, to the westward of a line connecting the north end of Formosa and the southwestern extremity of Japan, there is no flow of tropical waters to the northward; but, on the contrary, a cold counter-current filling the space between the Kuro-Siwo and the coast of China. As far as this cold water extends off the coast the soundings are regular, and increase gradually in depth; but simultaneously with the increase of temperature in the water, the plummet falls into a trough similar to the bed of the Gulf-stream."

According to this quotation, the sea of Japan is traversed from N. to S. by a cold current coming from the coast of Kamtchatka and the Kuriles; entering by the strait of Tsugar, and leaving by Corea channel, thus opposing the entrance through the strait of Corea of the warm waters of the equatorial and Kuro-Siwo currents, which tend to NE. to the northward of the Loochoo islands; but we would add that the foregoing assertion is too positive, in proof of which we cite the following passage from the instructions of Captain Legras:

"The speed of the currents is at times very slow; at others, very considerable. It is only known that a branch of the Kuro-Siwo is directed usually (although with numerous variations in its strength, direction, and breadth, and greatly influenced by the wind,) toward the NE., *after leaving the strait of Corea, and enters into the Pacific by Tsugar strait.* It is also a known fact that, in the autumn, another current is found, generally setting to ESE. in La Perouse strait, and to SW. in summer along the coast of Manchuria."

We will finish with an extract from the Coast Survey Report of 1867, which shows, in a more general manner than that of Lieutenant Bent, the existence of a cold current setting to the southward in the Japan sea:

"Between the Kamtchatka current and the Asiatic coast and islands is a cold polar counter-current coming from the Behring sea. It follows the coast of Kamtchatka, the trend of the Kurile islands, gives rise to the currents flowing west into the south part of the Okhotsk sea, and strikes the northern and eastern part of the sea of Japan. A small amount of the water of this current passes into the Japan sea through Tsugar strait, but the greater part keeps along the east coast inside, and probably underruns the great Japan current, the northwestern ledge of which is strongly marked by a sudden depression in the temperature of the water."

§ 37. THE KURO-SIWO OR JAPAN CURRENT.—One branch of the northern equatorial current, after having passed the Marriana islands, flows toward the eastern coast of Formosa, in a WNW. direction. It makes a sharp turn toward north while passing between Formosa and the Meiaco-Sima islands. Next flowing toward NE. it makes the circuit of the Loo-Choo islands; then it passes between the islands of Kakai-Sima, and Ou-Sima to the southward; and the large island of Kiusiu to the northward. It bears off along the coast of Niphon, passing by the bay of Yedo, and opens out, in a fan-shaped manner, toward the different points of the compass between NE. and E.; after leaving South island and Bayonnaise rock, situated nearly on the meridian of the bay of Yedo, the current occupies nearly the whole sector, which extends from 40° N. to Moor island.

This current is called *Kuro-Siwo* by the Japanese, that is *black current*, on account of its dark-blue waters, and presents many analogies to the Gulf stream of the Atlantic.

Its breadth, between the islands of Formosa and Majico-Sima, is hardly 100 miles; but it rapidly grows wider after the current has doubled the Loo-Choo islands. Between the Bonin islands and the coast of Niphon it attains a width of about 500 miles.

Its average speed appears to be about 1.5 knots. From the strait of Formosa to the coast of Japan it increases, and reaches its maximum between the meridians of Kiusiu and the bay of Yedo.

In this part of the current a speed of from 72 to 80 miles per day is sometimes observed. On the parallel 35° N., and at a distance of 200 miles from the coast, a current setting to ENE. has been proved to exist, with a speed of 48 miles per day; and on the same parallel, at a distance of only

75 miles from the coast, the speed has been found to be 72 miles.

King states that in the same latitudes there is a current of 5 knots per hour. During the winter (November) the currents on the coast of Japan have a more northerly direction; in summer (July) they incline more to the eastward.

The mean maximum temperature of the *Kuro-Siwo* is about 86°. The northern edge of the Kuro-Siwo is separated from the coasts of Yesso and Niphon (to the N. of Yedo bay) by a cold current coming from Kamtchatka and the Kurile islands, (vide the end of § 36.) This cold current is analogous to the one which lies between the Gulf-stream and the coast of the United States; its temperature is about 16° or 20° below that of the Kuro-Siwo. The limit of the two currents is marked by the sudden change in the color and temperature of the waters.

As in the Atlantic, eddies, bad weather, and thick fogs are here found. It is not so easy to determine the southern limit of Kuro-Siwo. The change of color in the water is nearly imperceptible; the change in temperature cannot be more than 7° or 9°, and is only gradually felt. Near 146° or 147° E., between Moor island and 40° N., the *Kuro-Siwo* current divides into two parts. One, called the Kamtchatka current, (vide § 38,) flows in a NE. direction, having for its axis a line drawn through 151° E. and joining 40° N. with Behring strait. The other branch, which is by far the larger, crosses the Pacific in a general easterly direction, in the same manner as the Antarctic drift-current in the southern hemisphere.

This main branch of the Kuro-Siwo is generally called the Japan, or *Tessan* current. The waters of this stream flow toward E. and SE., after passing the meridian 152° E., until they reach 172° E., between the tropic of Cancer and 40° N. The current bears toward E. between 172° E. and 163° W., and particularly in the zone to the westward of the Sandwich islands, included between 20° and 24° N. On the meridian of 163° W., and N. of the tropics, as far as 44° N., the direction of the current is nearly NE.; the southern part of the current is bounded by a line passing through 163° W., and joining the tropic, and 40° N. at 148° W. On this meridian of 148° W. the current flows NE. between 40° and 50° N., it then tends to E. and SE., and unites with the current of the coast of California, (vide § 39.)

The temperature of the current of Tessan has been found to be 81°.5 at 27° N. and 177° E.; only 61° at 36° N. and 128° W. This shows the cooling of the waters during their passage through 55° of longitude. The temperature has been found to be from 77° to 79° at 21° N. and 163° W.; in other words, at this point on the southern limit of the Tessan current the water was at least 3°.5 warmer than in the northern equatorial current, which flows in an opposite direction, a short distance farther south.

§ 38. THE KAMTCHATKA AND BEHRING CURRENTS.—The waters of the *Kuro-Siwo* separate into two branches, as has been stated above, near 146° or 147° E., between Moor island and 40° N. The least known, as well as the least important, of these branches takes the name of the *Kamtchatka current;* it flows toward NE., having for axis a line passing through 151° E., and joining 40° N. and Behring strait. This current, about which little is known, passes to the west of the Aleutian islands, at a distance of about 150 miles from the coast of Kamtchatka. In August and September the temperature has been found to be 52° off Petropaulski; its speed is about 0.3 of a knot per hour.

It may be here stated that there exists a cold polar counter-current (vide end of § 36) between the current above mentioned and the coast of Kamtchatka. It comes from Behring sea, follows the coast of Kamtchatka and the direction of the Kurile islands, and gives rise to currents which cease in the southern part of the sea of Okhotsk.

Behring current. Behring current is a stream which runs at a rate of about half a knot, and appears to issue from Behring strait, and bends first toward SSE., afterward passing to the eastward of St. Lawrence island. It then flows S. and SSW., toward the Aleutian islands, passing to the eastward of St. Mathew's island.

§ 39. THE CURRENTS OF THE COASTS OF CALIFORNIA AND MEXICO.—*On the coast of California*, from about 50° N. to the mouth of the gulf of California, 23° N., a cold current, 200 or 300 miles wide, flows with a mean speed of 0.7 of a knot, being generally stronger near the land than at sea.

Usually it follows the trend of the land, that is nearly SSE., as far as point Concepcion, (S. of Monterey,) when the current begins to bend toward S., SW., and then to WSW. off capes San Blas and St. Lucas. The temperature off Monterey is not more than 55°.5 or 57° and only 59° at 30° N.

Currents on the coast of Mexico.

On the coast of Mexico, from cape Corrientes (20° N.) to cape Blanco, (gulf of Nicoya,) there are alternate currents extending over a space of more than 300 miles in width, which appear to be produced by the prevailing winds, (vide § 24.)

During the dry season, January, February, and March, the currents generally set toward SE. During the rainy season, from May to October—especially in July, August, and September—the currents set to NW., particularly from Cocos island and the gulf of Nicoya to the parallel of 15°.

§ 40. THE CURRENTS OF THE BAY OF PANAMA.—After leaving cape San Lorenzo, or the equator, a current is found along the coast of South America, sixty miles in width; it follows the direction of the land, and, entering the bay of Panama, makes a complete circuit of that gulf. After meeting the western coast of the bay of Panama the current turns to the S. and acquires considerable velocity, especially during the dry season, from December to April, when the winds are frequent from ENE., (vide § 23.)

In the bay of Panama, and at its entrance, the currents are far from being regular, and, under certain circumstances, are quite strong. Eddies and a short, chop sea are particularly noticed in the SW. part of the mouth of the gulf. Farther out, near Malpelo island for instance, very rapid currents are found; these have been observed to set in entirely opposite directions, sometimes toward ENE., at others toward SW..

§ 41. THE CURRENTS OF THE COASTS OF CHILE AND PERU.—It has been stated, § 33, that the Antarctic drift current is separated into two branches, between 42° and 47° S. One of these branches flows toward NE., in the direction of Valdivia and Valparaiso.

This stream follows the various sinuosities of the coasts of Chile and Peru, and forms the important current, the existence of which was first noted by Humboldt.

The principal characteristic of this current is its relatively low temperature.

Abreast of Valparaiso 52° have been noted; of Coquimbo, 57°; of Arica, 64°.5; of Callao, 65°.5; of Truxillo, 69°; and off Cape Blanco, 66° to 73°.5. The general direction of the waters between Pisco and Payta is toward NNW. and NW.

Near cape Blanco the current leaves the coast of America, and bears toward the Galapagos islands, passing them on both the northern and southern sides. Here it sets toward WNW. and W. The breadth of the bed, on the meridian of the Galapagos, is from 400 to 500 miles; beyond this it widens rapidly, and the current is lost in the equatorial current near 108° W.

The breadth of the current near the coasts of Chile and Peru and as far as cape Blanco is about 150 miles. Its mean speed is nearly 15 miles. Between Payta and the Galapagos, where it obtains its maximum rate, it has been known to run on rare occasions as fast as 50 miles in a day; but on the coast, from Valparaiso to Callao, the greatest speed appears to be 24 miles per day; sometimes it is only 3 miles in 24 hours; nor is it very rare to find currents setting to the southward. These act in the most unforeseen manner, but usually last only for a short time.

As often happens in similar cases, the existence of a counter-current has been proved on different occasions. This sets toward the S. with a maximum speed of 0.5 of a knot per hour, is very irregular, and extends only a little distance from shore.

§ 42. THE CAPE HORN CURRENT.—In the "*Navigation of the South Atlantic ocean*," page 131, will be found a description of the cape Horn current, to which the reader is referred, and which will be here completed.

The waters of the Antarctic drift-current (vide § 33) divide, between the parallels 42° and 47° S., into two branches, one of which flows ESE. and SE., in the direction of the gulf of Peñas and the strait of Magellan, and forms the cape Horn current.

The stream follows the indentations of the west coast of Patagonia, though with a general SE. and ESE. direction, being about 150 or 200 miles in width. The current runs around Terra del Fuego, toward E.; then flows to NE., passing through the strait of Lemaire, and by Staten island.

The swiftness of the current is uncertain, on account of the great variation of the winds in these quarters. However, the following observations can be considered as forming a reliable average.

On the meridian of 80° W., and between 55° and 60° S., the current flows to SE. and ESE. at a rate of from 12 to 24 miles per day. On the meridian of 75° W., and about 120 miles to the southward of cape Pillar, the current flows to SE. at about 12 miles. On the same meridian, between 55° and 60° S., its speed is from 18 to 24 miles in an easterly direction. At 70° and 71° W. and 58° S., its direction is easterly, and rate about 30 miles per day. South of cape Horn, on the meridians of 68° and 69° W., it often attains a speed of 30 miles. To the southward and eastward of the cape the stream begins to curve to the northward and eastward, while between cape Horn and Staten island its speed near land is about 24 miles, toward NE. The regular currents are, however, affected by the tides running through Lemaire strait.

The flood is from east, on the northern coast of Staten island, and from north in Lemaire strait; according to Horsburgh it runs at about 2 knots per hour, and is influenced by the wind; the speed of the ebb is never more than 1 knot. King and Fitz-Roy state that the flood runs as fast as 5 or 7 knots through Lemaire strait.

Between cape Horn and Staten island the flood flows toward NE. Thus, at the full and change, high water occurs at 3^h 50^m at cape Horn; and at about 4^h 30^m at Staten island. The flood tide flows toward NE., with a speed of 3 knots or more per hour, between cape Horn and the strait of Lemaire. These currents are scarcely felt S. of Staten island.

Therefore, vessels entering the strait of Lemaire with favorable winds can easily pass through during the flood, and afterward take advantage of the ebb tide which sets to SW., between the strait and cape Horn.

§ 43. Icebergs.—In the northern Pacific there is no danger of meeting floating ice below 50° N., but it is not advisable to pass that parallel in making passage between Japan or China and California. In the southern Pacific floating ice has been encountered at all seasons, and often in quite low latitudes. Icebergs are, therefore, always to be feared, especially during the southern winter, as the nights are then long. They constitute a real danger, and the principal difficulty in making a passage from Australia, New Caledonia, New Zealand, or Tahiti, to cape Horn.

Icebergs have been encountered in March and April. They have also been frequently seen from April to August, particularly between 90° and 160° W., and as far north as 41° 45′ S. Finally, icebergs have been found in great numbers, especially between these same meridians, from September to January. There is, however, danger of meeting them at all times of the year.

In the following table will be found the latitude-limit of floating ice, on each meridian, for every 10°, or even for every 5°, when the changes are great. Though vessels may run *south* of this limit without encountering ice, they should always keep a bright lookout while in or near these localities.

Latitude-limit of floating ice.

Longitude.	Latitude-limits.	Longitude.	Latitude-limits.	Longitude.	Latitude-limits.	Longitude.	Latitude-limits.
°	° ′	°	° ′	°	° ′	°	° ′
60 W.	53 15	110 W.	42 45	170 W.	40 10	150 E.	51 30
65 W.	56 30	120 W.	42 20	175 W.	45 50	140 E.	47 15
70 W.	57 15	130 W.	41 45	180 W.	49 30	130 E.	45 20
75 W.	57 00	140 W.	41 20	170 E.	51 15	120 E.	45 00
80 W.	55 10	150 W.	40 45	160 E.	53 00	110 E.	45 00
90 W.	47 00	160 W.	40 30	155 E.	52 45	100 E.	44 40
100 W.	44 00						

A few general principles will now be given which should govern the navigator of the South Pacific. For information on this subject we are indebted to Messrs. Towson, Weddel, Boulton, Scoresby, etc.

From the 1st April to 1st October floating ice is hardly ever found to the northward of 50° S., and is rarely met even as far south as 50° or 53° S., except between 148° and 93° W. During this season it may be expected between 53° and 60° S., from 158° W. to cape Horn. From 1st April to 1st October.

5 N

From 1st October to 1st April.

From the 1st October to 1st April, icebergs are more numerous, stray bergs sometimes drifting as far to the north as 40° S., though they are rarely observed to the northward of 50° S. But navigators should be very vigilant after passing 50°, as there is very little space between 50° and 60° where icebergs have not been observed. They are particularly abundant beyond 52° S., and from the meridian of 173° W. to that of 88° W. Many are also found to the eastward of 70° W., and to the southward and eastward of cape Horn.

The direction and rate of their drift.

The icebergs of the Pacific generally drift toward E. by N., with a speed of 10 miles per day. But to the eastward of cape Horn they drift first to NE., then change their direction more to the eastward in approaching 40° S. On this parallel, and between 25° and 15° W., they set toward E., rarely advancing more than 1 mile per day. Then they bear away toward ESE. and SE.

Icebergs between the cape of Good Hope and Australia.

Between the cape of Good Hope and Australia it is impossible to state the parallel on which icebergs may not be encountered by a vessel making passage to the eastward; one year icebergs are found on one parallel, the next on another.

If, however, the parallel 52° S. is passed, the chances of meeting icebergs are undoubtedly increased. It is therefore best to keep *to the northward of* 51° S.

Between 70° and 80° W.

Icebergs appear to be more numerous between 70° and 80° W. and 56° and 58° S. than farther to the southward. This statement does not, however, hold true for the region east of 70° W.

Signs indicating their approach.

The following signs indicate the approach of an iceberg: A peculiar light, known as "ice-blink," which is sometimes seen at a great distance, even on a dark night. On coming close to the iceberg, this light has the effect of a white cloud settling over the rigging. The most certain indication is the falling temperature of the water, and its comparison with that of the air; it is therefore important to make frequent observations of the temperature of the water. This fall often amounts to 3°.5 and even 5°.5.

The proximity of icebergs is also known by the noise of the waves breaking over them; this is sometimes heard at a great distance, and resembles breakers on the shore.

With fair winds it is best to pass to windward of large icebergs, for then there is less danger of encountering the detached masses of ice, which always drift faster to leeward. These small icebergs are the most dangerous; they hardly show above the water, and cannot be seen when the sea is rough; they are often very deeply submerged, and if run into will cause bad leaks. Pass to windward of them.

The "ice-blink" of which we have spoken is, as a rule, observed only above large masses of ice, which are more or less flat and covered with snow. But this luminous appearance does not show above icebergs whose surfaces present a rugged appearance, or above those which have been capsized. Ice-blinks when seen.

It cannot be too often repeated that watchfulness is necessary. The man on lookout should be often relieved. He should keep his eye continually on the black line of the horizon ahead, and if he discover a white spot or even a light streak on the horizon, he may know it is ice. If a good watch be kept in this manner there is no danger of coming upon ice unexpectedly. It is also well to have another man on the watch for small icebergs, which are only to be seen at short distances. A good look-out necessary.

As has been said above, icebergs are most numerous during the southern summer. This is fortunate, as the nights are then shorter. Out of 550 icebergs, nearly one-half were seen in November, December, and January, while there were only 5 found in June and 3 in July. One-fifth of the whole number of icebergs seen in a year were observed in December alone. There are more from January to April than from August to November; thus the number in March and April is to the number in September and October as 5 to 3. Numerous during the southern summer.

According to Scoresby, the fall in temperature of the sea and air foretells icebergs, but it is extremely imprudent to trust to this warning. It is best to rely chiefly on the vigilance of those whose duty it is to keep a "sharp lookout." Have your yards nicely trimmed, and be ready at any moment to go about or haul up on either tack, as may be necessary. The reader should also refer to § 104, where further information on this subject will be found. Warnings.

PART II.

OBSERVATIONS ON THE PRINCIPAL ROUTES IN THE PACIFIC OCEAN.

CHAPTER I.

ROUTES FROM SOUTH TO NORTH, ON THE WESTERN COAST OF AMERICA.

§ 44. ROUTE FROM CAPE HORN OR THE STRAIT OF MAGELLAN TO VALPARAISO.—In the *Navigation of the South Atlantic* it was stated that vessels bound to Valparaiso from cape Horn should not cross the parallel of 50° S. farther west than the meridian of 80°. A summary of instructions for doubling cape Horn was also given, (p. 186 to 194.) Finally in a paragraph on the passage of the strait of Magellan, (§ 16 of the same work,) it was stated that it is preferable to enter the Pacific by cape Pillar, during *the southern winter;* while during *the southern summer* vessels can take the lateral channel and come out by the gulf of Peñas.

I. ROUTES AFTER LEAVING CAPE HORN.—The passage around cape Horn is usually made by good staunch sailing-vessels, or vessels with auxiliary steam power. The intention is generally to economize coal without prolonging the voyage. They cross 50° S. at about 80° W., that is between 78° and 82° W., and more or less to the westward according to the weather and winds they may have encountered. After having doubled the cape, and while making to the northward, it is a good plan to make the westing so as to cross 50° S. rather to the westward than to the eastward of 80° W.; but this parallel may be crossed at 79°, or even 78°, though it is not desirable to do so. After passing 50° S., keep to the northward, following 80° W., as nearly as the sea and dominant winds from NW. to SW. will allow. Beyond 45° S. it is best to keep between the meridian 78° 10′ and 79° 40′ W. Cross the parallel 40° S., near 78° 10′ or 78° 40′ W., then continue toward the north till you reach 37° 30′ or 37° S., thence make your easting and cross 35° S., between 74° 40′ and 75° 10′ W. From this point you can proceed without difficulty toward Valparaiso, merely making allowances for the leeway and current, which usually set to the northward and are apt to cause errors in the course.

Though it will generally be found easy to follow this route, as above described, we will give the difficulties to be encountered at each season of the year, and the course it is best to pursue in each case.

In January. In January it is advantageous to cross 50° S. a little to the westward of 79° 40′ W., as winds from NW. and NNW. rarely prevail in that locality. But between 45° and 40° S. it is preferable to keep to the eastward of that meridian, where the winds are often more favorable. Calms and baffling airs are frequent during this month.

In February. In February it is still more important, if it be possible, to cross 50° S. to the westward of 79° 40′ W. It may be generally stated that vessels will make this passage from cape Horn to Valparaiso much more rapidly if they cross the parallel of 50° S. even much farther to the westward. Thus quick passages can be made under canvas when 50° S. is crossed at 80° 40′ W. or to the westward of that meridian; 45° S. at between 80° 40′ and 81° 10′ W.; 40° S. between 79° 40′ and 80° 10′ W.; then 35° S. at about 74° 40′ W. In crossing the parallels at these points do not hug the NW. wind too closely, as there is danger of being driven to the eastward of the advised route. If the wind afterward shift to W. and SW., it is evident that you would not voluntarily lay as high as NW. to regain the points of crossing. You should merely try not to get too far from them.

In March. In March the best route is, on the other hand, to the *eastward* of 79° 40′ W.; it will therefore be sufficient to cross 50° S. just far enough to the westward to avoid the land, in case of several successive squalls from the W. This point should be left to the discretion of the captain: a sailing-vessel might pass near 78° 40′ or 78° 10′ W., while a vessel with auxiliary steam power would find no difficulty in crossing at 77° 40′ or 77° 10′ W. The winds are generally steady and favorable; yet there are always chances of winds from NW.; in case the wind does come out from NW., keep on the port tack as long as it can be done with prudence. On the one hand, be very careful not to get too close to the shore; on the other, remember that there is more chance of finding favorable winds near shore than out at sea in the neighborhood of 80° W.

Thus 50°, 45°, and 40° S. may be successively crossed at the meridian of 77° 10′ W.; the rest of the passage is easily made, crossing 35° S. at about 75° 10′ or 74° 40′ W. Calms and light airs are the only causes of detention north of the 35th parallel.

In April the same general route, given for the whole year, should be followed. In April.

We would merely remark, that at this time, between 50° and 35° S., the wind prevails more from NW. than during the first months of the year. This should therefore be borne in mind, to profit by any favorable wind, in order to keep as much as possible to the westward of 79° 40′ between 50° and 45° S.; and to the westward of 78° 40′ between 45° and 40° S. With this precaution it is possible to continue on the port tack when the wind comes out from NW., cross 35° S. near 75° 40′ W., and 34° S. between 74° 10′ and 74° 40′ W. North of 35° S., on nearing Valparaiso, calms are less frequent than earlier in the year, the winds blowing steadier from NE. to N. and NW. (Vide § 21.)

In May, as in February, it is best to make the northing to the westward of 79° 40′ W., when the winds permit. May is not a bad month if the route prescribed for February be followed. Near Valparaiso this is the winter season, (vide § 21.) In May

In June, as in February and May, it is advisable to cross 50° S. as far as possible to the westward. Between 50° and 45° S. the winds are more favorable to westward than to eastward of 79° 40′ W.; besides it is best, and nearly always possible, to profit by all the fair winds, in order to cross 45° S., and particularly 40° S. near 80° 40′ W. This advice is considered important, for in June the winds very often blow from NNE. to NNW. between 40° S. and Valparaiso. A vessel will not really be in a very good position to fetch this port under sail if she steer to the northward of NE. In June.

It is well known that W. and SW. winds are to be found between 40° S. and Valparaiso, and that in all cases the current is favorable. In § 21 will be found a description of the system of the winds in this region.

In July and August the same observations hold true. The only remark to make is, that during August there is no necessity of crossing 50° S. very far to the westward, as the wind is generally favorable, between 50° and 45° S., to both the eastward and westward of 79° 40′ W. But it is prudent, and always possible, to cut 45° S. between 79° 40′ In July and August.

and 80° 40′ W.; and afterward 40° S. at the same longitude as has been stated for June.

In September. In September it is easier to make the passage—between 50° and 45° S.—by sailing to the westward of 79° 40′ W., as here the winds prevail from W. and SSW., whereas to the eastward of this meridian they are from NW. After passing 45° S. the general route, given for the whole year, may be followed without difficulty. 40° S. should be crossed between 77° 10′ and 79° 10′ W.; thence the winds are fair and steady.

In October. In October, as in September, it will be more advantageous to cross 50° S. to the westward of 79° 40′ W. Settled and favorable breezes will nearly always be found in crossing 45° S. between 80° 10′ and 81° 10′ W.; 40° S. at 79° 40′ W.; and 35° S. at 74° 40′ W.

In November. In November the passage is slow and tedious. The general remarks which are given on the proper route to be followed during the whole year, apply particularly to this month. It is only to be noted that at this time the NW. winds are frequent from the parallel 50° to 35°, and both to the E. and W. of 79° 40′ W. Still it is better to cross 50° S. to the westward of 79° 40′ W. when it can possibly be done. As a general rule advantage should be taken of every favorable slant in the wind that may help the vessel to the northward and *westward*, for then a long stretch may be taken on the port tack when the wind comes out from NW. If possible cross 40° S. at about 78° 40′ W., as the wind is better to the eastward of 79° 40′, between 40° and 35° S. Beyond this point the voyage will be easily made.

In December. In December the same observations as for November. The winds will be found favorable and from W., SW., and S., after leaving 40° S., if that parallel be crossed to the eastward of 79° 40′ W., or even at 78° 40′ or 78° 10′ W.

II. ROUTES AFTER LEAVING THE STRAIT OF MAGELLAN.—The route through the strait of Magellan is always taken by steamers, or auxiliary-steam vessels, when the consumption of coal is a secondary consideration, (vide *The Navigation of the South Atlantic*, pages 194 to 198.) Consequently the passage from cape Pillar, or the gulf of Peñas, to Valparaiso, presents no serious difficulty.

Steamers can make their northing, merely taking care to

keep at a safe distance from land, that they may not be driven ashore in case of bad weather. As for vessels with auxiliary steam-power, they first make to the westward, steering on a WNW. or NW. course, either under steam or sail, if the winds allow. No other advice need be given for the beginning of this route, except to profit as much as possible by every favorable wind, so as to get away from the coast, when the fires may be hauled and the voyage continued under sail. The more the coast is left, (as far as 79° 40′, if it be possible,) the more chance there is of making a quick passage to the northward.

When vessels with auxiliary steam-power, leaving cape Pillar, shall have gone far enough to the westward and made sail, their route will not differ greatly from that of sailing-ships coming from cape Horn. This route, however, will be 2 or 3 degrees more to the eastward, and will be made under *less favorable* circumstances, the wind being more ahead, as may be seen by referring to the observations at the beginning of the present paragraph. This observation is still more applicable to vessels, with both steam and sail power, entering the Pacific by the gulf of Peñas, as their route is still farther to the eastward, and as it is advantageous, during nearly every month of the year, to keep *well* to the *westward* after passing cape Horn under sail.

The most prevalent winds, and those which are most to be feared as head-winds, come from NNW. and NW. It is therefore advisable to take advantage of every favorable breeze, after leaving the strait of Magellan, in order to make to the westward and to the northward. If this be done, a long stretch can afterward be made on the port tack without fear when the wind comes out from NW. And we repeat, there will be no serious difficulty—to a ship that can, if it become necessary, start ahead under steam and fore-and-aft sail—in taking the NNW. wind four points on the bow. The use of sail and steam combined has simplified all these routes, and there are no special points of crossing to be recommended for auxiliary steamers.

As an illustration of a voyage between cape Pillar and Valparaiso, we will first give an extract from the *Hydrographic Annals*, (vol. 12, p. 322.) This extract is compiled from the log of the *Rattlesnake*, captain Henry Trollope:

"The *Rattlesnake* made cape Virgins the 10th of May,

1853, and the same day was taken in tow by the steamer *Vixen*. They anchored seven times in eight days, and at noon on the 18th the tow-lines were cast off, at 20 miles to the westward of cape Pillar.

"The barometer rose until ten o'clock in the morning, when it commenced to fall very rapidly. The wind sprang up; and the *Rattlesnake* made good way under royals and port studding-sails, the wind being from eastward.

"Fitz-Roy recommends to run as far as 79° 40′ W., if you are bound to the northward; our voyage proved him right. The wind, which would have been ahead if we had not shaped our course as we did, came out favorable. It remained steady from the eastward until the 19th; but the weather was overcast, and the barometer fell $1^{in}.34$ during the eighteen hours included between the 18th at 6 p. m. and noon of the 19th. About one o'clock in the afternoon, after a calm lasting about one hour, and during a furious squall from NE. and ENE., the wind jumped around suddenly to NW., and blew from this direction with extraordinary violence. We lay-to on the starboard tack under the main topsail and storm-stay-sail. The barometer commenced to rise an hour after the commencement of the squall, and this was the time of its greatest strength; the wind, however, continued to be violent for 10 or 12 hours.

"Afterward we had a series of smaller squalls with a heavy sea, coming especially from W. and WNW. As we were well off shore, the wind was favorable and allowed us to steer N., without being set to the eastward of that point, until we struck 36° S. and 74° 40′ W. During all this time we encountered squalls and violent gusts.

"In approaching Valparaiso we nearly ran past the harbor, though we hove to when we judged we were 10 or 12 miles to the southward of point Curaumilla."

The following is an account of a voyage from cape Pillar to Valparaiso, made in 1860 by the ***Duguay Trouin***, (sail and steam power,) the flag-ship of Rear-Admiral Larrieu, commanding the Pacific Squadron, (***Ann. Hyd.***, vol. 18:)

"We doubled Tamar on the 20th March, about 1.30 p. m., and made little way during the rest of the day, the weather being alternately clear and foggy. About 8.30 p. m., the wind having shifted, we were in very good position for leaving the strait, being from 3 to 4 miles from Westminster Hall, bearing SW. At 8.30 p. m. the course was set at WSW., so as to pass through mid-channel, but at 9 o'clock, the wind having hauled ahead, it was changed to WSW. ½ W., the fore-and-aft sails being still carried. About 10.30 we sighted cape Pillar, bearing SSW., and very close aboard; this gave us a little uneasiness, as the wind had freshened, and we were already as close to the wind as we could lie with safety. Fires were lighted under another boiler, but we doubled the cape rapidly, and at midnight were out of all danger, having passed the meridian of the Judges and Apostles, which lie to WSW. of cape Pillar. Experienced strong head-winds and sea till we reached the parallel of cape Tres Montes. Having doubled cape Pillar, and wishing to make westing, I was obliged to steer to the southward on account of the wind, which varied from NNW. to W.

"At noon on the 26th March, we were at 56° 28′ S. and 83° 55′ W. On the evening of the 26th, the wind having shifted to WSW., we were able to take the port or northerly tack, but were compelled to lie-to all night under fore-and-aft-sail, as the wind was violent and the sea carried away our starboard quarter-boat. The *Duguay Trouin* labored greatly, notwithstanding her excellent qualities. The wind having moderated on the morning of the 27th, I set the maintop sail close reefed. After this we kept on a northerly course, the winds varying from NW. to SW. and S., with a regular barometer. As the barometer fell the wind remained steady from NW. and WNW., and I kept on the starboard tack; when it died away, I went about, the wind soon springing up from S., and blowing with extreme violence. About 8 o'clock on the evening of the 31st March, being at 48° 23′ S., we could only scud under a close-reefed maintop sail and a double-reefed foresail; about this time the barometer fell very rapidly to 28^{in}.54.

"We could not keep on our course, as our 'latteen' sails were not able to stand the heavy squalls; but, luckily, these violent squalls did not last long, moderating as they shifted toward SW. The weather became better after the 1st April, and we arrived at Valparaiso on the morning of the 7th."

Volume 22, "*Annales Hydrographiques*," contains the following observations on the voyage from cape Pillar to Valparaiso: "After a vessel has reached 72° 40′ or 77° 40 W.,* she can steer rapidly to the N. The winds are here generally from SW.; they are interrupted, according to the season, by calms and northeasterly gales. The weather becomes better when the wind hauls to S.; it shifts ordinarily from right to left. After a violent squall from SE. the wind may be expected to return to S., and haul again to SW., with better weather; thus following the laws of cyclones. The passage from cape Pillar to Valparaiso is sometimes made in 6 or 7 days. The mean length of the passage is about 12 days."

We will also quote an abstract from the log of the *Assas*, a screw-corvette, commanded by Captain DeKergist, (*Ann. Hydr.*, vol. 26:)

"Leaving cape Pillar on the 9th July, I kept to the westward, under steam, until the morning of the next day at 7.30 o'clock. Having made 105 miles, the fires were hauled. During the night the breeze had freshened and hauled to SW. and WSW. I then came-to on the port tack, with 2 reefs in the topsails, and at noon found a current of 16 miles, setting S. 60° E. The weather was bad during the following night, the wind strong from WNW.; went about on the starboard tack, to make to the westward as far as possible. On the 11th the barometer fell rapidly, and we lay-to under the main-topsail. On the evening of the 12th ship rolled 35° to leeward and 12° to windward. The barometer, already indicating a storm, went still lower, and I decided to light one fire to keep the ship from being driven to leeward. We kept on in this manner until the 14th, the wind varying from NW. to WNW. On the 14th July the barometer stood at 28^{in}.03. After a calm the wind slowly changed to SW. An observation obtained on the 14th put us 48′ more to the northward and 91′ more to the westward than the dead-reckoning for the four preceding days. The engine, which had been in motion for forty hours, had assisted the ship more than I had presumed. The squalls were not, however, over; those from the NW. having died away, the SW. ones began, shifting to WSW. and W. These soon raised a heavy sea, which made the

* This is probably an error, and should read, 77° 40′ or 82° 40′ W., as the longitude of cape Pillar is about 74° 40′ W.

vessel labor extremely and deadened her way, as we had to take it abeam. At daylight on the 15th, the sea having moderated, I bore up; the ship rolled 35° to one side and 28° to the other.

"On the 16th, the wind being from NW. and N., hauled up on the starboard tack. During the night of the 16th and 17th the barometer read: At midnight 28in.74; at 1 a. m. 28in.54; at 3 a. m. 28in.31; at 5 a. m. 28in.27. The wind suddenly failed, and we had scarcely time to set enough canvas to keep the ship steady against the broken sea, when a moderate breeze set in from S., freshening by degrees, and hauling to SW. After this the weather improved, and on the 21st we were able to cast loose the guns, and dry the deck under the carriages. On the evening of the 23d we thought we sighted land at Bucalemo, and on the morning of the 24th made Valparaiso light, after a 15 days' passage from cape Pillar."

Passage of the *Venus*—Captain Roy—from cape Pillar to Valparaiso:

"We passed through the straits in four days and six hours. The thermometer only once fell as low as 44°.5. We had splendid weather for a landfall at cape Virgins, and very little wind in the strait. The fog was quite thick and rendered it difficult for us to see the land at the western end of the strait; but this was compensated for by a smooth sea and a light breeze from N. and NW., which lasted until the following day, enabling us to make 60 miles to westward, under steam. On the 13th February, at 6.30 p. m., the breeze came out from WNW.; we then hauled fires and made sail. On the 15th the wind hauled to NW. and N., and increased, obliging us to lie to. On the 16th the wind fell a little, returning toward WNW. and W., from which points it blew freshly, until the 19th, when it shifted to SW. and SSW., blowing a gale with a heavy sea, but giving us good way. On the 21st, being at 41° S., 81° 40′ W., the barometer, which had fallen to 29in.37 with the wind from N., rose to 30in.47 with winds from SW.; the wind then abated and shifted to SSE. and SE., from which points it blew with great strength the following day. It gradually diminished until the morning of the 24th, when it died away altogether, after having brought us 90 miles from point Coronilla. Lighted fires and steamed. The horizon was very foggy, and, contrary to my expectation, we did not, sight land before dark. At 10 p. m. we sighted Valparaiso, where we dropped anchor on the 25th February, at 9 a. m. In this passage we were fortunate enough to lose nothing in latitude after leaving cape Pillar, and were able to make our westing so as to cross 50° S. at 78° W., after which we bore up toward our destination."

We will conclude with the account of the cruise of the iron-clad *Belliqueuse*, flag-ship of Admiral Penhoat, (*Ann. Hydr.*, vol. 30:)

"After doubling cape Pillar, (a sharp-pointed sugar-loaf, united to the land by a low isthmus,) on the 6th of March, we stood on our course to the westward. My intention had been to pursue this route to 80° W., but about 2 a. m. the breeze which had set in from W. at midnight, freshened considerably, and, the sea increasing, we steamed only about 3 knots. The wind then shifting to WSW., the fires were banked, and we hauled up on the port tack, on a NW. by N. course.

"On the 7th of March we were at 50° S. and 78° 03′ W. The breeze continued to freshen, the gusts became heavier and more frequent, soon blowing a gale. We lay to under two close-reefed topsails, two reefs in the fore trysail, and the storm stay-sail.

"On the evening of the 8th the wind moderated, and finally died away, though the sea still ran very high. The wind soon came out strong from NNW.; we then went about on the starboard or west tack, with two reefs in the topsails. About 3 a. m. it blew a violent squall from NW. to SW.; but shortly after the wind died away, and the weather improved.

During these gusts we had been carried considerably to the eastward, but had gained in latitude. In the evening we were 60 miles west of the middle of Campana island, with a light breeze from SW., but the sea still so rough that the vessel could make no headway under sail. The fires were lighted, and we continued on our course to the west. On the 11th of March a good breeze set in from SW., and the sea became considerably smoother. Hauled fires and made sail. The same day we crossed 45° S. at 78° 40′ W.; from this point we had favorable breezes from SW., shifting to S. and SE. On the evening of the

15th we were 45 miles from Valparaiso, where we anchored the 17th of March."

§ 45. ROUTE FROM CAPE HORN OR THE STRAIT OF MAGELLAN TO THE "INTERMEDIATE PORTS" OF COQUIMBO, MEXILLONES, ISLAY, IQUIQUE, AND ARICA.—The route to be followed to reach these ports is nearly the same as that prescribed for making Valparaiso, (vide § 44.) The vessel should be steered so as to make land about 28° or 30° S., (unless the destination be Coquimbo;) thence follow the coast with a steady wind from S. and SE., (vide § 22,) keeping between 5 and 15 miles from shore. The position can always be obtained by sighting the land, and a good landfall can be made, which the winds and currents might prevent if the off shore were followed.

If it is not deemed advisable to approach the land so far to the southward of the destination, steer for the island of Juan Fernandez. Here the chronometers can be corrected, and the main land afterward approached with greater assurance.

§ 46. ROUTE FROM CAPE HORN OR THE STRAIT OF MAGELLAN TO CALLAO.—Ships bound to Callao, from cape Horn or the strait of Magellan, will first steer as if they were going to Valparaiso. Their route will only be different after leaving about 40° S.; from this parallel they will, if possible, steer a little more to the northward. We will, however, give a few observations, relative to this passage, for each season of the year; these will complete those given in § 44, to which the reader should first refer. In January, after having crossed 50° S. at 79° 40′ W., or even a little to the westward of that meridian, if possible, ships coming from cape Horn should shape their course so as to cut 45° S. at 79° 40′. Between 45° and 40° S. it is preferable to keep to the eastward of 80° W., where the winds would appear to be steadier and more favorable than farther to the westward. Cross 40° S. at 77° 40′ W., and make the northing from 40° to 30° S., between 77° and 78° W., as the winds near those meridians are more favorable. A light SSW. to SSE. wind will probably be found near 30° S. and 77° W., which will shift to SE. beyond 25° S.; thence the passage to Callao will be easily accomplished. In January.

In February, after having crossed 40° S. near 79° 40′ W., there is no better rule than to sail immediately to the northward. In case winds blow from N. or NW., which rarely In February.

happens, keep the sails full. The wind is sure to shift soon to its usual quarter, WSW. and SW. Between 40° and 35° S. the wind is more favorable and steadier east of 80° W. than to the westward of this meridian; the reverse is the case between 35° and 30° S. Beyond 30° S. the wind will always be favorable. The 30th parallel should be crossed between 78° and 80° W.

In March. In March, beyond 40° S. and 78° 10′ or 78° 40′ W., the winds will be generally favorable, though as far as 30° S. they frequently blow from NW. Make as much northing as possible without hugging the wind too closely, and keep between 76° and 80° W.

When there are leading winds it is well to keep as close as possible to the meridians 79° and 80°, so as to prolong the port tack in case of a series of winds from NW. or NNW.

It is also important not to cross 30° S. to the westward of 80° W.; this parallel should be crossed between 77° and 80° W. The winds will be constantly favorable from this point to Callao.

In April. In April, after having reached 35° S., near 76° W., (vide § 44,) steer to the northward. The winds will be generally favorable, though frequently from NNE. to NNW. It is advantageous to cross 30° S., between 73° 10′ and 74° 40′ W. From this point steer so as to cross 25° S. at 74° W.; the winds are here nearly always steady between SSE. and SSW.; thence the winds are favorable. Though this route is more easterly than those for the preceding months, still it is advisable to follow it, as NW. winds are common between the parallels 30° and 25°, and west of 75° W. This unfavorable direction of the wind is even more noticeable farther to the westward.

In May. In May, if it is possible, cross 40° S. between 78° 10′ and 79° 40′ W. From 40° to 30° S., the winds usually vary from NNE. to NW. and WNW., occasionally interrupted by winds from SW.

Unfavorable northerly winds are still more frequent, and the breeze generally less steady to westward of 80° W. It is therefore well to keep between 78° and 80° W. in making to the northward, and to choose that tack which will help most to run down the latitude. After having crossed 30° S. near 79° W., the dominant winds are from SE.; they will become steadier as the vessel makes to the northward, though there is always a chance of finding a breeze from NNE. to NW. as far as 20° S. This month corresponds to the beginning of the winter season, (vide § 21.)

In June, 40° S. should be crossed to westward of 80° W., if possible at 81° or 82° W. The parallel of 35° should be crossed if it can be done between 80° and 83° W.; here the winds blow nearly as often from N. and NNW. as from SW. To the eastward of 80° W., the chances are not so good. Beyond 35° S. the winds are generally favorable to 30° S., which you should try to cross between 78° 10′ and 80° 10′ W. Afterward continue to steer to the northward, keeping to eastward of 80° W.; the winds will be throughout favorable. In June.

In July, cross 40° S. and 35° S. between 80° and 82° W.; 30° S. between 78° and 80° W., and 25° S. between 77° and 80° W., taking care, however, not to go farther to the westward. These crossings, or rather this large belt, would seem to be the most favorable for vessels making passage to the northward under sail. However, N. and NW. winds are always to be expected at this season, and often keep vessels jammed on the wind until they have passed 25° S. In July.

August is also an unfavorable month; the following crossings are, however, the best: 40° S. between 80° and 82° W.; 35°, 30°, and 25° S., crossed successively between 77° and 80° W. Between 40° and 35° S. there is apt to be a series of winds from NNE. to NW., while southerly winds prevail between 35° and 25° S. Beyond 25° S. the winds are generally favorable. In August.

In September, make as much northing as possible between 40° and 30° S., keeping, as much as circumstances will allow, between 77° and 79° W. The winds often blow from SW., but there will sometimes be a series of NW. winds, though they are less common than during the preceding months. Beyond 30° S. the wind is favorable, and from S. to SE. In September.

In October, cross 40° S. near 79° 40′ W. if possible, (vide § 44.) After passing this parallel, make to the northward, keeping between 79° and 82° W. In this part of the route the winds are almost always favorable, varying generally from WNW. to SW., S., and SE. Cross 30° S. at about 80° W.; thence the winds predominate from S. to SE. In October.

In November. In November, it will be found better to cross 40° S. east of 80° W., as the winds are here a trifle better than to the west of this meridian; besides, farther to northward the NW. winds become rarer, and are inclined to shift to SW., S., and SE. After crossing 30° S., between 77° and 80° W., southerly and southeasterly winds predominate.

In December. In December, the same route is advisable, with this difference, that it is still more important to cross 40° S. east of 80° W., between 78° and 80° W., for example. You will generally find favorable winds from some point between WSW. and S. NW. winds are rare. The chances are even more favorable north of the parallel 35°; here the wind frequently blows from SSW. to S. and SSE. Beyond 30° S., which should be crossed between 77° and 78° W., steady S. and SE. winds prevail.

§ 47. ROUTE FROM CAPE HORN OR FROM THE STRAIT OF MAGELLAN TO PAYTA AND GUAYAQUIL.—The instructions given in paragraphs 44 and 46 should be followed. SE. trades are met at 30° S.; afterward steer directly for the port, with favorable winds; or sight "Lobos de Afuera" and point Aguja, afterward passing close to cape Blanco if the destination be Guayaquil. Additional instructions will be found in § 56.

§ 48. ROUTE FROM CAPE HORN OR THE STRAIT OF MAGELLAN TO PANAMA.—The observations given in paragraphs 44 and 46 are applicable as far as 30° S. Here steady trades will be encountered from S. and SE. The route then passes about 30 or 40 miles from point Aguja and cape Blanco. Thence head to the northward, with leading winds, so as to make cape San Francisco if need be. In this neighborhood the land is high, bluff, and covered with large trees; the sea-breeze blowing from S., the land-breeze from SSE. To the southward of the line, say 5° S. of the equator, the winds are nearly always favorable; still, be on the watch for northwesterly winds, as they sometimes blow from February to June. During these months it is advisable to pass about 100 or 150 miles from cape Blanco, so as to be able to make a good stretch on the port tack; rap full if you happen to have a NW. wind. When the wind blows from NNE. or NE., which, however, seldom happens except in May, keep well full on the starboard tack. North of the line the winds are generally favorable, except in January, February, and March, when they frequently blow from the northward, (vide § 23.)

For full directions on the latter part of this voyage, vide § 54.

§ 49. ROUTE FROM CAPE HORN OR THE STRAIT OF MAGELLAN TO ACAPULCO, SAN BLAS, AND MAZATLAN.—Captain Sherard Osborn, R. N., makes the following notes on this passage: "Supposing a vessel, bound for the western coast of Mexico, safely round cape Horn, and running before the southerly gale which almost constantly blows along the shore of South America, she ought to shape a course so as to cross the equator in about 98° or 99° W. long., so that when she gets the NE. trade she will be at least six or seven degrees to the eastward of her port, San Blas or Mazatlan, and have at the same time a sufficient offing from the Galapagos islands to avoid their currents and variable winds.

"We crossed in 105° W. long., having been recommended to do so by some old merchants at Valparaiso, and were consequently, although a remarkably fast-sailing ship, a lamentably long time making the distance. Our track led us to be exactly in the same longitude as our port when we got the trade, and it hanging well to the northward, we were constantly increasing our distance until in the latitude of San Blas, when an in-shore tack of course shortened it. But by the course I have recommended the *first* of the NE. trade will drive the vessel into the meridian of her port, and she will thus daily decrease her distance.

"Care must be taken, in standing in for land, not to get to leeward of San Blas, as there is a strong southerly current along the coast, especially off cape Corrientes."

These instructions are meant for vessels bound to San Blas or Mazatlan. Those bound to Acapulco, Istapa or Realejo should cross the line more to the eastward, where, however, calms are common, particularly near the Galapagos islands in January, February, and March. It will therefore be advantageous to cross the equator between 86° and 88° W., if running for one of the eastern ports of the Mexican coast, or about 98° W. if bound to one of the western. Steamers can, of course, run through the "doldrums" whenever most convenient.

In paragraph 55 will be found an account of an easterly passage between Callao and San Jose de Guatemala, (*Frigate Havana.*) We will also give in § 60 the remarks of Captain Wood on the passage between the Galapagos islands and cape St. Lucas. Information on the winds and currents of this passage may also be found in paragraphs 1, 24, and 39.

We would also state that instructions have been given on the route between cape Horn and the equator in § 44 and § 46.

In January, February, and March.

From a careful study of the wind-charts the following facts are deduced concerning the voyage from the line to the port of destination. In January, February, and March, ships bound to Acapulco, San Blas, and Mazatlan should cross the line at about 100° W. They should steer, the winds being strong and steady from SE. and S., so as to reach 10° N. near the meridian of their port. If they are going to Acapulco, they should cross 10° N., between 101° and 103° W. Thence, till they make the land, they will have from 6 to 7 per cent. of calms, and NW., N., and NE. winds. Ships going to San Blas and Mazatlan should cross the parallel of 10° between 105° and 107° W. Beyond this point they are liable to about 5 per cent. of calms and variable winds from NE. to NW.

During this season, vessels bound to Istapa and Realejo should cross the equator between 85° and 87° W. They will then have about 5 per cent. of calms, and variable SE. and S. winds. The parallel of 10° N. should be crossed to the eastward of the meridian of Istapa or Realejo, as the case may be; as NE. and NW. winds are common north of 10° N., ships keeping to the westward of 78° or 80° W. are very likely to be becalmed after passing 10° N.

In April, May, and June.

In April, May, and June, vessels bound to Acapulco should cross the line in the neighborhood of 97° or 98° W. Thence they will be able to make quick time to the northward, with the wind quite steady from SE. and S.

They should reach 10° N. at about 97° W., and will be exposed to more and more calms as they sail to the north; especially if they run to the westward of the meridians 98° or 100° W. The southerly winds die away on approaching Acapulco, and come out from E., NE., and N. This is a very long and tedious passage for sailing-vessels.

Vessels bound to San Blas and Mazatlan should cross the equator at about 98° W. They should then steer so as to cross 10° N. between 100° and 102° W.; the SE. breeze dying away toward the parallel of 10°. North of 10° it is usually calm, though light airs are found, especially from NE. to N., as far as 20° N.; beyond this parallel steady W. and NW. winds generally prevail. Ships are sometimes becalmed for a long time, between 10° and 20° N.

Ships going to Realejo and Istapa should cross the equator between 85° and 87° W. From this point they can easily steer for their destination with southerly winds, varying at times from SW. to SE.

In July, August, and September the route for vessels bound to Acapulco is shorter than that of the preceding months. They should cross the line at about 96° or 97° W., and steer with the variable southerly and southeasterly breeze, so as to reach 10° N. at 98° W. Thence they will be exposed to more and more calms as they approach and pass 100° W. In the neighborhood of 15° N., variable NW. and SE. winds may be expected. In July, August, and September.

Ships bound to San Blas and Mazatlan should cross the line in the neighborhood of 100° W. The SE. trades are here prevalent, and will be kept as far as 10° N. This parallel should be crossed at about 107° or 108 W. Beyond 10° N. there are about 6 per cent. of calms, and the prevailing winds are westerly, varying from NW. to SW. If possible cross 15° N. at about 108° or 109° W. Beyond 15° N., NW. winds are common as far as Mazatlan or San Blas. The chance of meeting calms will be at first 6 per cent.; it will afterward increase, and reach as high as 11 per cent. beyond 20° N. If the above crossings be followed the last part of the passage will be rendered easier, for the course can be laid at NNE. on the port tack.

Ships bound to Istapa and Realejo should cross the equator at about 85° W. They will find very favorable southerly winds, varying from SW. to SE., until they reach their anchorage.

In October, November, and December vessels going to Acapulco should cross the line near 96° or 97° W., and then, with favorable winds from S. to SE., head for 10° N. at about the same longitude. In October, November, and December.

Beyond 10° N. both easterly and westerly winds are frequent, with 5 per cent. of chances of calms. On approaching 15° N., which should be crossed between 98° and 99° W., the chance of meeting calms and NW. winds becomes much greater. The latter part of this route can be greatly shortened by vessels with steam-power. We do not think that even sailing-vessels need follow a more westerly route; for west of 100° W., and north of 10°., there are 9 per cent. of chances of calms, and NW. and NE. winds.

Vessels going to San Blas and Mazatlan are obliged to make their northing in the unfavorable belt above mentioned. They should cross 10° N. near 106° or 107° W., where they will have steady winds from SE. They should then try to get as far as possible to the northward, taking advantage of the NW., N., and NE. breezes, and going about whenever they can make to the northward.

Vessels bound to Realejo or Istapa will make good headway with the southerly winds, from the line—between 86° and 88° W.—to 10° N., crossed a little to the westward of the port of destination. The wind becomes variable after passing this parallel; in fact westerly winds are the commonest, the usual directions being NNE., NNW., NW., and sometimes WSW.

§ 50. ROUTE FROM CAPE HORN OR THE STRAIT OF MAGELLAN TO SAN FRANCISCO.—Captain Osborn says: "I would cross the equator in about 100° W. longitude, cross the NE. trade with a topmast studding-sail set, and thus pass into the limit of the westerly winds about 300 miles to windward of the Sandwich islands, and once in them take good care to keep to the northward of any port, for as you approach the shore the wind will draw round north, and the current to the southward increase."

This advice is good; with the exception of the point at which the line should be crossed.

We give below a summary of Maury's instructions, to which we shall add a few remarks on the best points of crossing, taken from information furnished by the late Superintendent of the Naval Observatory at Washington:

"The California-bound vessels should aim to enter the SE. trade-wind region of the Pacific as far to the west, provided they keep on the eastern side, say, of 118° W., as they well can; they should not fight with head winds to make westing, nor should they turn much from the direct course when the winds are fair. But when winds are dead ahead, stand off to the westward, especially if you be south of the trade-wind region. Having crossed the parallel of 35° S. and taken the trades, the navigator, with the wind quartering and all sails drawing, should now make the best of his way to the equator, aiming to cross it between 105° and 120°, according to the season of the year and the directions and tables hereinafter given.

"In urging upon California-bound vessels the importance of making westing about the parallel of 50° S., I do not mean that they should expose themselves to heavy weather, or contend against adverse circumstances, in order to get west on this part of the route. I simply mean that, if a vessel, after doubling the cape, can steer a WNW. course, as well as a NW., or a NW. as well as a NNW., or a NNW. as well as a N. course, that she should on all such occasions give preference to the course that has most westing in it, provided she does not cross 50° S. to the westward of 100° or thereabouts, nor 30° S. to the westward of 115°, nor enter the SE. trade-wind region to the west of the last-named meridian. This is the western route, and is to be preferred by all vessels at all seasons.

"Between the equator and 10° or 12° N., according to the season of the year, the California-bound navigator may expect to lose the SE. and to get the NE. trade winds. He will find these last nearest the equator in January, February, and March; but in July, August, and September he will sometimes find himself to the north of the parallel of 15° N. before he gets fairly in the NE. trades. And sometimes, especially in summer and fall, he will not get them at all, unless he keeps well out to the west. Having them, he should steer a good rap-full, at least, aiming, of course, to cross the parallel of 20° N. in about 125° W., or rather not to the east of that, particularly from June to November. His course, after crossing 20° N., is necessarily to the northward and westward until he loses the NE. trades. He should aim to reach the latitude of his port without going to the west of 130° W., if he can help it, or without approaching nearer than 250 or 300 miles to the land until he passes out of the belt of the NE. trades, and gets into the variables, the prevailing direction of which is westerly.

"The Farallones, seven small islands, about 30 miles from San Francisco, are in the fair-way to the harbor. They afford a fine landmark, and should be made by all inward-bound vessels. The course from the south Farallone to the mouth of the harbor is about N. 73° E.; *true* distance 27 miles. The fort on the south point of Alcatraz island is said to be the best course in. Vessels, upon approaching the heads of San Francisco, especially in the winter months, are liable to be beset by fogs, and delayed for many days.

"Between the northwest coast and the meridian of 130° W., from 30° to 40° N., the prevailing direction of the wind in summer and fall is from the northward and westward; whereas, to the west of 130°, and between the same parallels, the NE. trades are the prevailing winds for these two seasons. There is a marked difference in the directions of the winds on the opposite sides of the meridian of 130° W. in the North Pacific.

"Vessels bound to San Francisco should not, unless forced by adverse winds, go any farther beyond the meridian of 130° W. than they can help.

"Supposing that vessels generally will be able to reach 30° N. without crossing the meridian of 130 W., the distance per great circle from cape Horn to its point of intersection with that parallel is about 6,000 miles.

"And supposing, moreover, that California-bound vessels will generally, after doubling cape Horn, be able to cross the parallel of 50° S. between the meridians of 80° and 100° W., their shortest distance in *miles* thence to 30° N., at its intersection with the meridian of 130° W., would be to cross 40° S. in about 100° W.; 30° S. in about 104°; 20° S. in about 109°; the equator in 117° W.; and 30° N. about 130° W., (126° if you can.) By crossing the line 10° farther to the east or 10° farther to the west of 117°, the great circle distance from cape Horn to the intersection of 30° N. with 130° W. will be increased only 150 miles.

"Navigators appear to think that the turning-point on a California voyage is the place of crossing the equator in the Pacific. But the crossing which may give the shortest run thence to California may not be the crossing which it is most easy to make from the United States or Europe; and it is my wish to give in these Sailing Directions the routes which, on the average, will afford the shortest passages to vessels that have doubled cape Horn and are bound direct to California."

Such is the principal advice given by Maury on the passage between cape Horn and San Francisco. We will add the following remarks.

After leaving cape Horn it is well to make as much westing as possible. Vessels are quite certain to be delayed, not only by the wind, which often blows from W., but by the strength of this wind, and the heavy swell and rough sea it causes. It should, therefore, be understood that the advice to make westing merely means to neglect no opportunity of getting to the westward whenever it can be done without making anything to the southward. In § 44 and § 46 detailed instructions will be found on the part of the passage between the cape and 30° or 25° S.; that is on the route to be followed to reach the SE. trades.

After entering this zone it is perfectly easy to steer for the equatorial crossing. But it is an open question at which point it is best to cross. We shall, therefore, pursue this question still further with the view of determining its best solution.

We will first give the following table, (extracted from the Sailing Directions,) which gives for each month the average of the shortest passages which were made prior to 1854:

Mean of the best passages—prior to 1854—from 50° S. to San Francisco.

Months.	No. of passages from which the mean has been taken.	Longitudes west at which the parallels have been crossed.						Days of passage.		
		50° S.	40° S.	35° S.	30° S.	25° S.	0°.	From 50° S. to the line.	From the line to San Francisco.	From 50° S. to San Francisco.
		°	°	°	°	°	°			
January	9	80	83	87	90	92	111	22½	21	43½
February	8	82	85	88	90	93	111	25	20	45
March	13	82	85	88	89	93	110	25	24	49
April	9	83	87	86	89	92	109	24½	30	54½
May	12	82	85	87	87	90	109	24¾	30½	55
June	11	82	84	86	89	91	110	27	28	55
July	6	82	88	90	92	95	115	23	28	51
August	8	84	86	85	87	90	108	25	31	56
September	4	82	86	87	87	90	111	21	24	45
October	12	80	82	84	86	89	110	24	23	47
November	11	83	85	84	84	88	108	24	23	47
December	10	83	83	84	87	91	113	22	21	43

After an examination of this table, we determined to see if there would not be some advantage, at certain seasons of the year, in crossing the equator on a more westerly meridian. With this intention we constructed the following table:

For each month, we have placed on the *first line* the mean point of crossing, corresponding to the most rapid passages which have been made *between the equator and California*. These points of crossing, as well as the number of days at sea from the line to San Francisco, have been taken from the tables in Maury's work. The number of days of passage from 50° S. to the line is placed in the first line, and represents the means of the monthly tables of routes.

The second line for each month contains analogous information, taken from the preceding table, and which corresponds to the best passages made before 1854.

Thus the *first line* is reserved for the new or westerly routes; the *second line* for the best routes prior to 1854.

In comparing these two lines, it is important to note that the total number of days of passage for the old route (*second lines for each month*) corresponds to the observed minimum.

But it is not so for the first lines of each month (*western routes*,) for which the number of days of passage from the line to San Francisco only represents the minimum. We have not been able to determine the minimum number of days from 50° S. to the line, and have confined ourselves to putting the monthly mean in its place.

Comparison between the eastern and western routes, from cape Horn to San Francisco.

Months during which the line was crossed.	Number of passages of which the mean was taken.	Meridians between which the line was crossed.	Number of days of passage— From 50° S. to the line.	From the line to San Francisco.	From 50° S. to San Francisco.
		Longitude W.			
January	4	Between 115° and 125°	23	From 18 to 19.	41½
	9	At 111°	22½	21	43½
February	1	Between 115° and 120°	22	18	40
	8	At 111°	25	20	45
March	13	Between 110° and 115°	31	25	56
	13	At 110°	25	24	49
April	7	Between 115° and 120°	29	23	52
	9	At 109°	24½	30	54½
May	13	Between 110° and 120°	30⅔	28	58⅔
	12	At 109°	24¾	30¼	55
June	21	Between 110° and 120°	30	32	62
	11	At 110°	27	28	55
July	7	Between 115° and 125°	31	From 29 to 30.	From 60 to 61.
	6	At 115°	23	28	51
August	3	Between 105° and 110°	33	25	58
	8	At 108°	25	31	56
September	5	Between 115° and 125°	25	26	51
	4	At 111°	21	24	45
October	16	Between 110° and 120°	25¾	26	51¾
	12	At 110°	24	23	47
November	14	Between 110° and 120°	26	24	50
	11	At 108°	24	23	47
December	3	Between 115° and 120°	25	18	43
	10	At 113°	22	21	43

Notwithstanding the unfavorable manner in which this table was necessarily constructed, it will still be seen that the total length of the passages, on the first line for each month, is less than the total length of those on the second line in January, February, and April.

This result shows that there is some advantage in following a western route, at least during the above-mentioned months.

Findlay also states that the mean passage from 50° S. to San Francisco, is 53½ days for ships that cross the equator between 115° and 120° W.; while it is 53.8 days for those which cross the line between 110° and 115°. This fact again shows that there is an advantage to be gained (however slight) by taking a westerly route.

In conclusion we would state that the line should never be crossed, by sailing-vessels, to the eastward of 110° W.

In January, February, and March.

In January, February, and March cross 10° N. between 120° and 123° W. To the southward of this point the trades blow; but soon after leaving 10° N. the wind will come out steady from NE.

The change will generally take place without an intervening calm. Vessels will be able to steer a little free through the NE. trades; as 20° N. will not have to be crossed to the eastward of 128° W., nor 30° N. to the eastward of 133° W. Beyond 30° N. the wind will become variable, first hauling to ENE. Make as much to the northward as possible; but do not attempt to make any easting until the region of westerly winds, or what is generally the same, the latitude of San Francisco, is reached.

In April, May, and June.

In April, May, and June, cross the line between 118° and 123° W.; here the wind prevails from SE., and the course can be laid to reach 10° N., between 123° and 125° W. Thence keep the sails well full, with the NE. trades, which will often haul to N. and stand on the same tack to 30° N., between 133° and 138° W.

The last part of the passage will be the same as that for the preceding months; and the winds will not generally become favorable, or westerly, south of 37° or 39° N.

In July, August, and September.

In July, August, and September, the best point to cross the line is about 125° W.; and the parallel 10° N. at 130° W. The SE. trades will be carried across the line and as far as this latter parallel; but from 10° to 20° N., about 7 per cent. of calms and prevailing NE. breezes may be looked for; SW. winds have also been found in this locality. Cross 20° N., between 133° and 136° W.; beyond this parallel the steady NE. trades will be found, and the attempt should be made to cross 30° N. near 140° W. After this the wind may be very variable in direction, though probably from the northward and eastward as far north as 34° or 36° N., where the first puffs of the westerly wind will be felt. To make a quick passage run well up into the zone of west winds, and make the easting to the northward of the parallel of San Francisco.

In October, November, and December the equator should be crossed near 113° W.; and 10° N. between 118° and 120° W. The SE. trades will be carried to nearly 10° N., vessels usually running from one set of trade-winds into the other without any intervening calms; then keep a point or two free, and cross 20° N. between 127° and 129° W. It will, however, be necessary to lay a little closer when between 20° and 30° N., so as to reach the latter parallel at 133° or 134° W. As the wind becomes variable north of 30°, advantage should be taken of every favorable slant. The parallel of 38°, or perhaps 40°, once reached, bear to the eastward with the west wind, taking care, however, to cross the 130th meridian north of the parallel of San Francisco; as northerly winds in that locality are common. In October, November, and December.

§ 51. ROUTE FROM VALPARAISO TO THE INTERMEDIATE PORTS AND CALLAO.*—*Remarks by M. Lartigne*, (vide S. Pacific Directory, page 911.) "The navigation of the Peruvian coast is very easy in summer; the land-breezes are moderate; the weather, which is generally clear, allows the latitude to be observed nearly every day, and to recognize, by this means, the part of the coast opposite to which you may be. There is then no inconvenience in keeping a moderate distance off, so as to meet with fresh breezes, and thus shorten the passage.

"The weather, which is often cloudy in winter, will not allow of observations to be taken every day, and you must then direct your course by your dead reckoning, or from the more remarkable objects lying on the coast. Those which are met with between the Quebrada Camarones and the valley of Tambo may be made out at a considerable distance, so that when between these two remarkable points you may proceed by keeping 20 or 25 miles off the land. At this distance the sea-breezes keep up through the greater part of the night.

"The only objects at all remarkable that are to be met with between the valley of Tambo and that of Quilca, are the points of Islay and Cornajo, but these cannot be made out at more than 10 or 12 miles off, for when farther off

* Intermediate ports ("Intermedias") is the name given to the harbors along the west coast of S. America between Valparaiso and Callao. —*Translator.*

they appear confounded with the high land of Peru. It seems that in this season, you must continue to fix your position by the sight of the land, and so follow the coast at less than 10 or 12 miles distant; but as you then only find light airs, interrupted by calms, which may last for several days together, you run the risk of being carried too near the land by the heavy swell which is felt on all its extent. The depth off it is considerable, and the quality of the bottom very bad; it is only at the opening of the valleys that you can hope to find, at 2 or 3 miles from the shore, less than 30 fathoms water, over a bottom of mud or fine sand. The only advantage that will be gained by sailing so near the land will be to profit by the slightest breeze to get an anchorage, and to be seldom exposed to the chance of overrunning it; but these advantages, as will be seen, are not of a nature to compensate for the inconveniences, or rather the dangers, to which a ship is exposed.

"It would, therefore, be better to sail farther off the land, keeping at 20 or 25 miles' distance; as when between the Quebrada Camarones and the valley of Tambo, the swell is not felt at this distance, and the winds will be fresh; but the currents, which constantly run to NW., cause the reckoning to be very erroneous, and you may be carried to leeward of your port or the anchorage you may be seeking. Beyond this, this inconvenience is without danger, and cannot occasion more than a hindrance; for in returning to the required destination, sailing to the southward, you must run to the offing, bear up to the wind, and then, approaching the land, reach the port which has been overrun. It is, notwithstanding, necessary, following the general rule, to make an exception, which in some circumstances may shorten the passage. We have said that the breeze was sometimes tolerably fresh, and that then the counter-current, which runs to the south along the land, extends some miles in the offing; it is evident that it would be better to work in this counter-current, at all times when the force of the wind allows it, and you have not overrun your port more than 5 or 10 miles; but if you should have done so to a greater distance, it will be preferable to take directly the first course, and profit by this breeze to get away from the land.

"It will be advantageous to maneuver thus, every time you are on any portion of the coast which is described.

"What has just been said, relative to the mode of navigating and running along the coast of which we have just been speaking, applied to the portion comprised between the valleys of Quilca and Ocoña. But it is necessary to observe that the valley of Camana, which is as easy to be made out at 20 or 25 miles' distance as that of Quilca, has the inconvenience, as well as the latter valley, of not being perfectly recognizable until it bears to the NE., when it is passed, and you cannot reach the anchorage on that tack.

"In winter, as in summer, you must always be particular to approach the land to the south of the intended port, and then range the land at a short distance. The breezes being fresher in summer and the sea smoother, the land is made more easily than in winter."

Captain Fitz-Roy says: "When going to the northward, along the coast of Chile, steer direct to the place, or as nearly so as is consistent with making use of the steady winds which prevail in the offing. Little difficulty is found in going to the northward along the coast of Peru; a fair offing is all that is required to insure any vessel making a certain port in a given number of days."

Captain Chardonneau gives the following instructions for the coast of Peru:

"As fogs are frequent, vessels should not get within 4 miles of the coast, yet in order to reach their destination easily they should not, on the other hand, run out to sea for more than 15 miles; that is, the prominent headlands should be kept within sight, as no confidence can be placed in the dead reckoning on account of the strong and variable currents."

2d. From Valparaiso to Callao.

This is an easy passage, and we will confine ourselves to a few observations on the voyage, taken from Capt. Basil Hall's log while on this coast:

"*From Valparaiso to Callao, 27th January to 5th February, 1821.*—The wind on this passage is always nearly the same, viz, SSE. It sometimes hauls a point or two to the eastward, but the passage is always practicable. The only precaution to be attended to is to run well off the land in the first instance, say 150 miles, on a NW. course, and then steer direct for San Lorenzo, a high and well-defined island forming the eastern side of Callao bay. It is usual to make the land of Morne Solar, which lies 10 miles to the southward of Callao, and then run into the roads by the Boqueron passage, or proceed around the north end of San Lorenzo. In entering by the Boqueron great attention must be paid to the lead and the bearings, and an anchor kept ready to let go.

"It is generally calm in the mornings, and sometimes foggy; but about 11 o'clock it clears up and a breeze springs up from the southward, which enables ships to reach the anchorage generally without a tack, after rounding the north end of Lorenzo, so that upon the whole this outer route, which is entirely free from danger, is preferable to the other, at least for a stranger."

Valparaiso to Callao, touching at the intermediate ports.

"*Passage made between the 27th May and the 24th June,* 1821.—From Valparaiso we steered at the distance of about 60 miles from the coast as far as lat. 22° 30′ S., when we hauled in; afterward coasted along in sight of the shore, at the distance of 20 or 25 miles, as far as Arica. The winds being light from SSE., it was not till the 7th June that we anchored there. Thence we coasted along by Quilca, Sama point, and Ilo to Mollendo, the winds being generally from the eastward, and drawing off shore at night; calm in the mornings, and hauling in from the sea in the day; the weather invariably fine. From Mollendo to Callao we had a fresh breeze off shore about SE. On approaching Solar point the wind fell light, and we were obliged to tow the ship through the Boqueron passage into Callao roads."

Valparaiso to Callao, touching at Coquimbo, Huasco, Copiapo, Arica, and Mollendo.

"*Passage made between the 15th November and the 9th December,* 1821.—The winds during these passages along shore are always light and from the southward, hauling in from sea during the day and freshening from off the land in the night.

"Between Mollendo and Callao there is a pretty steady breeze from ESE., with a drain of current along shore—a remark which applies to the whole coast from Valparaiso to Lima.

"A remarkable increase of the great SW. swell is observable at the full and change of the moon, on the coast, especially from Arica to Huacho inclusive, a circumstance which renders it difficult, and sometimes impossible, to land at those places."

§ 52. ROUTE FROM VALPARAISO TO SAN FRANCISCO.—After leaving Valparaiso, a ship will generally find no difficulty in steering NW. and reaching the region of the SE. winds. After striking the trades, the course should be shaped so as to cross the equator at the point mentioned in § 50.

§ 53. ROUTE FROM CALLAO TO PAYTA AND GUAYAQUIL.—As the wind in this locality always prevails from the southward and eastward, the passage from Callao to Payta and Guayaquil may be made without any difficulty.

Capt. Basil Hall gives the following example:

"Voyage from Callao to Pacasmayas, Payta, and Guayaquil, 17th to 25th December, 1821.—The winds between Callao and Guayaquil are moderate from the southward; at night hauling to the southeastward, and in the day from SSW.

"This is the period at which the rains are expected to set in, and the heavy, threatening aspect of the clouds over the hills gave us reason to expect that we should not escape; but none fell during our stay."

Vessels coming from Callao should steer to double point Aguja, which is long and level, terminated by a bluff about 160 feet high.

Vessels bound to Payta will find, "after leaving Foca point, a line of cliffs, about 130 feet high extending as far as Payta point, which is 9 miles distant, N. by E. Between these points and 1½ miles from the coast is a group of hills called the Silla, or Saddle of Payta, thus described by Capt. Basil Hall: "The Silla is sufficiently remarkable; it is high and peaked, forming three clusters of peaks, joined together at the base, the middle being the highest; the two northern ones are of a dark-brown color, the southern is the lowest and of a lighter brown. These peaks rise out of a level plain, and are an excellent guide to vessels bound for the port of Payta from the southward." Bound to Payta.

There is no danger in entering the harbor of Payta; after rounding the outer point with a signal-station on its ridge, False bay will be opened; this must be passed, as the true bay is round Inner point. That point ought not to be hugged too closely, for there are some rocks at the distance of a cable's length, and the wind baffles often. After rounding Inner point a vessel may anchor where convenient, in quiet, still water, in about 6.5 fathoms, over a muddy bottom. The holding-ground is excellent, and notwithstanding the fresh breezes, there is nothing to apprehend. These winds are constant every day, setting off the land from 10 a. m. to sunset, but they raise no swell, as they blow over the high land. In entering the roadstead it is better to shorten sail before rounding the Signal point, as heavy gusts sometimes occur in Colorado bay as well as off that point. It is seldom that the anchorage is reached in one tack, but there is plenty of room for working.

Vessels bound to Guayaquil.

Vessels bound to Guayaquil should make the land at point Picos, which is easily known by its sand dunes.

A few miles farther north is the low, wooded point of Ma pelo.

Ten miles to the northward and westward of this point the depth is about 41 fathoms, sand and mud. Vessels should pass about 5 miles to the southward of Santa Clara island, with a depth of between 15 and 20 fathoms, and then steer N. 59° E. for 25 miles, passing between point Arenas an the S. buoy on Mala bank.

A pilot will generally be found at point Arenas; if not, it is easy to reach the anchorage at Puna. The channel W. of Mala bank is the best; Mala hill being a good landmark, vessels drawing 18 feet can, at high water, clear the bar north of Puna (with a pilot) and ascend the river to Guayaquil, a distance of 80 miles.

§ 54. Route from Callao to Panama.—Paragraph 48 contains the proper instructions for reaching Panama from 30° S. The reader should keep in mind, while running through the SE. trades that it is advisable to make point Aguja and cape Blanco before standing up the coast for cape San Francisco with the favorable wind. The wind is usually fair to northward of the line, except occasionally in January, February, and March, (vide § 23.)

Captain De Rossencoat gives the following excellent instructions on this route:

"After leaving Callao and doubling the Pescadores and Pelado islands, steer for point Aguja, where the wind is generally fresher than at other points on the coast. Keep about 20 or 25 miles from the coast between point Aguja and cape San Francisco, and then head for the Pearl islands, where the land-fall is always uncertain, as the currents at the mouth of the gulf are always more or less affected by the prevailing winds.

"During the fine season, when the NE. wind is well established, that is after the beginning of January, the currents set to W. During the winter, on the contrary, the prevalence of the SW. winds makes an easterly current. In both cases their strength is sufficient to make grave errors in navigating.

"Apart from this uncertainty of the currents the gulf of Panama is easy of access. As soon as the Pearl islands are sighted steer for Otoque islands, which are high, and visible at quite a distance; there are two principal ones and between them a small islet. On reaching this point, Taboga island, where ships generally anchor, will be in sight.

"To reach the anchorage pass between Taboguilla and Urana, being careful to keep sufficiently close to Urana to avoid a ridge of rocks which is only awash at very low water, and is situated half way between point Urana and a small round island, about 1 mile SSE. of Taboguilla. If this passage cannot be made without going about, it will be better to keep to the eastward of Taboguilla; vessels can anchor very near land in this roadstead with 11 or 13 fathoms of water.

"Steamers ply regularly between Taboga and Panama. To reach the anchorage off the town, steer so as to pass to the eastward and quite close to the Farallon of San José. When this bears S. 11° E. by compass, and the steeples of the cathedral N. 60° W., drop anchor in 5 fathoms of water, muddy bottom."

Captain Harvey, of H. M. S. Havana, remarks as follows on this voyage, (*Naut. Mag., Nov.*, 1860:)

"We left Callao on the 14th May, crossed the equator on the 20th, in 82° 37′ W. With the exception of some variable weather on the 26th, we took the wind with us up to the Pearl islands, making Galera at 3h a. m. on the 28th, and anchored off the town of Panama on the following day. A ship bound for Panama should make her way up the bay on the eastern side of it and work up inshore between the Pearl islands and the main, where there is good anchorage, should it fall calm or the current prove too strong against her. During our stay of six weeks we had the usual sultry weather, with rain and thunder storms."

Captain Fitz-Roy says:—"Sailing-vessels bound to Panama should endeavor to get within 3 or 4 miles of Chepillo island, especially between December and June, and so have all the advantage of the prevailing wind. From this position Ancon hill will be seen, and should be kept a little on the port bow, as the wind hauls to the westward on approaching Panama.

"The passage from the southward into the gulf of Panama is easily made during the greater part of the year, by keeping about 60 miles from the coast north of Guayaquil, and, after crossing the line, shaping a course for the Galera islands, at the same time taking care, especially in the dry season, to stand inshore with the first northerly winds. By so doing vessels will most probably have the current in their favor along the coast, whereas by keeping in the center or on the western side of the gulf, a strong southerly set will be experienced.

"After making Galera and clearing the San José bank, the navigation between the Pearl islands and the main is clear and easy, with the advantage of being able to anchor should the wind fail or the tide be against the vessel. As a rule, this passage should be taken; but, with a strong southerly wind, the navigator is tempted to run up the bay, in which case he should still keep on the western shore of the Pearl islands, where anchorage and less current will be found should the wind fail, an event always to be expected in these regions."

Observations of Commander James Wood:—"The passage from the southward to Panama bay is easily made during the greater part of the year; but in the fine season, when within the influence of the northers, the following plan should be adopted: Make short tacks inshore, as there is generally a set to the northward found within a few miles of the land, and where that is interrupted, a regular tide is exchanged for a constantly contrary current farther off. Between Chirambira point and cape Corrientes the land is low and faced with shoals, caused by the mouths of the numerous rivers which have their outlets on this part of the coast; but after passing cape Corrientes, it may be approached pretty closely, except off Francisco Solano point, where some rocky-shoal patches extend to seaward, as the coast is in general bold-to. Care, however, should be taken not to run into the calms caused by the highlands, as it is difficult to get off into the breeze again, and the swells set inshore, where it frequently happens that no anchorage is to be found till close to the rocks.

"In beating up the bay of Panama, in the fine season, the eastern passage, or that between the Islas del Rey and the main, is to be preferred, as, with one exception, it is free from dangers. The water is smooth, and a regular tide enables you to make more northing than it would be possible to do, in nine cases out of ten, against the strong current and short, high sea which at this season prevail in the center or on the western side. During the rainy season a straight course up the bay is preferable to entangling yourself with the islands, the current generally following the direction of the wind."

§ 55. ROUTE FROM CALLAO TO GUATEMALA AND MEXICO.—After leaving Callao and entering the SE. trades, first steer for point Aguja, (vide § 54.) Then if bound to Realejo or Istapa, shape the course so as to cross the equator between 85° and 88° W. But if the destination be Acapulco, San Blas, or Mazatlan, it is best to pass to the southward and westward of the Galapagos islands with the prevalent SE. winds. In this case the line should be crossed between the meridians 96° and 100° W.

The last part of the voyage, from the equator to the port of destination, is the worst. Instructions were given in § 49, on the route from the line to the principal ports of Guatemala and Mexico, to which the reader should refer. The following example of a passage between Callao and the coast of Guatemala is taken from the remarks of Captain Harvey, of H. M. S. Havana, (Naut. Mag., Nov., 1860:)

"Leaving Callao, we stood well out from the land, current in our favor. In 7° 30′ S. and 83° W., we had as much as 36 miles N. 72° W.; and in the same afternoon we were running through patches of brick-colored water. At noon on the 29th, in lat. 1° 6′ N., long. 86° 54′ W., we ran through a strong tide rip, extending NW. and SE. as far as could be seen. It was most distinctly marked, the water to the southward having a greener color. The temperature before entering it was 72°; when 500 yards farther, and inside or north of the line, it was 78°; at half past twelve it was 80°. Up to this we had had a current to the NW. by W. of more than a mile an hour; but now we had less, about half a mile an hour, and in a more northerly direction. The next day the wind shifted to WSW., and we lost the trade in 3° 19′ N. and 87° 34′ W. amid thunder-storms and squalls of rain.

"On the 1st May Cocos island was seen from the mast-head, bearing ENE.; and tropic birds and black fish came about the ship. After dallying with calms and westerly winds from the 1st to the 6th, we were treated with a fine easterly breeze and a current to the W. by N. of 38 miles. We concluded that it was the end of a papagayo, and for several days after we had the usual Central American weather. On the 11th we found ourselves inshore looking about for Istapa. The Guatemala peaks could not be seen, and the shore showed us nothing but one unbroken line of beach and trees, with a heavy surf; but in the evening we had the satisfaction of making out El Agua, or Water volcano, the east center peak of the range. The next morning, on standing in, we observed three vessels at anchor to the westward, and running down to them came to in 13½ fathoms, thinking that we had reached Istapa. The first visitors informed us of our mistake and that we were really at San José de Guatemala."

§ 56. ROUTE FROM CALLAO TO SAN FRANCISCO.—The passage from Callao to the equator can be accomplished with a leading or fair wind. The line should be crossed west of the Galapagos islands, though it is not necessary to keep as far to the westward as advised in § 50.

In January, February, and March. During these months the equator should be reached near 108°, and 10° N. near 114°, or 116° W. Up to this point vessels will have the SE. trades, and only 4 per cent. of calms north of the line. After striking the NE. trades, at about 10° N., steer a trifle free, and cross 20° N. about 127° W., as farther to the eastward, and near 120° W., the wind is sure to haul to the northward. Nothing will therefore be gained by hugging the wind, as eventually more westing will result, and the voyage be lengthened.

The last part of the voyage is the same as that described in § 50.

In April, May, and June. In April, May, and June strike across the prevalent SE. trades, and reach the equator between 108° and 113° W., and 10° N. at about 120° W.; thence continue according to the directions given in § 50.

In July, August, and September, sailing-ships leaving Callao should follow the same route as vessels coming from cape Horn; that is, they should not cross the equator to the eastward of 125° W. By keeping on this route they will have steady SE. trades until they reach 10° N., which should be crossed at 130° W. The remainder of the passage should be made as described in § 50.

In July, August, and September.

In October, November, and December cross the line near 108°; and 10° N. between 118° and 120° W., with steady SE. trades. After this keep a little free with the NE. wind. Ships generally run from one set of trades to the other without any intervening calm. This passage should also be finished according to the instructions given in § 50.

In October, November, and December.

§ 57. Route from Payta or Guayaquil to Panama.—We recall this route only from memory. In § 54 instruction will be found for making the passage from Callao to Panama, which also apply for the voyage from Payta or Guayaquil to the same destination.

§ 58. Route from Payta or Guayaquil to San Francisco.—It will be advantageous at all seasons to keep to the southward and westward of the Galapagos islands. Cross the equator in the same manner as described in § 56. Complete instructions for the route north of the line will be found in both § 56 and § 50.

§ 59. Route from Panama to Mexico.—It is exceedingly difficult to determine which is the best route between Panama and the Mexican ports. Calms (vide § 1) and light baffling airs often make the passage excessively long; and, coasters excepted, ships without steam should, if possible, avoid making this passage.

According to Fitz-Roy's instructions vessels can only go up the coast by keeping near land, after doubling cape Mala, and by making use of the land and sea breezes. But the voyage should not be attempted at all, except by good ships with a large crew, unless the destination is one of the ports of Central America, and then only because it is the only one to take.

Commander James Wood says: "If bound to the westward during the season of *northers*, a great deal of time may be saved by keeping close inshore, and thus taking advan-

From December to April.

tage of them; they will carry you as far as the gulf of Nicoya. When past the Morne Hermoso, 'papagayos' may be looked for, and with them a course should be steered for the gulf of Tehuantepec, when it will depend on the port you are bound to whether, after crossing the gulf by the aid of one of its gales, you should keep in or off shore. If bound for Acapulco, keep in and beat up; but if bound to the westward, you cannot do better than make a W. course as nearly as the winds will allow you.

From May to October. "The passage to the westward from Panama during the rainy season is a most tedious affair; calms, squalls, contrary winds, and currents, a heavy swell and extreme heat, as well as an atmosphere loaded with moisture and rain, are the daily accompaniments. It often occurs that 20 miles of westing are not made in a week, and it is only by the industrious use of every squall and slant of wind that the passage can be made at all. Opinions are divided among the coasters as to the propriety of working to the southward, and trying to get rid of the bad weather, or beating up within a moderate distance of the land. My experience would lead me to prefer the latter, as the strong winds and frequent squalls, which so often occur near the land, sometimes allow a good long leg to be made to the northwestward, while, farther off, this advantage is sacrificed for only a shade finer weather."

These instructions will be completed by referring to §§ 1, 23, and 24. §§ 60, 61, and 62 should also be consulted.

§ 60. Route from the Galapagos islands to Cape San Lucas.—Commander Wood gives the following instructions for this passage: "The trade-wind seems to possess no steady influence to the eastward of a line drawn from cape San Lucas, in 23° N., to the Galapagos islands on the equator. Among these islands the southeastern trade-wind is steady during nine or ten months of the year, and it is only in January and February, and sometimes in March, that they are interrupted by long calms, and occasional breezes from north and northwest, but these are never of any strength. To the northward of them the eastern limit of the trade seems to depend upon the time of the year. In the early part of April I have found it between the parallels of 8° and 13° N., 900 to 1,000 miles farther to the eastward than at the end of June; and in the intermediate months, either more or less to the eastward, as it was earlier or later in the season, but in no case that I have met with has a steady or regular trade been experienced until the above line has been reached. It is this circumstance, and the prevalence in the intermediate space of westerly winds, calms, and contrary currents, that makes the passage from Panama to the westward, as far as this line, so tedious. I have been 40 days beating from the entrance of the bay, in 80° W., to the eastern edge of the trade, in 111° W., a distance of less than 2,000 miles, or, on an average, about 40 miles per day."

§ 61. ROUTE FROM PANAMA TO REALEJO, AND FROM REALEJO TO ACAPULCO.—Below will be found Captain Basil Hall's observations relative to the passages from Panama to Realejo and Acapulco. We would also refer the reader to § 59 for further information on this subject:

"On leaving Panama for Realejo, come out direct to the northwestward of the Pearl islands; keep from 60 to 90 miles off the shore as far as cape Blanco, (gulf of Nicoya;) and on this passage advantage must be taken of every shift of wind to get to the northwestward. From cape Blanco hug the shore, in order to take advantage of the northeasterly winds which prevail close in. If a *papagayo* (as the strong breeze out of that gulf is called) be met with, the passage to Realejo becomes very short.

"From Realejo to Acapulco keep at the distance of 60, or, at most, 90 miles from the coast. We met with very strong currents running to the eastward at this part of the passage; but whether, by keeping farther in, or farther out, we should have avoided them, I am unable to say. The above direction is that usually held to be the best by the old coasters.

"If, when off the gulf of Tehuantepec, any of the hard breezes, which go by that name, should come off, it is advisable, if sail can be carried, to ease the sheets off, and run well to the westward, without seeking to make northing; westing being, at all stages of that passage, by far the most difficult to accomplish. On approaching Acapulco the shore should be got hold of, and the land and sea breezes turned to account.

"This passage in summer is to be made by taking advantage of the difference in direction between the winds in the night and the winds in the day. During some months, the land winds, it is said, come more off the land than at others, and that the sea-breezes blow more directly on shore; but in March we seldom found a greater difference than four points; and to profit essentially by this small change, constant vigilance and activity are indispensable. The sea-breeze sets in with very little variation as to time, about noon, or a little before, and blows with more or less strength till the evening. It was usually freshest at two o'clock, gradually fell after four, and died away as the sun went down. The land-breeze was by no means so regular as to its periods or its force. Sometimes it came off in the first watch, but rarely before midnight, and often not till the morning, and was then generally light and uncertain. The principal point to be attended to in this navigation is to have the ship so placed at the setting in of the sea-breeze that she shall be able to make use of the whole of it on the port tack, before closing too much with the land. If this be accomplished, which a little experience of the periods renders easy, the ship will be near the shore just as the sea-breeze has ended, and then she will remain in the best situation to profit by the land wind when it comes, for it not only comes off earlier to a ship near the coast, but is stronger, and may always be taken advantage of to carry the ship off to the sea-breeze station before noon the next day.

"These are the best directions for navigating on this coast which I have been able to procure; they are drawn from various sources, and, whenever it was possible, modified by personal experience. I am chiefly indebted to Don Manuel Luzerragui for the information they contain. In his opinion, were it required to make a passage from Panama to San Blas, without touching at any intermediate port, the best way would be to stretch well out, pass to the southward of Cocos island, and then run in with the southerly winds as far west as 96° before hauling up for San Blas, so as to make a fair wind of the westerly breezes which belong to the coast. An experienced old pilot, however, whom I met at Panama, disapproved of this, and said the best distance was 50 or 60 miles all the way. In the winter months these passages are very unpleasant, and it is indispensable that the whole navigation be much farther off shore, excepting only between Acapulco and San Blas, when a distance of 30 or 35 miles will be sufficient."

§ 62. ROUTE FROM PANAMA TO SAN FRANCISCO.—The following is a resumé of Maury's instructions for this route:

"The passage under canvas from Panama to California, as at present made, is one of the most tedious known to navigators.

"From the bay of Panama make the best of your way south, keeping near 80° W., until you get between 5° N. and the equator. After crossing 5° N., make a SW. course if the winds allow; if the wind be SW., brace up on the starboard tack; but if it be SSW., stand west, if it be a good working breeze. But if it be light and baffling, with rain, know that you are in the doldrums, and the quickest way to clear them is by making all you can on a due south course."

Speaking of the barometer in the Pacific, Maury says:

"The mean height in the equatorial calms is less than the mean height in the trades on either side. This difference does not probably exceed one-tenth of an inch (0.1 inch.) But close attention to the barometer in and about these calms will often enable the navigator to decide whether the winds he may have be really trade-winds or not; for, after having been fighting these calms, if you get the wind from NE. or SE., as the case may be, and the barometer *rises*, then you may be sure that you have the trades. In the calms of Cancer and of Capricorn there is a descending instead of an ascending current of air; therefore the barometer ranges higher, on the average, within those two calm-belts than it does anywhere else. The difference, however, does not exceed the tenth of an inch. Close attention to this instrument will often enable the navigator to decide, when he has crossed this belt and got into the region of trades, even before he gets the wind from the trade-quarter. He determines this by the fall of the barometer, when he enters the trades from the calms of the "horse-latitudes," but by its rise when he enters the trades from the equatorial calm-belt.

From June to January.

"Suppose that after crossing 5° N. you have got to the west of 85° without having crossed the equator. Now, if the time of the year be in that half which embraces July and December, the prevailing winds will be between SE. and S. inclusive, and the course is west as long as there is a breeze; as soon as the breeze dies, and you begin to fight the baffling airs, conclude that you are in the vicinity of the doldrums that are often found here, either between the NE. and SE. trades, or between one of these trades and the system of southwardly monsoons that blow north of the line, and between the coast and the meridian of 95° W. These belts of doldrums lie east and west, and the shortest way to cross them is by a due north and south line. Having crossed 95° W., stand away to the northward and westward with a free wind, and after reaching 100° W., aim to pass some distance from Clipperton.

From January to June. "If the passage from Panama be attempted in January, February, March, April, May, or June, time will probably be saved by going south of the equator, for at this half of the year the NE. trades and the equatorial doldrums are often found between 0° and 5° N. Between the meridians of 80° and 85° W., in this part of the ocean, these winds and calms are found even in the mouths of July and August. The navigator should therefore run south of the Galapagos islands, and not cross the equator again to the northward before he reaches 105° W. Aim to cross 10° N. at 120° W., when the NE. trades will probably be found.

"West of longitude 100°, and between the parallels of 5° and 10° N., the winds in the months of November and December are variable between NE. and S. by way of the east. In January, February, and March they are quite steady as NE. trades. In April they are variable. The doldrums are generally found between those parallels in this month. During the rest of the year the winds are all the time between SE. and SW.

"It will be well to cross the parallel of 10° N. at least as far west as the meridians of 105° or 110° W. Here between the parallels 5° and 10° N. the winds in November are steady from SSE. and S. December, April, and May, are the months for the doldrums in this part of the ocean. Having crossed the parallel of 10° N. between 105° and 110° W., the navigator is then in the fair way to California.

"In making the west coasts of Mexico and the United States, the kelp is said to form an excellent landmark. This weed is very long, and grows in the rocks at the bottom. When, therefore, on approaching the coast, you come across lines or swaths of tangled kelp, its being tangled or matted is a sign that it is adrift. But when you come across it tailing out straight, it is then fast to the rocks at the bottom, and it is dangerous to get among it.

"Vessels with steam-power should steam whenever necessary, after leaving Panama, and aim to cross the line near 85° W. Thence the route lies either to northward or southward of the Galapagos, according to the season. The remainder of the voyage can be easily made under canvas."

Such are Maury's instructions. The following abstract is also eloquent upon the same subject:

"*Passage of the Havana, Captain Harvey, from Panama to San Francisco.*—July 27th, sailed from the anchorage off the island of Taboga, for San Francisco. On the 1st August, in lat. 2° 30′ N., the land obliged us to tack off. On the 6th we passed four miles to the southward of the position of Rivadera shoal, continued westward, although forced northward, until the 10th, when we tacked to make southing.

"On the 21st sighted Clipperton island, bearing W. by N. ½ N. Hauled up to pass south of it, and stood along the island, trying for soundings, but no bottom at 150 and 180 fathoms, two miles distant. It was covered with myriads of birds, abundance of large drift-wood and pieces of wreck. On the north side the sea was much less, and landing was apparently easy in whale-boats. The island is correctly stated as being visible between 12 and 15 miles, but it is a formidable danger, and a wide berth should always be given to it at night. On the 24th August, in lat. 14° 11′ N., long. 114° 18′ W., picked up the trade-wind after only a few hours of variables. In lat. 34° 30′ N., long. 140° 06′ W., we tacked to make easting, as the wind was hauling more northerly; and made the Farralon light at three in the morning of September 15th, and were soon after in the midst of fog. Anchored in San Francisco harbor at four p. m., having been forty-nine days twenty hours on our voyage.

"This passage is at all times a trying one, lasting frequently 60, 70, and even more than 100 days. Vessels formerly took the inshore track, which may occasionally succeed. But by following Maury, the heavy rains, excessive heat, and doldrums certain to be met with inshore are avoided, perhaps entirely, but certainly to a very considerable extent. The U. S. frigate *Independence* arrived on

October 1st, fifty-nine days from Panama. The shortest passage on record is 45 days."

§ 63. ROUTE FROM MEXICO TO SAN FRANCISCO.—*Captain Sherard Osborn's instructions* :—"A vessel making the passage northward from San Blas had better make an inshore tack until she reaches the latitude of, or sights, cape San Lucas, as she will then get the true wind, which blows almost without intermission along the line of coast from the northward. A west, or may be *south* of west course will only be first made good, but as the offing is obtained, the wind will be found to veer a little to the eastward. However, it will always be the object to make headway, and get out of the tropic without any reference to the longitude, as a strong NW. wind will soon in lat. 25° or 28° run off the distance, provided you have sufficient northing.

"The attempt to beat up inshore amounts to perfect folly, if it does not deserve a worse name, a strong current accompanying the wind; and the latter must be taken into consideration when running in for your port with westerly winds."

Commander James Wood's instructions :—"When once within the influence of the trades, a passage is easily made either to the southward, westward, or northward; but it must be borne in mind that the eastern verge of this trade seems, in these parts, to be influenced by the seasons, (vide § 60.) Thus in June and July, I found it fresh from NNW., and even at times NW., as far out as the meridian of 125° W., whereas in March and April it was light from NNE. to E. and ESE. from our first meeting it in 98° W. till past the meridian of cape San Lucas in 110° W., where I picked up a good steady breeze from NNE.

"As a general rule the wind is found to haul more to the eastward as you get farther from the land, and I did not find this rule affected by the latitude, as, although, as I have stated, the wind hangs to the northward, and even at times to the westward of north, near the eastern limit of the trade, from the tropic of Cancer to the variables near the equator, I found it about the meridian of the Sandwich islands as far to the eastward on and near the line as it was in 35° N., in which latitude the westerly winds are in general met with."

The following remarks may be added to the advice given by Captains Osborn and Wood:

A vessel sailing from one of the ports of Mexico or Guatemala, should gain the NE. trades as soon as possible. When they are reached keep the ship a little free on a NNW. course, until the brave west winds are found to northward of the parallel of San Francisco.

Vessels leaving Mazatlan during these months will have quite fresh winds varying between N. and W.; with these they can easily pass to the northward of the Revillagigedo islands. In January, February, and March.

Ships leaving San Blas can also fetch to the northward of these islands, by making up the coast on the land tack if necessary. After they have crossed the meridian of 120° W. between 20° and 24° N., they will enter the region of steady NE. trades.

During the same season, vessels leaving Acapulco will have at first 6 per cent. of calms and variable NE. and NW. winds. They should make as much to the westward as possible, and will meet the NE. trades beyond and near 110° W. After this they should run a little free on the starboard tack and end the voyage as already described.

Ships leaving Istapa and Realejo should steer as nearly as possible SW., with the prevalent NE. wind, and thus reach 8° or 7° N. near 93° W. Thence the course should be due west to 108° W., the wind being quite fresh from SE. After crossing this meridian vessels should begin to bear to the northward, and cross 10° N. near 111° or 112° W. The NE. trades will be found in this locality.

Ships coming from San Blas or Mazatlan during these months will do well to beat up against the NW. winds, as advised by Sherard Osborn, as far as cape San Lucas. They should then keep close to the wind—which will be variable and from N. to W., mainly from NW.—standing well out to sea on the starboard tack. Nothing will be gained by crossing 120° W. to the northward of 20° N., as beyond this parallel the wind hauls more to N., while to the southward it inclines toward NE. In all cases the wind will become more and more favorable to the westward, enabling vessels to gradually bear away to the northward for the region of west winds. In April, May, and June.

Vessels leaving Acapulco at this season are liable to be detained by numerous calms, (vide § 1,) and NW. to NE.

winds. They should therefore run to the southward and westward for the SE. trades, near 10° or 8° N. Good way may be made to the westward, near these parallels, and to 118° W.; when, bear to the northward and cross 10° N. near 120° W. Here vessels usually run into the trades without meeting any calms.

The course from Istapa or Realejo to 8° or 7° N. and 92° W. is SSW. or S. Vessels will here strike the southerly or south-easterly wind and then should steer due west, the SE. wind freshening as they go to the westward. Near 118° W. they should begin to make northing, probably meeting the NE. trades at 10° N. and 120° W.

In July, August, and September.

Ships sailing from Mazatlan or San Blas, during July, August, or September, should take the route given for the preceding season; the circumstances attending the passage will, however, be more unfavorable, the calms more numerous, (vide § 1,) and the baffling airs from NW. more frequent. Vessels with steam-power will therefore have a great advantage, as they can steam to cape San Lucas.

During this season it will be greatly to the advantage of sailing-vessels leaving Acapulco, Istapa, and Realejo, to make for the SE. trades, to the southward of 10° N. Steamers, while they should take advantage of every puff of wind that may help them on their course to W. or NW., should keep up steam until they reach the NE. trades.

In October, November, and December.

Ships starting for Mazatlan or San Blas during this season will find almost the same weather as that for the preceding three months; it will, however, be a little better. Those leaving Acapulco should steer to the westward, as in January, February, and March. They will first have light variable winds from NE. to NW., and be liable to about 9 per cent. of calms. The NE. trades blow beyond 110° W., and become steadier and steadier to the westward of that meridian. Ships leaving Istapa or Realejo should first make to the southward and westward, in search of the SE. trades, to the southward of 10° N. From the beginning of the passage they will generally have gentle, though variable, winds to help them toward the trade-wind region. After meeting the SE. wind, good way can be made to the west on the parallel of 8° N. Vessels should commence to bear to the northward at 113° W., and cross 10° N. between 116° and 118° W. After this the NE. trades will be found, when the ship can be hauled up on the starboard tack.

§ 64. Route from Monterey to San Francisco.—The following instructions are from the "*U. S. C. S. Reports*:"

"Sailing-vessels bound to the northward from Monterey during the summer season should stand well off shore, not too close-hauled until about 200 miles from the land, when they will be beyond the influence of the southerly current, and in a situation to take advantage of a slant of wind, which frequently occurs from the WNW. They would do well not to approach the land, unless favored by the winds so as to enable them to lay their course, or nearly so, until up with the latitude of the destined port. During the summer.

"Steamers should follow the coast from point to point as nearly as possible, always keeping within 15 miles of the land. They will by this means shorten the distance, and frequently avoid the strong NW. wind, as they will often find it quite calm close in with the shore, when there is a wind to seaward.

"Vessels bound to the northward in the winter season should keep as close along the land as practicable, and take every advantage of all southerly winds to make latitude. They should always endeavor to make the land at least 20 or 30 miles to the southward of the destined harbor." During the winter.

§ 65. Routes from San Francisco to Vancouver.—In the last part of § 24 instructions relating to the prevailing winds between San Francisco and Vancouver will be found.

From November to April, or during the bad season, the passage should commence by putting well out to sea. This will generally be easy to do, as the wind is oftenest from NW. When far enough from land to have nothing to fear from SW. or NW. squalls, make as much to the N. as possible. Beyond the parallel of cape Mendocino the SW. winds prevail, enabling vessels to finish the voyage without difficulty. From November to April.

From April to November, or during the good season, the wind almost invariably blows from the northward, between NW. and NE. NW. is the favorite quarter, though SW. and SE. winds have been known in this locality. After leaving San Francisco run about 100 or 150 miles off shore, and then make to the northward, profiting by every shift in the wind, and always standing on the most favorable tack. From April to November.

CHAPTER II.

ROUTES FROM NORTH TO SOUTH ON THE WESTERN COAST OF AMERICA.

§ 66. ROUTES FROM VANCOUVER TO SAN FRANCISCO AND MONTEREY.—In § 24 the reader will find information concerning the winds to be expected on this passage.

The following instructions are taken from the U. S. C. S. Reports:

"If bound to the southward keep the coast in sight, and take advantage of either tack upon which the most latitude may be made, always making the land to the northward of the port in summer, and to the southward in the winter season.

"Bound to San Francisco or Monterey, use every opportunity to observe for latitude and longitude, so as to know the vessel's position up to the latest moment, as fogs and haze, preventing observations, prevail near the land. Allow generally for a southerly set of ½ a mile per hour, until within about 50 miles of land; after which, at times, it is not appreciable. With these precautions vessels may steer boldly on, shaping a course for the south Farallon, an islet about 250 feet high and a mile long, having 14 fathoms water, and good holding-ground on the SE. side."

From November to April.

From November to April, or during the bad season, vessels are liable to have head winds from SW., between Vancouver and the parallel of cape Mendocino. However, these winds are variable, blowing quite often from NW.; the currents, moreover, are generally favorable, and consequently there will be no serious difficulty to overcome. To the southward of 40° N. the NW. winds will become more frequent, and consequently the weather more propitious.

From April to November.

From April to November, or during the good season, the NW. wind prevails from Vancouver to Monterey, and even beyond this point. This is therefore an easy passage to be made at this season as the current is also favorable. Keep between 50 and 100 miles of the coast, thus avoiding the fog and the anxiety caused thereby. The wind frequently blows from SW. during the morning, and from NW. during the afternoon.

§ 67. ROUTE FROM SAN FRANCISCO TO MEXICO.—Ships leaving San Francisco for Mazatlan or San Blas will, at all seasons, have the wind and current favorable. The former is generally from the northward and westward; and from December to June is well settled as far as the port of destination.

To the southward of 30° N. the winds usually become lighter from July to December. During this season the chances of becoming becalmed off the coast of Mexico are greater. It will be a good plan to run in close to cape San Lucas, and to be careful that the vessel is not set to leeward of her port by the current.

Vessels bound to Acapulco will experience the same wind and weather. Beyond cape Corrientes they are liable to be delayed by calms, particularly between June and October. During these months, which correspond to the bad season, the winds usually blow from NW., (vide § 24,) but they are generally so light, and last for such a short time, that patience only is necessary. Auxiliary steam-vessels will, therefore, have an immense advantage during the latter part of this voyage, as well as on all the coast of Mexico and Colombia. During the winter, or the good season, the chances of calms are less, and the descent of the coast can be made with more facility, from cape Corrientes to Acapulco, or even to Istapa and Realejo.

§ 68. ROUTE FROM SAN FRANCISCO TO PANAMA.—Maury says: "Vessels out of San Francisco, intending to touch at Panama or any of the ports south, should stand out well from the Mexican coast. Information as to the best route for these passages is wanting. But I should, with such information as I at present have with regard to this navigation, feel disposed, were I bound from San Francisco to Panama, to steer straight for the line somewhere about 104° W., and stand on S. until I could, with the SE. trades, run in on the starboard tack for the land."

This *off-shore* route is undoubtedly the best for sailing-vessels, at least during the rainy season of the coast of Mexico, (from May to October.) Vessels keeping near the coast at this time of the year may find themselves, at any moment, becalmed or headed off by the SE. winds, which are quite frequent. Auxiliary steam-vessels can, if need be, use their engines, and make good headway by crossing 20° N. near 108° or 109° W.; and 10° N. at about 98° W. To the southward of 10° N. they will find quite settled winds from S. and SE., which, at times, shift to SW., to the eastward of 88° W.

From May to October. From May to October, as has been already said, sailing-vessels leaving San Francisco should stand well out to sea, with the favorable NW. winds. They should steer so as to cross 20° N. near 118° or 120° W.; then head S. or SSE., with the NE. or NW. winds. If these grow light it should be remembered that the wind is more settled to the westward, and that the chance of meeting calms is greater near the coast. To the southward of 10° N. settled SE. trades will be found. Here haul up on the port tack, and stand on until sure of reaching Panama, on the starboard tack, when go about. The Galapagos islands can be passed to the northward, and 90° W. near 4° N. To the eastward of this proposed point of crossing the winds will generally be found to be settled from SE. and S.; however, they often veer as far as SW.

From November to April. The dry season on the coast of Mexico lasts from November to April, when vessels should also follow the off-shore route; they should not, however, go quite so far to the westward at this season, though it will probably be found necessary to make a little more southing. From San Francisco to 20° N. the winds are generally favorable from NE. to NW.; and this route passes some distance to the westward of the Revillagigedo islands. Below 20° N. the NE. trades blow with such regularity that 10° N. can be crossed at 110° W. As soon as the wind begins to grow light steer S. for the SE. trades; with which stand on on the port tack until certain that the vessel will fetch to southward of the Galapagos on the starboard tack. The starboard tack will bring the ship near cape San Francisco, and the last part of the voyage will be identical with that described in § 54 and § 57.

Vessels with auxiliary steam-power should cross 20° N. at 109° or 110° W., and go down the Mexican coast with the prevalent NW. winds, steaming when becalmed. It will be to their advantage to cross 10° N. in the neighborhood of 89° or 91° W.; but they will afterward encounter SE. and S. winds; these being often variable and light, steam will have to be used to reach Panama, (vide § 73.)

§ 69. ROUTE FROM SAN FRANCISCO TO CALLAO.—This is one of those routes which Maury has studied, and for which he has given good instructions. We can do no better than quote below the principal passages:

"The best route is still undecided.

"Many very clever navigators give a decided preference to the eastern passage from California; but while they judge, for the most part, each by his own individual experience, I have the experience of them all to guide me in my judgment. I think it not at all unlikely that the opinion expressed by Captain Shreve, of the Cleopatra, may be found, on further investigation, to hold good for a part of the year. He says:

"'I would advise all captains leaving San Francisco for Callao in the months of August, September, and October to take the inner passage, that is, being in the long. of 110° W., lat. 8° N., steer along the equator by the wind, passing either side, or between the Galapagos islands, as the wind will permit. Had I taken this route instead of crossing the SE. trades, it would have shortened my passage one month, which has been proved by the West-Wind and several other ships during the above months. I inquired of several disinterested captains as regards the passage to Callao; all advised crossing the SE. trades. It may do when the sun is far north. This passage is little understood as yet. I had no difficulty with my ship (steady trades) in beating from Callao to the Chincha islands in three days; therefore, what difficulty can exist in beating from the equator to Callao?'

"Individual cases may be cited in favor of each route, but upon the whole, and with such lights as I have, I am inclined to give the preference to the western or off-shore route as the one which for most of the year and on the long run will give the shortest average passage, and which average, when the route comes to be properly understood and followed, will probably be brought down as low as 50 or 52 days the year round.

"Most vessels on this voyage make a mistake, especially in summer and fall, in the passage across the belt of NE.

trades. Being anxious to get to the east, they edge along, aiming to lose these winds in 90° or 100°, as the case may be. Then they encounter the southwardly monsoons that are found at this season of the year between the systems of trade-winds in the Pacific off the American coast, as they are along the African coast in the Atlantic. The vessels taking this course, and being so baffled, have now to make a sharp elbow and run off 8 or 10 or even more degrees to the westward before they clear this belt of calms and monsoons and get the SE. trades. Of course the voyage is greatly prolonged by this.

"The route which, as at present advised, I would recommend is that navigators steer the same course from California that they should if bound to the United States, until they pass through the SE. trades and clear the calms of Capricorn. Therefore I say to the Chincha-bound trader, when you get your offing from the 'heads,' steer south, aiming to cross the line not to the east of 115°, for the rule is the farther east the harder it is to cross the equatorial doldrums in the Pacific, as well as it is in the Atlantic.

"When you get the SE. trades, crack on with topmast studding-sail set, until you get the 'brave west winds' on the polar side of the calms of Capricorn. Now turn sharp off from the route around cape Horn, and run east until you bring your port to bear to the northward of NE., when you may 'stick her away.' Now, by this rule, the Chincha-bound navigator may sometimes, before he gets these westerly winds, find himself as far south as 40° or 45°, and as far west as 120° or 125°. Let him not fear, but stand on until he gets the winds that will enable him to steer east, or until he intercepts the route from Australia to Callao, when he may, without fear of not fetching, take that.

"In the summer and fall of the northern hemisphere—June to November—the calm belt of Capricorn will be cleared generally on the equatorial side of the parallel of 30° S.; at the other seasons you will have frequently to go 6° or 8° farther.

"On this voyage, navigators, as soon as they leave the SE. trades, are often tempted by puffs and spirts of westerly winds to stand east; and thus time is lost by running east with a 4 or 5 knot breeze in the calm belt of Capricorn. They should stand south until they clear it, preferring, as a rule, to take the chances of better winds and the certainty—which is some compensation—of shorter degrees of longitude beyond."

After these statements Maury gives a series of "tables of passage," containing for each month a certain number of *west* and *east* routes.

In examining these tables the first point that strikes the eye of the reader is, that the general mean for the whole year is shorter by 9 per cent. for the west route than that obtained for the other route.

The following results were deduced from a comparative examination of the two routes for each month:

In January, the mean from San Francisco to the line by the western route is 21 days; and from San Francisco to Callao, 55.8 days. By the eastern route, 41 days were consumed in reaching the line, and 81.5 days in the whole passage.

The advantage in favor of the western route is therefore 26 days.

In February and March, the total length of the voyage by the W. route is 57 days. Examples of the E. routes are rare; if it were possible to obtain them they would undoubtedly be longer.

In April, 25 days to the line by the W. route, and 54.7 days for the whole passage; 22 days to the line by the E. route, and 89 days for the whole passage. By keeping to the westward 34 days were therefore saved.

In May, 26 days to the equator by the W. route, and 61 days from San Francisco to Callao; by the E. route, 31 days to the equator, and 66 days to Callao. Advantage in favor of the W. route 5 days.

In June, 22 days to the line by the W. route, and 50.5 days for the whole passage; by the E. route, 31 days to the line, and 62.3 to Callao. Gain of 12 days by W. route.

The next four months show opposite results.

In July, 26 days in reaching the line, and 57 days for the whole passage; by the E. route, 29 days to the line, and only 48 to Callao. Gain of 9 days by the E. route.

In August, 26 days to the line by the W. route, and 57.6 days for the whole voyage; by the E. route, 33 days to the line, and 57 days in all. A very small advantage on the side of the E. route.

In September, 25 days to the line, by the W. route, and 58.8 days to Callao; by the E. route, 23 days to the line, and 48.5 to Callao. Ten days' gain by the E. route.

In October, 27 days to the line, by the W. route, and 71 days for the whole passage; by the E. route, 28.2 days to the line, and only 57 days to Callao. Gain of 14 days in favor of the E. route.

In November, 27 days to the line, by the W. route, and 55.5 days for the whole voyage; by the E. route, 30 days to the line, and 55.5 days, or in other words the same length of voyage to Callao.

In December, 23 days to the line, by the W. route, and 58.8 to Callao. No example of the E. route given.

From which we conclude that nothing will be gained by taking the E. route, except during the months of July, August, September, and October; while by the W. or trade-wind route the passages will be much shortened, especially between January and June.

We would therefore advise sailing-ships, leaving San Francisco, to regulate their course in accordance with the instructions given in § 68 and the principles laid down by Maury: that is, to endeavor to cross the equator near 118° W., between May and October, and near 113° W. between October and May.

Auxiliary steamers, frigates, and corvettes, for instance, should follow the route given in the preceding paragraph, (§ 68,) that is, keep nearer to the Mexican coast, and make the coast of S. America near cape San Francisco.

Vessels taking the trade-wind route should keep on the port tack until Callao bears to the northward of NE.

They may often be compelled to go to the southward of 30° S. before they meet the W. winds. With these they should steer to the eastward until they are in good position to make their northings.

In *January*, 30° S. will generally be made, on the port tack, between 118° and 122° W., then run to the east on the parallel 33° until near 98° W.

In *February*, *March*, and *April*, cross 30° near 118° W., and then steer to the eastward on 35° or 36° S.

In *May*, cross 30° in the neighborhood of 123° W., and make to the eastward between 32° and 33° S.

In *June*, 30° S. can be crossed between 123° and 128° W.; the west winds prevail at 35° S.

In *July*, cross 30° S. between 118° and 123° W.; and as the west winds are to be found a little farther north, run down the easting between 33° and 34° S.

In *August*, the SE. trades will be very variable, and 30° S. should be crossed between 118° and 128° W., more or less to the east or west, as circumstances allow. The easting should be made on the parallel of 32°.

In *September*, 30° S. cannot be crossed to the eastward of 128° W.; the easting should be made on 33°, or near that parallel.

In *October*, 30° S. can be crossed a little farther to the eastward than during the preceding month, and the easting made on 32° S.

In *November*, after having crossed the line near 113° W., and 30° S. between 118° and 122° W., steer to the eastward near 33° S.

In *December*, cross 30° S. near 118° W., then follow 32° S. to the eastward.

Auxiliary steam-vessels, taking the E. route, will be enabled to cross 10° N. between 89° and 91° W. They will then experience variable winds till they reach cape San Francisco. These winds blow quite often from S. and SE., with 5 per cent. of chances of calms.

Steam will have to be used about half the time in this locality. The passage from cape San Francisco to Callao will be tedious, as both wind and current are from the southward. If this part of the voyage be made under canvas it will last at least 25 days. For further observations on this subject consult § 75.

§ 70. ROUTE FROM SAN FRANCISCO TO THE INTERMEDIATE PORTS.—This voyage should be made in exactly the same manner as that described in § 69; that is, keep always to the west, or SE. trade-wind route. Stick to the port-tack until you strike the west winds to the southward of 30° S. Afterward the course should be to the east until the port of destination bears to the N. of NE. Thus, if the port of destination be Iquique or Arica, 92° W. should be reached before bearing away to the northward.

§ 71. ROUTE FROM SAN FRANCISCO TO VALPARAISO.—According to Maury, the route to Callao is longer than that to Valparaiso; owing to the fact that sailing-ships have to make the parallel of Valparaiso before striking to the northward for Callao, consequently the mean of the voyages from California to Peru being about 56 days, vessels can count on a mean passage of about 50 or 55 days from California to Valparaiso.

If the reader will refer to § 68 and § 69 he will there find all the necessary instructions relating to this passage.

Thus, as has been stated, the equator should be crossed very near 118° W. from May to October, and at about 113° W. from October to May.

Make a long port tack rap full through the SE. trade-belt.

In January, vessels can make good headway to the eastward after reaching 34° or 35° S. In February, bear away to the east, between 35° and 36° S. In March, at the same latitude, or a little farther south. In April and May, bear to the eastward near 34° or 35° S. In June, steer east, when south of 35° S. In July, August, and September, near the parallel of 34° S. In October, November, and December, at 36° S. In no case begin to steer directly for Valparaiso before reaching 83° W.

§ 72. Route from San Francisco to cape Horn.—Nothing especial is to be added to what has been stated in the preceding paragraphs, particularly in § 69.

After standing well full on the port tack, through the SE. trades, steer south until the region of W., NW., and SW. winds is reached, or in other words until between 35° and 40° S. Then bear away around cape Horn on a great circle route, as nearly as the wind will permit.

Information relating to icebergs will be found in § 43. The reader should also consult § 107, § 108, and § 109.

§ 73. Route from Mexico to Panama.—Reference should first be had to § 63, where will be found Captain Wood's instructions. According to this officer, vessels leaving Mexico will have no difficulty in going either north or south after they have once penetrated the NE. trade region. He also shows how the eastern limit of the trades changes at different seasons.

Observations of Captain Basil Hall: "The return passages from Mexico to Panama are always easy. In the period called here summer, from December to May, a distance of 90 to 150 miles from the coast insures a fair wind all the way. In winter it is advisable to keep still farther off, say 300 miles, to avoid the calms, and the incessant rains, squalls, and lightnings, which everywhere prevail on the coast at this season. Don Manuel Luzurragui advises, during winter, that all ports on this coast should be made to the southward and eastward, as the currents in this time of the year set from that quarter."

Captain Hall's instructions seem to be completely indorsed by a study of the prevailing winds in these quarters.

December to May.

From December to May vessels can easily keep along the coast as far as 10° N., or even all the way to Panama, the wind generally varying between NE. and NW., but there is a chance of meeting south-easterly and southerly winds in the offing, south of the 10th parallel.

After May, calms will render the passage particularly difficult, and it will be advisable to keep farther from the coast. Vessels coming from Mazatlan or San Blas should cross 12° or 13° N. near 100° W.; then 10° N. at 92° or 93° W.

From May to October.

From May to October the direction of the wind is generally favorable as far as 103° W.; prevailing SE. winds will then be found, becoming steadier and stronger to the southward and eastward of that meridian. The last part of the passage will ordinarily be long and irksome. Vessels leaving Acapulco during this season should at first make as much to the southward as possible, and not attempt to work to the eastward until they are in the neighborhood of 12° S.

We will finish with a quotation from Fitz-Roy:

"Vessels bound to Panama from northward should make the island of Hicaron, which lies about 50 miles westward of Mariato point, and from this endeavor to keep under the land as far as cape Mala. If unable to do this, they should push across for the opposite side of the continent, when the current will be found in their favor. On getting eastward of cape Mala the safest plan is to shape a course for Galera island, and to use the eastern passage. At the same time, if tempted up the gulf by a fair wind, vessels should endeavor to get on the western coast of the Pearl islands, which have the advantages already explained."

§ 74. Route from Mexico to Guayaquil.—The principal points of this route, which is usually long and tedious, will be found in § 69, under the head of the E. route from

California to Callao. The reader should particularly refer to the assertions of Captain Shreve, quoted by Maury. Useful information will also be found in §§ 73 and 75, which contain extracts from Basil Hall and Rosencoat.

We shall therefore confine ourselves to the following advice (from Findlay) relative to the best manner of approaching Guayaquil:

"Coming from the northward Santa Clara island may be made, which is visible about 16 miles, and at first appears like three hummocks; and Zampo Palo, the high range on Puna island, will generally be seen at the same time.

"Santa Clara should not be approached nearer than 2 miles, or within the depth of 12 fathoms, the best track being about 5 miles to the southward of it, in from 20 to 15 fathoms water, whence a NE. ½ E. (by compass) course for 25 miles will lead toward Arenas point, (vide § 53.)

§ 75. ROUTE FROM MEXICO TO CALLAO.—Auxiliary steam-vessels should always approach the land near cape San Francisco. Their steam-power will help them both in the first part of the voyage and while beating up to Callao beyond capes San Francisco and Santa Elena.

Sailing-vessels and auxiliary steamers desiring to economize coal, will find it generally advantageous (at least from May to December) to take the west route.

The first part of this route is nearly identical to that given in § 68 and § 73, while the latter part will be the same as that described in § 69. The west route seems to be undeniably the best for vessels leaving San Blas, Mazatlan, or Acapulco. If the point of departure is farther S., Istapa or Realejo for instance, there may be some doubt as to which is the best route; but even then it would seem that the westerly passage is the one to be preferred. The SE. trades will be found to the southward of 10° S.; when haul up a little free on the port tack, and pass a trifle west of the Galapagos islands.

Below are the instructions given by Captain Basil Hall; it will be seen that they do not absolutely conform to our own advice on this passage, for we still hold to the opinion that the W. route is the preferable one for sailing-vessels:

"If it is required to return direct from San Blas to Callao, a course must be shaped so as to pass between the island of Cocos and the Galapagos, and to the south-eastward, till the land be made a little to the southward of the equator, between cape San Lorenzo and cape Santa Elena. Then work along shore as far as point Aguja, in lat. 6° S., after which work due S., on the meridian of that point, as far as 11° 30′ S. and then stretch inshore. If the outer passage were to be attempted from San Blas it would be necessary to run to 25° or 30° S., across the trade, which would be a needless waste of distance and time."

Observations on the route from San José de Guatemala to Callao, by Captain Rosencoat.—"Following the information I had on this route, I decided to make for Cocos island, where I presumed I would find westerly winds, to carry the ship a little to northward of the gulf of Guayaquil and thence to the Peruvian ports by the usual route. We followed this route, but instead of the westerly winds near Cocos island found a light though steady breeze from E. and ESE. The wind held from this quarter till we reached Chatham island, (the most eastern of the Galapagos,) and instead of dying away, as is usual in this locality, freshened and still kept steady from the eastward. I was therefore compelled to run through the SE. trades and then make my easting with the variables near the tropic of Capricorn, in order to reach the prevalent winds of the coast of Peru.

"Vessels deciding to cross the SE. trades should, however, run boldly through the region of SE. winds, no matter how much they may at first be set to the westward. Those, on the contrary, deciding to make the passage along the coast, should, from January to April, keep within 15 miles of land, and take advantage of the Mexican current, which sets ESE. with a speed of one knot and a half per hour. The WNW. and SSW. winds which we found during these four months, on the coast of Central America, would have given us a quick passage to cape San Francisco, or even to a more southerly point on the coast, as the prevalent wind is SE. in the gulf of Panama at this season; except near Gorgona island, where frequent calms are said to exist.

"Vessels taking the inshore route can hardly expect to make the passage from San José de Guatemala to cape Blanco in less than 15 days; and the whole voyage to Callao in less than 35 days. We took 40 days, but our unsuccessful attempts to make to the eastward at the beginning, probably lengthened the whole voyage. If we had steered boldly for the SE. trades at first, we probably would not have been delayed by calms north of the equator.

§ 76. Route from Mexico to the intermediate ports, Valparaiso and cape Horn.—It was stated in the preceding paragraph that the W. route seems to be the best for sailing-ships making passage from Mexico to Callao. We have also stated that the E. route has its partisans, and that vessels leaving the southern ports, such as Istapa or Realejo, may have quite a short passage.

But when vessels, bound to a port south of Callao, leave the Mexican coast, it seems impossible that they should prefer any other to the W. route. Therefore ships leaving San Blas, Mazatlan, or Acapulco should profit by every breeze that will set them to the southward, always choosing, if the wind comes out ahead, that tack by which they can make SW. or SSW. in preference to the one which will take them to the southward and eastward. Vessels leaving Realejo should steer SSW. as nearly as possible. By following this advice they will find the SE. trades near 10° N. No other recommendation is needed, except to keep "rap full" on the port tack, and never hug the wind. If bound to one of the intermediate ports, do not go about in the trades until your port bears N. of NE. But if the destination be Valparaiso, keep on the port tack to the region of prevalent westerly winds, near 32° or 34° S., (vide § 71.) Ships bound around the Horn should steer S. until they have settled and steady west winds, when they can gradually bear away to the eastward, (vide § 72, § 107, § 108, and § 109.)

§ 77. Route from Panama to Guayaquil, Payta, and Callao.—After leaving Panama the end in view should be to make as much as possible to the southward, and thus reach the SE. trade-region.

This advice, given by Fitz-Roy, is certainly the best, but sailing-ships cannot always follow the most favorable route, especially in a locality of calms, squalls, and light breezes.

Referring to James Wood, we find the following: "But the great difficulty at all times consists in getting either to the southward or westward of Panama. The passage to the southward is made in two ways—either by beating up the coast against a constantly foul wind and contrary current, or by standing off to sea till sufficient southing is made to allow you to fetch your port on the starboard tack. Both plans are very tedious, as it frequently takes twenty days to beat up to Guayaquil, whilst six or seven days are an average passage down." (Vide Maury's instructions, § 62.)

In short, it will be to the advantage of ships leaving Panama for Guayaquil to keep close to the land. If they are bound farther down the coast, they should, after doubling cape Blanco, follow the instructions given in § 79.

§ 78. ROUTE FROM PANAMA TO THE "INTERMEDIATE PORTS," VALPARAISO, AND CAPE HORN.—As stated in the foregoing paragraph, make to the southward as much as possible, taking advantage of every favoring breeze, and thus reach the SE. trades by the most direct route. Below 5° N., and especially to the westward of 80°, settled southerly winds will be found, ranging between the SE. and SW. points of the compass.

The passage may be made along the coast to cape San Francisco, or even to cape Blanco, though if the wind shows a tendency to come from the southward and eastward, it is well to stand off-shore on the port tack before reaching San Francisco. If bound to any of the intermediate ports, the inshore route is preferable, as a long stretch on the port tack is liable to set vessels too far to the southward and westward of their destination. We again repeat the rule, that vessels should never go about in the SE. trades until the port of destination bears north of NE., (vide §§ 71, 72, 76, and 84.)

§ 79. ROUTE FROM GUAYAQUIL AND PAYTA TO CALLAO.—Fitz-Roy gives the following instructions on this route:

"On leaving Guayaquil or Payta, if bound to Callao, work close inshore to about the islands of Lobos de Afuera. All agree in this. Endeavor always to be in with the land soon after the sun has set, so that advantage may be taken of the land-wind, which, however light, usually begins about that time; this will frequently enable a ship to make her way along shore throughout the night, and will place her in a good situation for the first of the sea-breeze.

"After passing the above-mentioned islands, it would be advisable to work upon their meridian until the latitude of Callao is approached; then stand in, and if it be not fetched, work up along shore, as above directed, remembering that the wind hauls to the eastward on leaving the coast. Some people attempt to make this passage by standing off for several days, hoping to fetch in well on the other tack; but this will generally be found a fruitless effort, owing to a northerly current which is often found on approaching the equator. The mean passage for weatherly vessels is 15 or 20 days from Guayaquil to Callao."

Capt. Basil Hall gives the following instructions:

"The passage from Guayaquil to Callao requires attention, as may be seen from the following directions, which I obtained from Don Manuel Luzurragui, captain of the port of Guayaquil:

"'The average passage in a well-found and well-managed ship is twenty days; eighteen is not uncommon; and there is an instance of a schooner doing it in twelve. From the entrance of the river as far as Punta de Aguja, (in lat. 6° S.,) the shore must be hugged as close as possible, in order to take advantage of the changes of wind, which take place only near the shore. In this way, by due vigilance, slants may be made every day and night. On reaching Punta de Aguja, work to the southward, as nearly on the meridian of that point as may be, as far as 11° 30′ S., and then strike inshore for Callao, and if it is not fetched, creep along shore, as formerly directed.

"'Persons accustomed to the navigation between Callao and Valparaiso are tempted to stand boldly out in hopes of making their southing with ease, and then run in upon a parallel. But this is not found to be practicable, and indeed the cases have no resemblance, since the passage to Valparaiso is made by passing quite through the trade-winds and getting into the variables; whereas Callao lies in the heart of the trades; accordingly, a ship that stretches off from Guayaquil comes gradually up as she stands out, and finally makes about a south course; when she tacks again, the wind shifts as she draws in, and she will be fortunate if she can retrace her first course, and very often does not fetch the point left in the first instance.

"'To work along shore with effect, the land must be kept well on board, and constant vigilance be bestowed upon the navigation, otherwise a ship will make little progress.'"

Capt. Andrew Livingston, well known in the nautical world, makes the following remarks on navigating to windward from Huanchaco to Callao:

"The most intelligent, experienced persons with whom I

conversed generally recommended standing off shore during the night, and inshore during the day; but advised that any person in charge of a vessel beating thus to windward should take care to be pretty close to the shore by sunset, to take advantage of the wind, which about that time generally draws rather off the land, though not sufficiently to deserve the name of a land-breeze."

On the above I remark, that on account of the land trending so much to the eastward, if you stand twelve hours off shore and twelve hours inshore, at the same rate of sailing, and have gained any southing of consequence, you will still be a considerable distance off shore when your twelve hours are completed standing in; and I think that it will be found in general most advisable to stand off only about ten hours and in for fourteen hours; as, even if you get inshore rather too soon, you can, by making a short tack or two, be sure of being near the shore at sunset, when you may expect the wind rather to favor you for gaining southing with your port tack on board.

On the off-shore tack you will generally find that the vessel comes up more and more as you stand off, but do not let this persuade you to stand off too far, even should the vessel head up S., or S. by E. by compass, as you will lose more on the inshore tack, when you must be headed off in proportion as you have headed up on the off-shore tack. The inspection of the chart will at once convince any person of this fact, even if there is no northerly current. Of course, bringing that directly on or abaft a vessel's beam must sag her to leeward.

On the coast of Peru the water is frequently of a dirty-brown color, and sometimes quite red, as if mingled with blood.

§ 80. ROUTE FROM GUAYAQUIL AND PAYTA TO THE INTERMEDIATE PORTS.—Vessels should strike through the trades on the port tack, as if bound to Valparaiso, and go about when their port bears to northward of NE. They will consequently leave the trades, and make their ea ting near 30° or 32° S. (Vide §§ 70, 76, 78, and 81.)

§ 81. ROUTE FROM GUAYAQUIL AND PAYTA TO VALPARAISO AND CAPE HORN.—Fitz-Roy's instructions on this route may be condensed as follows:

"Sailing-vessels bound from Guayaquil to Valparaiso

should stretch out to sea, crossing the Peruvian current before passing the meridian of 92° W. From this they should push to southward, not caring about being driven to the westward if southing can be made, as they will have no difficulty in making their easting on the parallel of Valparaiso. This passage is generally made in 37 days."

We have nothing to add to these general observations; besides, detailed instructions will be found in §§ 71, 72, and 84 on the proper manner of steering through the trades and prevalent W. winds in order to reach Valparaiso or cape Horn.

§ 82. Route from Callao to the Chincha islands.—We will cite the instructions on this passage given by Captain Chardonneau, in his work on the coast of Peru, (pages 22 and 94:)

"Going from Callao to the Chincha islands or Pisco, it is best to keep at a distance of from 25 to 40 miles from the coast, until SW. of Cerro-Azul. Then stand in to within 10 miles of the land, as, from this point to Pisco, there is nearly always a light northerly breeze during the morning.

"It frequently happens that after finding a calm in the morning abreast of Cerro-Azul the ship will at night be anchored off Pisco. The current sets steadily to WNW. in this locality.

"It is advisable to keep away from the shore during the night, and near it during the day, until beyond the 13th degree of latitude, then work along at a distance of 4 or 5 miles from the land.

"Captain Harvey's advice for the autumn months is to run 26 hours to seaward and 22 to landward, thus being at the end of 48 hours to windward of San Gallan."

Fitz-Roy also states that it is best to beat up near land, between Callao and the Chincha islands. The same rule should be followed as he has given in § 79, for the passage from Guayaquil or Payta to the Lobos de Afuera islands.

Lastly, *Maury* cites, in the *Sailing Directions*, an extract from the log of the Hornet, Captain Knap, which we quote below:

"The passage from Callao to the Chincha islands offers no especial peculiarities. I would simply state that it had better be made in the region of steady trades, that is, clear of the land, and out of the influence of calms and baffling airs, as these make vessels lose at least twelve hours out of the twenty-four. I think you cannot count on land-breezes near the shore, at any rate not in the autumn.

"I reached the Chinchas by two tacks, one of twenty-six hours off shore, the other of twenty-two hours, which brought me to San Gallan, fifteen miles to windward of the group. Forty-eight hours in all from Callao. I observed the same rule while descending the coast from cape Blanco to Callao, that is, I kept an offing of three or four degrees, thus getting away from the influence of the coast calms, an influence which I had unfortunately experienced while sailing from point Santa Elena to cape Blanco."

The above quotations do not absolutely agree. It is, however, certain that long and rapid passages have been at times made by ships keeping near the land, and again by others which have given the shore a wide berth; and the successful captains have never failed to praise the respective routes which they have followed. We believe, however, that as a general rule it is best to keep a good distance from the land at the beginning of the passage. But we would also add that it is not advisable to run too long off shore on the port tack, as the wind may haul toward ESE. toward the end of the off-shore tack, and the ship may be set back to the northward while approaching the coast on the starboard tack.

Fitz-Roy's instructions—to approach land during the afternoon and thus be in readiness to take the off-shore tack at sunset—would seem to be the best. Stand on this tack until the morning at 9 o'clock, when go about.

By proceeding in this manner navigators need not fear (if the wind haul ahead) that they will be as far from the Chinchas at the end of twenty-six hours as they were at starting.

§ 83. ROUTE FROM CALLAO TO THE "INTERMEDIATE PORTS."—After stating that vessels bound to Valparaiso should run off shore, Fitz-Roy remarks:

"But for the intermediate ports the case is different, (excepting Coquimbo,) as they lie considerably within the trade-wind, and must be attained through that medium alone. A very dull sailer might indeed do better by running through the trade, and making southing in the offing, so as to return to the northward along the coast, than by attempting to work to windward against a trade-wind, which never varies more than a few points.

"It may be recommended to work along shore, in a good sailer, as far as the island of San Gallan, whence the coast trends more to the eastward, so that a long leg and a short one may be made (with the land just in sight) to Arica, or to any of the ports between it and Pisco.

"From Arica, the coast being nearly north and south, vessels bound to the southward should make an offing of not more than 45 or 60 miles, (so as to insure keeping the sea-breeze,) and work upon that meridian, till in the parallel of the place to which they are bound, but on no account is it advisable to make a long stretch off; for as the limit of the trade-wind is approached, it gradually hauls to the eastward, and great difficulty will be found in even fetching the port from which they started."

Captain Basil Hall makes the following remarks on this passage:

"There is no difficulty in making a passage along the south coast of Peru from the eastward; but from the westward a great deal of vigilance is requisite to take advantage of every occasional shift of the wind, since by this means alone can a passage be made. The best authorities are, I think, against standing out to sea, to the southwestward, in the hopes of fetching in upon the starboard tack. The Constellation, American frigate, tried this passage, but she lost a great deal of time thereby, being at least three weeks in going from Callao to Mollendo. The San-Martin, bearing Lord Cochrane's flag, made the passage from Callao to Arica, which is considerably farther, in thirteen days, by keeping inshore and taking advantage of the changes which take place, with more or less regularity, every evening and morning.

"As the weather along the south coast of Peru is invariably fine, ships are not otherwise incommoded at the various anchorages, than by a high swell, which always rolls in at the full and change of the moon. Arica is the only place having any pretensions to the name of a harbor."

Commander Chardonneau gives the following advice:

"Vessels bound from Callao to Iquique, Arica, and Islay should work along the coast to Morro de Chala. The boards should be short and close under the land, so as to take advantage of the fresh wind in the offing during the day, and the land-breeze during the night. Between San Lorenzo and San Gallan islands it is advisable to keep clear of Asia or Cerro-Azul bight, as vessels are liable to be becalmed in that locality and the eddy currents are strong and uncertain. The best plan is to keep to seaward of a line drawn from San Lorenzo to the northernmost of the Chincha islands.

"To southward of Morro de Chala vessels should keep from 30 to 100 miles from the coast, as beyond the latter distance the wind hauls to the eastward, and renders it difficult to fetch the port when bound in on the land tack.

"The parallel of their port once reached, vessels should run in for the land, being careful not to lose anything that they have made in latitude. If the landfall be made a few miles to windward the anchorage can always be reached, even during a calm, as the current and 'catspaws' are sufficient for working up to the port."

§ 84. ROUTE FROM CALLAO TO VALPARAISO.—*Fitz-Roy's instructions:* "For a sailing-vessel bound from Callao to Valparaiso, there is no question but that by running off with a full sail the passage will be made in much less time than by working inshore, for she may run quite through the trade, and fall in with the westerly winds which are always found beyond it. The average passage is about three weeks. Fast-sailing schooners have made it in much ess time; and there is an instance of two men-of-war sailing in company, having gone from Callao to Valparaiso, remained there two days, and re-anchored at Callao on the twenty-first day. But these are rare occurrences, and only to be done under most favorable circumstances, such as meeting with a *norther* soon after leaving Callao."

Captain Basil Hall's passages; Callao to Valparaiso, 28th February to 18th March, 1821.—The return-passage from Peru to Chile requires some attention, and may generally be made by a man-of-war in less than three weeks; it has been made in less than a fortnight by a frigate, which, however, on the next occasion, took twenty-eight days. The point which contributes most to the success of this passage is keeping well off the wind after leaving Callao, and not having any scruples about making westing, provided southing can also be gained. The SE. trade-wind, through which the greater part of this course is to be made, invariably draws to the eastward at its southern limit, and therefore eventually a ship can always make her southing. The object, however, being to get past the trade and into the westerly winds, which lie to the southward, a ship ought to keep the wind at least abeam, while crossing the trade. In winter, that is when the sun is to the northward of the equator, the trade-wind blows steadier, and its southern extreme lies 4° or 5° to the northward of its summer limit, which may be taken at about 30° or 31° S.

"*Chorillos* (*near Callao*) *to Valparaiso*, 10*th to* 28*th August*, 1821.—This being what is called the winter passage, we lost the trade-wind in lat. 25° S., after which we had the winds to the SW. as far as lat. 27° S., long. 88° W., when they shifted to the NW. and W., and so to the SW. and S., as far as lat. 33° S., long. 78° W. We were much embarrassed by calms, light winds, and heavy rains, after which the wind came to the northward and NNW., with thick, rainy weather. We made the land to the southward of Valparaiso on the 27th, and got in next day by the wind coming round to the SW.

"At this season of the year, when northerly winds prevail, with heavy rain, and unpleasant weather, it does not seem advisable to make the coast to the southward of the port. Neither ought a ship, I think, to run into Valparaiso in one of these gales, since the wind frequently blows home, and is attended by a high swell. During the winter the barometer, the threatening aspect of the weather, and the rising swell generally give sufficient warning. Previous to a *norther*, also, the land of Concon, and that beyond it to the northward, are seen with unusual sharpness and distinctness.

"This passage in eighteen days may be termed short. Formerly thirty days was usual; it afterward sunk to twenty-five days, and, at the period of our arrival, three weeks was considered good. Sir Thomas Hardy, in His Majesty's ship Creole, made the passage from Huacho in something less than fourteen days, the distance being more than 2,200 miles. This was early in May, 1821; and it is well worth attending to, that the trade-wind was crossed, with a foretop-mast studding-sail set, no regard being paid to any object but getting through the trade-wind as fast as possible. The same ship, however, in February and March of the following year was twenty-eight days making the passage; but this is unusually long for a man-of-war."

Commander Chardonneau's observations.—"A ship leaving one of the ports of Peru for Chile should make a good offing, work to the southward and westward, and keep ½ or 1 point free. The wind will become fairer and fairer as the ship leaves the coast and makes to the southward. Many captains brace their yards so that the foretop-mast studding-sail will just 'touch,' and then steer by it. This seems to be a good plan. Stand on in this manner for the variable winds, always to be found south of the tropic, and between it and the parallels of 28° or 30° S. These winds are often fresh and blow from SW., S., and NW., with squalls and rain. Then steer to the southward and eastward—without seeing land—as far as the parallel of the port of destination, or a little to the south of that parallel, if it is summer, (from September to April.) Make the land to southward of your port.

"Vessels going to Valparaiso should head for the island of Juan Fernandez as soon as they strike the variable winds, and passing to the northward of the island and in sight of it, stand to the eastward until they sight point Curaumilla, if it is the summer season. But in winter (from May to August) steer for Valparaiso, taking care not to enter the port if the wind be from the N. and the barometer be falling, for the anchorage is then dangerous, and it is preferable to lie-to outside until the wind comes out from the west, when there will be no risk."

We will complete these instructions by giving, for each month, the principal crossings on the route from Callao to Valparaiso. By first tracing on the chart the indicated route, the navigator will be enabled to make the passage under approximative mean conditions. Still it should be understood that the route may be altered if the wind is contrary. As a general rule, the ships which keep a little free are the ones which make the quickest passages; consequently it may become necessary to make a longer circuit than that given, and yet have a quick voyage.

In January, cross 15° S. near 79°; 20° S. near 81°; 25°

S. near 82°; 31° S. near 82°; 32° 30′ S. near 80° W.; then head for Valparaiso.

In February the port tack carries vessels to 25° S. and 83° W.; beyond this parallel the chances will vary, and they may find winds from S. and SW.; 80° W. should generally be crossed between 32° and 33° S.

In March, 20° S. can generally be crossed near 81° W.; 30° S. near 84°: here the winds are from SE. to SW., and haul to the westward as the ship goes south. The parallel of 32° 30′ should be crossed near 80° W.

In April, cross 20° S. at about 81° W.; vessels can often reach 25° S. between 81° and 82° W.; 30° S., near 81° W.; 31° or 32° S., near 80°; and thence steer for their port.

In May, cross 20° S. at 81° W.; 25° and 30° S. at 82°, and then make for Valparaiso.

In June, cross 20° S. at 81° W.; 25° S. at 82°; 30° S. between 80° and 81°; and then shape the course so as to make the parallel of Valparaiso near 75° W.

In July, vessels will often be able to cross 20° and 25° S. at 81° W.; as the northerly winds are more frequent than during the preceding month they may fetch 27° 30′ S. at 79° W.; 30° S. at 77° W.; sailing directly for their port from this point.

In August, cross 15° S. in the neighborhood of 80° W.; 20° and 30° S. near 82° W.; 32° S. at 80°; and 32° 30′ S. at 75° W.

In September, cross 20° S. a little to the eastward of 81° W.; 25° and 30° S. between 81° and 82° W.; and 32° S. at 79° W.

In October, cross 25° S. between 82° and 83° W.; beyond 30° or 31° S. a fair or leading wind will generally be found.

In November, cross 20° S. near 82° W.; 25° S., between 82° and 83°; and attempt to reach 30° S. near 80° W.

In December, 15° S. should be crossed at 80° W.; 20° S. at 82°; 25° S. at 84°; from this parallel to 30° S. the wind varies from SW. to ESE.; beyond 30° S. the direction of the wind is favorable for running down the easting.

§ 85. ROUTE FROM CALLAO TO CAPE HORN.—We have seen in the preceding paragraph that ships, leaving Callao for Valparaiso, should keep a good full while passing through the SE. trades. This recommendation is just as important for ships bound around cape Horn. In this passage the region of westerly winds should be reached as soon as possible; no matter if the ship be carried very much to the westward in search of them. In short, the farther the ship crosses the parallel of 35° to the westward, the more favorable will be her position for doubling the cape with the NW. and SW. winds.

As we have stated in § 72, the great-circle route is the best for doubling the cape.

A very good point to cross 30° and 35° S., is between 88° and 93° W.

§ 86. ROUTE FROM THE "INTERMEDIATE PORTS" TO VALPARAISO AND CAPE HORN.—The following quotation is from Fitz-Roy's instructions for the navigation of the coast of Chile, and may be of use to vessels bound from the intermediate ports to the southward:

"If bound to the southward steer direct for the place, if fortunate to have a wind which admits of it; but if not, stand out to sea by the wind, keeping every sail clean full, the object being to get through the adverse southerly winds as soon as possible, and to reach a latitude from which the ship will be sure of reaching her port on a direct course. Every experienced seaman knows that, in the regions of periodic winds, no method is more inconsistent with quick passages than that of hugging the wind. When Rear Admiral Sir Thomas Hardy was on the coast, he used to cross the southerly winds with a topmast studding-sail set, his object being to get through them."

This advice is especially given to vessels bound to Valparaiso. We have nothing to add to what we have said on analogous routes in §§ 71, 76, 78, 81, and 84.

After leaving the intermediate ports it will be still more to the advantage of ships bound to cape Horn to keep a "clean full" while making passage through the zone of southerly winds; in fact, they should steer a trifle freer than if bound to Valparaiso. It is probably best to cross 35° S. near 83° W., or even farther to the westward; and it can be generally stated that the farther 35° S. is crossed to the westward, the sooner the cape will be doubled. The length of the route will certainly be compensated for by the increase of speed, caused by keeping away, both in the trades and in the belt of westerly winds.

§ 87. ROUTE FROM VALPARAISO TO CAPE HORN.—After

leaving Valparaiso southerly winds will be encountered, except during the rainy season, when they are interrupted by *northers*, (vide § 21.)

It will therefore be easy for ships to make to the westward as far as Juan Fernandez.

As a general rule 35° S. should not be crossed at less than 80° or 82° W. It will be advantageous—as stated in the foregoing paragraph—to cross this parallel as far to the westward as possible; that every advantage may be taken of the prevalent W. winds, which haul to northward and to southward as far as 40° or 42° S.

In January, cross 35° S. at 82° W.; 40° S., when possible, between 80° and 85° W.; 50° S. at 81° or 82° W.

In February, the same route.

In March, cross 35° S. at 80° or 80° 30′ W. Aim to cross 40° between 80° and 83° W.; after which bear to the southward.

In April, May, and June, vessels are sufficiently far to the westward if they cross 35° S. between 78° and 80° W.; and 40° S. between 80° and 83° W.

In July and August, cross 35° S. between 81° and 82° W.; 40° S. between 82° and 83° W.

In September, October, November, and December, cross 35° S. as far as possible to the westward. Thus, if it can be done, the parallels of 35° and 37° should be crossed between 82° and 85° W. The S. winds at this season prevail as far as 40° or 42° S. The west route is therefore much the best.

The crossings just indicated for the different seasons show approximately the route which the winds allow. Sailing-vessels should keep still farther to the westward if the wind permit. Auxiliary-steam vessels need not, of course, make such a long circuit. They can head, for example, for 35° S. near 78° W.; and 40° S. near 80° W., using steam if necessary.

We take from the *Ann. Hydr.* the following extract, relating to a voyage from Valparaiso to cape Horn, made by the sailing-frigate *Alceste*, Captain *Brosset:*

"In order to double cape Horn the proper route to be made, after leaving Valparaiso, is to keep 150 or 200 miles from the coast of Chile; thence to descend parallel to this coast until 50° S. is reached; after this, to make an oblique route around the cape.

"I left Valparaiso on the 14th December, and was favored with moderate winds from SSW., varying to S., SSE., and even SE. I made my offing by passing to the northward of Juan Fernandez islands.

"On the 18th December I was about 570 miles from the coast of America, lat. 33° 30′ S., long. 85° W., and in good position to make to the southward; but I crept along at a slow rate, and was off my course when close-hauled, being several times obliged to take the port or west tack in order to keep at a convenient distance from land. From the 18th to the 28th December the wind blew constantly from S. to W.; oftenest from SSW. and SW., varying in intensity from a light breeze to a moderate gale.

"At 50° S. and 86° W. I commenced to bear away toward E., and, assisted by the SSW. wind—which now became favorable—I crossed, on the 31st December, the 56th degree south latitude at 76° W. I supposed that, in all probability, I should quickly double cape Horn with the strong west winds which seem to reign perpetually in these localities, when the SW. wind hauled to S. and SSE., and dying away altogether was succeeded by a calm. Nor did I make Staten island until the 11th January. During the 12 days taken to double cape Horn (from 56° S. and 76° W. to Staten island) I constantly experienced winds from ESE., E., light breezes from NE., and calms and light airs from S. to SSE., with which I slowly made headway to the eastward.

"On both the 5th and 6th January I had a gale from the east.

"Although this gale only took place on the 5th and 6th January, the rotation of the wind commenced from W. on the 2d, and with the hands of a watch; that is, in a direction opposite to the usual one. According to the Dutch instructions this is nearly always a certain sign of a tempest."

§ 88. Route from Valparaiso to Concepcion.—Below will be found the instructions of Capt. Basil Hall on this passage, from whom we have had occasion to quote several times before:

"*Passage made in 20 days, October*, 1821.—As the prevalent winds along this coast are from the southward, it is necessary to take advantage of every slant that will allow of southing being made; and we were fortunate in meeting with a westerly wind on the third day after sailing, which carried us more than half the distance. The wind subsequently was S. by W., which made the rest of the passage to Concepcion almost a dead beat. We arrived at Talcahuana, in Concepcion bay, on the 8th. During the 9th it blew fresh from the northward. We afterward beat up to the bay of Arauco, and to the island of Mocha, in 38° 19′ S., having on this occasion been favored with a south-easterly breeze, and then a southerly one to stand in with.

"We endeavored to reach Valdivia also, but the wind came from S. by E., and blew so hard that we were obliged, for want of time, to give it up. On the return passage to Valparaiso, we had light NW. and W. winds, then SW., and so on the southward and S. by E., which is the most common wind.

"These particulars would seem to point out that a passage may always be made to the southward, for the winds are seldom steady for twelve hours, and by taking care to profit by every change, southing must be made.

"The passage from Valparaiso to Concepcion is generally made in ten days, which is also the usual time required for a passage to Callao. The distance, however, in the first case is 200 miles, and in the latter, 1,320 miles—a circumstance which points out very decidedly the direction of the prevalent winds."

We will also give an abstract of the instructions of Captain *Fleuriot de Langle*, (vol. 22 Ann. Hydr:)

According to this superior officer, the winds which prevail during the greater part of the year from SSE. to SSW., along the coast of Chile, render the passages from Valparaiso to the southern ports long and tedious. These winds are interrupted in winter by *northers* of short duration. Ships will consequently be obliged to beat up on the port tack; and will find enough wind and sea to compel them to take at least two reefs in the topsails. It would, at certain times, seem preferable to stand well out to sea, and not to go about until the latitude of the port is reached; at others it is best to beat up the coast. If bound to Maule or Talcahuana it is not advisable to stand on the port tack farther than to mid-distance between Valparaiso and Juan Fernandez.

If this advice be followed there is a chance of striking a

northerly breeze, even during the fine season. The weather is also better and the sea smoother than it is farther to the westward.

Between Juan Fernandez and the coast of Chile, it often happens that the wind prevails from SW. near the coast and from SE. farther out; or vice versa. Thus it is sometimes very difficult, after having made a great deal of latitude, to run down the longitude. By tacking under the land the passage will generally be shortened; the time from Válparaiso to Concepcion being only from 4 to 8 days by the coast route; while it is from 10 to 15 days by the off-shore route.

While beating up the coast the temperature of the water will show whether or not the ship is close to land, as the sea is colder near shore. According to M. Fleuriot de Langle, vessels bound to Valdivia or the Chiloé islands, and those bound around the cape, should keep on the port tack until they sight Juan Fernandez; they will usually make this island in three days. As the land is high, be careful not to approach too close to leeward, as ships are apt to lose the wind.

CHAPTER III.

Routes from the western coast of America, across the Pacific.

§ 89. ROUTE FROM VALPARAISO OR CALLAO TO AUSTRALIA, (*by the trades.*)—Vessels bound to Australia from the coast of Chile or Peru, should first strike for the SE. trade region. Those leaving Valparaiso will therefore steer immediately NNW. or NW.; while those starting from Callao should head W. The settled SE. trades being once found, the passage to the Marquesas islands is quickly made.

Fatu-Hiva, or Madeleine island, is a good land-fall, and is 8 miles long N. and S., and 4 miles broad. The land is high, with a peak rising for nearly 4,000 feet.

Between the coast of America and the Marquesas the wind will become steadier and haul more and more to the eastward as the ship goes to the westward.

Beyond the Marquesas constant trades can be relied upon, particularly from July to October; but they will not be so fresh or steady during the rest of the year, (vide § 4.) During the spring months, and after the sun has passed the summer solstice of the N. hemisphere, the voyage can be made to the northward of the Samoa group; passing close and to the southward of Uvea island, to the westward of the Fijis, the tropic can be crossed a little west of the meridian of the isle of Pines. During the rest of the year, from November to June, steer direct from Fatu-Hiva to Tonga-Tabu, and thence so as to cross the tropic near the same meridian, 166° E.

For further instructions on this route vide §§ 132 and 133.

Auxiliary steamers, starting from Valparaiso, can shorten this passage by passing the Paumota islands to the southward or by sailing through the group on the 20th parallel, (vide § 93.) After reaching about 141° W., they can leave 20° S., and steer so as to intersect the parallel of Tahiti near 158° W. They can then pass between the Samoas and Tongas. The last part of the passage is similar to that already given. But we think that it will be advantageous for sailing-ships to keep to the route we have indicated, especially from March to October; that is, to sight Fatu-Hiva and pass to the northward and clear of the Paumotas. It often happens that calms or westerly breezes are found south of the Paumotas; though that, of course, will be no inconvenience if steam-power is available.

§ 90. Route from Valparaiso or Callao to the Indian ocean, Saigon, Batavia, Melbourne, etc.—This passage can be made in two ways: *by the trades, and by the cape Horn route.*

These two routes are so different that it is difficult to decide which is the better. To make a choice between the two, one should take into account the qualities of the ship; the importance attached to making a passage more or less rapid, no matter what it may cost; the season of the year, etc.

The trade-wind route will be less boisterous, but even this is often dangerous; in certain cases it will be sensibly longer. The cape Horn route is severe and generally quicker, passing through higher latitudes.

Vessels bound to Melbourne should navigate in the SE. trades according to the instructions given in § 89. Once south of New Caledonia, they should approach the coast of Australia, in the manner described in § 138; the passage should be finished as stated in § 178. Sailing-ships are almost certain to have head winds and squalls between New Caledonia and Melbourne, (vide §§ 6 and 7.) It will be especially difficult to pass through Bass strait, as January, February, and March are the only months when occasional easterly winds may be expected. 1st. The trade-wind route.

During these months, and from November to March in general, vessels bound to the Indian ocean and Batavia should, after sailing over the route just described, pass to the southward of Australia. This part of the route will be found in detail in § 172.

It is possible to pass north of Australia at all seasons; the weather being better by this route. From March to September there will therefore be a choice between Torres strait, which can be traversed with the SE. monsoons, and the route by St. George's channel and the N. of New Guinea.

In either case pass north of the Samao islands, close to the southward of Rotumah island and north of the New Hebrides. For further information vide § 171. During the other season, from September to March, the NE. trade route can be taken, (vide § 92.) Make the Bashee islands between Formosa and Luzon, and run down the China sea with the NE. monsoon. Instructions relating to the latter part of the voyage will be found in § 166 and subsequent paragraphs.

The following quotations, referring to trade-wind passages made during the seasons of both monsoons, are from Horsburgh, (vol. 3, p. 764:)

"Several ships have made very rapid passages from the coast of South America to India. Captain Peircy left Valparaiso in January, 1814; after crossing the Pacific, he entered the China sea, passing through the Bashee group; crossing the strait of Malacca, he arrived at Bengal 86 days after leaving Valparaiso.

"The *Sherburne* left Copiapo for Calcutta on the 27th February, 1824; sailed between the Marquesas and Society groups, and sighted the most easterly of the Navigator islands. Instead of the prevalent SE. winds, she often experienced light and variable airs; she took the route through St. George's channel; thence sailed along the north coast of New Guinea, through Gillolo and Ombay channels, and entered the Indian ocean on the 15th June, after doubling the S. point of Sandal wood."

2d. The cape Horn route.

When the destination is Réunion, Bombay, Calcutta, Batavia, or even Saigon, quicker passages will undoubtedly be made by this route. It will be particularly advantageous for ships leaving Valparaiso and Callao, between the months of August and January. During this season the weather will not be so severe in that part of the passage made in high latitudes; the temperature will be warmer, and, the days being longer, the floating ice can be more easily avoided, (vide § 43.) Moreover, during this season the trade-wind passage will be more uncertain as bad weather and westerly winds are common in the western part of the intertropical zone, (vide § 4.)

During the rest of the year, and particularly when starting between March and June, the SE. trade passage can be made with steady winds; the weather will be fair, and the difference in time between the two routes will be little or nothing. It is well understood, however, that the only route at all seasons, for good and well-equipped vessels bound to Reunion or Bombay, is that by the cape.

One of the inconveniences of the southern route, especially during the winter, when the nights are very long, is the danger of meeting icebergs. The chances of running foul of them can, however, be diminished after leaving cape Horn, by steering so as to cross 40° S. near 41° or 42° W., when run down the easting; but this circuit will greatly increase the length of the voyage, and a more rapid passage will probably be made, by gradually shaping the course (after leaving cape Horn) for 40° S. to the eastward of 28° or 29° E. if bound to Reunion; for 40° S. at 37° or 42° E. if bound to India, or near 62° E. if the destination is Batavia.

Maury gives the following instructions for this route:

"I have advised a ship-master who consulted me by letter to go by way of cape Horn. The distance, by the way of the cape, to Calcutta is 10,500 miles; while the distance by the usual route west, or 'running down the trades,' as it is called, is 13,000 miles. The difference in time will be quite as great as this difference of distance would indicate. Indeed, in addition to distance, time is also in favor of the cape Horn route, for the winds are stronger and quite as fair.

"As one stands between the capes of the South Atlantic and looks north upon the chart, he sees a part of the ocean in the shape of the letter A, without the cross, which is untraveled, except by whalemen and sealers. The track to and around the cape of Good Hope forms one side of the letter; the track to and fro around cape Horn, the other. Between these two the ocean is a solitude. Among the many thousand logs on file in the observatory there is not one to show that any trader has ever performed the voyage from the offings of cape Horn to the offings of the cape of Good Hope.

"The way by the cape Horn route to India is to proceed from Valparaiso as though you were homeward bound around the cape, and then, with 'the brave west winds' which prevail there, to run east with flowing sheets, passing between the isles of South Georgia and Sandwich-land, keeping a bright lookout for icebergs. The route thence crosses the prime meridian in about 54° S., 20° E. in 50° S., 35° E. in 40° S., by which time the navigator will again find himself in the traveled thoroughfares, and will know how to proceed.

"Distance in miles from Valparaiso via—

	Cape Horn route.	Western route.
To Canton	11,500	10,800
To Shanghai	12,200	10,500
To Java-head	9,700	......

"In the southern summer the voyage from Valparaiso to Canton may, on account of the winds, be performed quite as quickly via cape Horn as it may be by the route west. If the 'brave west winds' will enable a ship by cape Horn to average only ten miles a day more during the voyage than she can in running down the trades west, time, which now is worth so much in navigation, would be somewhat in favor of the cape Horn route, even to Canton."

§ 91. Route from Valparaiso or Callao to New Caledonia and New Zealand.—General observations on this route will be found in § 89, to which we will add the following instructions:

After leaving Valparaiso, especially in an auxiliary steamer, and when the sun is in the S. hemisphere, the route lies through the southern part of the Paumota archipelago and along the parallel of 20° S., (vide § 93.) But sailing-ships will, we think, always do better if they steer for Fatu-Hiva, after entering the trades. Starting from Callao, they should certainly head for the Marquesas. After correcting their position, by sighting Fatu-Hiva, they should follow the instructions given in § 132 for vessels bound to New Caledonia and New Zealand.

As a general rule, it is always easy to reach Noumea, since this route lies entirely in the trade belt. The wind will always be found steady, and especially settled, from March to October, (vide § 4.)

From December to May. Ships bound to New Zealand during these months should steer, after leaving Fatu-Hiva, so as to pass to the west of Cook's group; thence S. of Nicholson's shoal; they should cross the tropic near 173° W.; and pass to the westward of the Kermadec isles.

From May to November. At this season it is preferable to pass to the northward and westward of the Tonga islands and to intersect the tropic near 180° or 178° E. From this west point there will be fewer chances of being set to leeward after running out of the trades and while heading for the Bay of Islands with the prevalent westerly winds, (vide § 132.)

§ 92. ROUTE FROM VALPARAISO OR CALLAO TO CHINA.* —The passage from South America to China may be made by either the NE. or SE. trades. These two routes are called respectively the *northerly* and *southerly*.

The southerly route is to be preferred by vessels leaving Valparaiso or Callao from February to July. The northerly route is the better one during the remainder of the year.

1st. *The southerly route.*—First head, as has been stated in §§ 89 and 91, so as to make Fatu-Hiva, thence steer to the southward of the Gilbert group and to the northward of the Pelew islands. After this shape the course—if the passage is being made between March and October—toward the strait of San Bernardino, (vide § 101,) and enter the China sea through the passage between Luzon and Mindoro. Here the SW. monsoon prevails. But from October to March, after leaving the Pelew islands, keep to the northward of the Philippines, and gain the China sea by passing through the Bashee group. Thence the passage to Hong Kong will be easy, the NE. monsoon being favorable.

There is also another route which has its advantages for ships coming from the coast of America from February to July. To follow it, steer, after sighting Fatu-Hiva, so as to intersect the line near 166° or 168° W. Then take a NW. direction, and pass north of the Marshall group. After crossing the line at 166° W., vessels can equally as well pass between Bonham island (Ralick group) and Mulgrave island (Radack group;) and then make to NW. between the two chains of islands. In both cases, after they have once passed to the northward of the Marshall group, they will make rapidly to the westward, keeping to the southward of Guam island, (Marianas;) after this make for San Bernardino strait, in the same manner as advised during the SW. monsoon.

2d. *Northerly route.*—After leaving Valparaiso or Callao, head, with the SE. trades, to cross the line near 138° W., and 10° N. near 143° W.; the NE. trades will be found near this parallel. The parallel of 18° N. should be crossed in

* Instructions for the "cape Horn route" will be found in § 90.

the neighborhood of 160° W. Thence follow the instructions given in § 101.

§ 93. ROUTE FROM VALPARAISO TO THE MARQUESAS AND TAHITI.—Vessels starting from Valparaiso and bound to the Marquesas islands should cross the tropic near 88° W., and 12° S. at 108° W.; and then lay a course straight for Fatu-Hiva, (or Magdelena;) from this island the anchorage in Taïo-Hae bay (Nukathiva) is easily reached. Sailing-ships bound to Tahiti will also find it to their advantage to use the same route. After sighting Fatu-Hiva they can run through the Paumota group, as stated in § 120; but it is shorter to cross the tropic near 98° W. and 20° S. near 118° W., and make the westing through the archipelago on this parallel, which is comparatively clear of dangers.

According to Wilkes a ship may keep on the 20th parallel until she reaches 141° or 142° W., and then head for Tahiti direct. On this route, Carysford island (Tureia) will be sighted to the southward at about 138° W., and Whitsunday island (Nukutavake) to the northward at about 139° W. Still keep on the 20th parallel and run between Barrow island, (Vanavana,) long. 139° 05′ W., and Byam-Martin island, (Pinaki,) long. 140° 25′ W. Thence pass to northward of St. Paul, (Hercheretue,) long. 145° 05′ W. All these islands are visible at a distance of from 7 to 10 miles.

Captain Richard Fay gives the following directions for making a landfall at the Society islands from the southward: "It is always dangerous to approach the Paumota group from the southward or southward and westward; pass west of Moorea, and then steer for Tahiti. However, it is an easy matter during the good season to make Tahiti from the east by passing between Anaa and Maitea islands, and very near the latter. Keep a good distance from the NE. coast of Tahiti, and make the landfall at point Venus;* but do not run in closer than 6 miles, on account of the reefs which extend for 4 or 5 miles off shore, especially between Mahena and an islet situated to the E. of point Venus.

"By following this last route you will be able to make to the northward on the starboard tack, with a free wind, which afterward hauls aft as you approach the land. When as high as point Venus you can run inside the reef by the Tanoa or Papiete passage, according to the appear-

* There is a light on point Venus, (vide § 184.)

ance of the weather. The pilot will, however, come on board and take the ship, by one or the other of the passes, to the anchorage of Papiete, where come to in 7, 8, or 9 fathoms of water; bottom, sand and mud.

"Vessels coming from the E. or Valparaiso, and bound to Nukahiva, should sight Magdelena island, situated to the southward of Nukahiva. They should then head for Taio-Hae bay or Comptroller bay. The latter is situated on the eastern coast of the island, and is larger than the former; but it is far from the French station and the residence of the chief at Taio-Hae."

Observations on this route by Mr. G. Biddlecombe, Master H. M. S. Actæon.—"On leaving the coast of Chile or Peru, run into the SE. trade-wind, or in latitude 20° S., as soon as possible, when you will generally have strong easterly winds and fine weather; you may then stand to the westward in that latitude till you bring Pitcairn island to bear about SW., when you should steer for it, taking care not to get to the westward of the island, as the current runs strongly to the westward, owing to the prevailing easterly wind, except about December and January, when a northerly or NW. gale sometimes sets in. From Pitcairn island you will be enabled to shape a course for the Marquesas, taking care then to keep to the eastward, as the SE. or SSE. trade blows through the islands."

§ 94. ROUTE FROM CALLAO TO THE MARQUESAS AND TAHITI.—After leaving Callao or any of the adjacent ports the course should be west to gain the trades, which will be found quite near land. Afterward, there are two routes, either of which may be followed. The first is a direct one to the Paumota islands; the second takes the navigator within sight of Fatu-Hiva, and then along the northern edge of the Paumotas.

Both passages are easy, and only demand attention while passing the low islands of the Paumota group. The trades grow fairer as the ship clears the coast. In the neighborhood of the Paumotas they often become squally and variable.

Apropos of the direct route we will cite Wilkes's remarks, (vide "U. S. Exploring Expedition:")

"Coming from the eastward, from Callao for instance, the most direct route is to follow the parallel of 18° 30′ S.

You will thus sight Clermont-Tonnerre island, (*Pukaruha*,) long. 136° 20′ W., and can pass either to the northward or southward of Serles island, (*Reao;*) thence shape the course to double Aki-Aki to the northward, and pass Harp island (*Hao*) to the southward. Afterward, sail to the southward of Dawhaidia, (or of the two groups, Ravahere and Marokau,) and head for Tahiti. The only danger by this route appears to be the Buyers, (*Reiotoua;*) but this group, supposed to be situated at 18° 20′ S., 143° 07′ W., has been looked for in vain by Lieutenant Parchappe, and has been erased from the latest edition of charts.

"There is still another route, on the parallel of 15° S. By this you will first sight Houden, (*Puka-Puka*,) from which make a straight course for Kotzebue, (*Aratika*,) or the northern part of King, (*Taiaro.*) Thence pass to the northward of Vincennes, (*Kauchi*,) and to the southward of Elizabeth (*Toau*) and Greig, (*Niau*,) and steer for Tahiti."

As we have stated, it is preferable to sight Fatu-Hiva or Magdelena before making for the Paumota group, (vide §§ 103 and 120.) This is the most frequented route from Callao or Payta to Taïo-Hae bay, (Marquesas.)

A westerly current may be looked for in this passage running at the rate of at least 10 miles per day. If no observations can be obtained, it will be well to count on a speed of 20 miles, especially if the course be for the Paumotas direct.

§ 95. ROUTE FROM VALPARAISO OR CALLAO TO THE SANDWICH ISLANDS.—After losing sight of the coast of South America, make for the SE. trade region, steering NNW. or NW. from Valparaiso, and W. from Callao. After reaching the trades head for the crossings mentioned below.

From June to October. From June to October cross the equator between 133° and 138° W., and 10° N. at 140° or 141° W., thus avoiding the constant calms which prevail to the eastward of 133° W. Ships generally carry the SE. trades to 10° N., and even beyond that parallel; passing from one set of trades to the other without intervening calms.

From October to June. From October to June cross the line between 132° and 134° W., and 10° N. at 138° W. The ship will lose the SE. trades between the equator and 10° N. and experience few or no calms.

Beyond 10° N., steer with the NE. trade-winds, so as to fetch well to windward of the port of destination, keeping well to the E. of Hawaii, and stand in for the land on a parallel north of the port. Navigators should remember that the equatorial current may set them to the westward and to leeward. If bound to Honolulu, they should make the island of Maui from the northward.

§ 96. ROUTE FROM PANAMA TO AUSTRALIA, NEW CALEDONIA, AND NEW ZEALAND.—Starting from Panama and bound to the W., it will be advisable to make as much southing as possible in order to reach the SE. trades, (vide §§ 62 and 77.) Vessels whose destination is Australia, should therefore head on the same route as that described for the passage to San Francisco or Callao.

It will generally be very advantageous to cross the line east of the Galapagos, near 83° W., when possible. Auxiliary steamers will, of course, always follow this easterly route. If compelled to pass north of the Galapagos, the chances are great that they will be detained by calms and light, baffling airs, except during October, November, and December, when the winds are more settled.

After the trades are once reached to the southward of the line, they should steer for the high lands of Fatu-Hiva, (Marquesas.) For details concerning the remainder of the voyage vide § 91.

§ 97. ROUTE FROM PANAMA TO CHINA.—First follow the routes given in the preceding paragraph, and that for making the passage from Panama to San Francisco contained in § 62.

If the point of departure be Central America or Mexico, follow the instructions given in § 63. An example of this passage will also be found in § 99. These instructions all apply until the region of steady NE. trades is attained.

Vessels should afterward steer (with the NE. wind) a course that will bring them to the meridian of the Sandwich islands between the parallels of 15° and 20° N., and then finish the voyage as described in § 101.

§ 98. ROUTE FROM PANAMA TO THE MARQUESAS AND TAHITI.—Instructions relating to this passage may be found in § 96. They can be abbreviated as follows:

After leaving Panama make to the southward as much as possible, until the SE. trades S. of the line are reached.

Then head for Fatu-Hiva, and finish the voyage as described in §§ 103 and 120.

§ 99. Route from Panama to the Sandwich islands.—Reach the SE. trades as soon as possible, following the instructions given in §§ 62 and 96. Afterward cross the equator (to the N.) near 128° W.; 10° N. near 138° W.; and finish the passage in accordance with § 95. Be careful to reach a parallel to the northward of the destination as soon as possible.

We will, in addition, give as an example of this passage an extract from volume 19, *Ann. Hydr.*, taken from the log of the frigate Havana, Captain Harvey:

"On the 30th June we left Herradura bay, (E. coast of the gulf of Nicoya.) We experienced light and variable winds, calms, squalls, rain, and often thunder-storms, until we reached 106° W.; and, indeed, the weather was always variable and squally until the last week of the passage. From the 28th July, lat. 15° N., long. 113° W., to the 9th August, lat. 14° N., long. 127° W., we had westerly winds, in accordance with Maury's instructions for this season. Between 125° and 140° W. the wind was from N. by W. to N. by E.; it then shifted to NE. We sighted the island of Maui on the morning of the 19th, and anchored the same evening at Honolulu in 8 fathoms; that is too close inshore, for the best anchorage is in 14 or 15 fathoms.

"According to Maury this passage can be more quickly made by pushing to the S. and SW., until you meet the SE. trades well over the line, between 2° and 5° N.; then run to the westward till between 115° and 125° W. Then bear to the northward, sooner or later, according to the season.

"In summer the last longitude (125°) should be preferred. After the zone of 'doldrums' you will reach the region of NE. trades. You will thus avoid the unfavorable weather of the Central American coast.

"The *Swift*, by taking this route, went from Panama to Honolulu in 47 days. But in 1848, the *Herald*, notwithstanding the fact that she was towed a distance of 1,000 miles from Panama, took 42 days to clear the 'doldrums,' crossing them on the parallel of 9° or 10° N.

"It is best at all seasons, when approaching the Sandwich islands, to give Hawaii a wide berth, and to make the land on the north coast of Maui. Do not anchor off Honolulu during the four winter months, (from December to April,) as the southerly squalls are dangerous and the holding-ground poor. During the rest of the year vessels can safely anchor in from 15 to 20 fathoms; bringing Punchbowl to bear NE. by E., and the point on Diamond head E. by S.

"Though the passage leading to the port is well buoyed, it is dangerous to pass through, even with a favorable wind, unless thoroughly acquainted with the reefs."

§ 100. Route from San Francisco to Australia, New Caledonia, and New Zealand.—We will first give Maury's instructions for the passage from San Francisco to Sidney:

"From California to Australia the route out of San Francisco should be down as soon as possible into the NE. trades, as though you were bound to China, India, or the Sandwich islands, crossing the equator anywhere between the meridians of 140° and 150° W., according as you prefer to run down your westing principally in the NE. or SE. trades.

"I give the preference to the latter, generally, because they are more steady, reliable, and certain than their congeners of the northern hemisphere; at least such is the rule. The distance by this route to Bass strait will be about 7,000 miles, and an increase upon this of the average distance to be sailed on the passage going, together with the distance returning, will not amount, as before stated, to more than six or eight hundred miles.

"Aim to cross 30° S., on the passage from California to Australia, in the neighborhood of 170° E. Thence the course is between Australia and New Zealand, direct for your port.

"In these passages, as on the California routes generally, navigators have to cross the calms of Cancer and Capricorn, as well as those of the equator, which last are found between the NE. and SE. trade-winds, but upon different parallels, according to the season of the year.

"It may therefore be remarked here, once for all—and which remark navigators, bound either from the United States or from Panama to California, are requested to bear in mind—that the barometer will often enable the navigator to tell when he has crossed these belts of calms and entered the trades."

The course should therefore be first down into the region of the NE. trades, as stated in § 69. After reaching these steady, settled winds, make the following crossings:

From January to July.

From January to July, cross 10° N. at 143° W.; and the equator at 148° W.

In January, February, and March, no calms, properly speaking, will be found between the NE. and SE. trades.

In April, May, and June there will be only about 2 per cent. of chances of calms in this region.

In July, August, and September.

In July, August, and September, cross 10° N. at 148° W., and the equator between 150° and 153° W. In this season, and if the precaution be taken not to follow a more easterly route than the one indicated, there will be only from 2 to 3 per cent. of calms between 10° N. and the line.

From October to January.

From October to January, cross 10° N. at 138° W., and the equator at 143° W. By following this route there is only from 2 to 3 per cent. chances of calms between the two trade-wind regions. Farther to the westward, at this season, the ship would be liable to meet more calms.

At all seasons cross 10° S. near 154° or 156° W., or anywhere between those meridians. For information regarding the termination of the voyage, vide §§ 91, 96, and 132.

§ 101. ROUTE FROM SAN FRANCISCO TO CHINA.—We will commence with Maury's instructions on this passage:

"The distance between California and China being nearly double the distance between the United States and Europe, a vessel navigating these waters has a wider range in latitude than one trading across the Atlantic has, in which to hunt good winds. All vessels going west from California will, almost of necessity, stand to the southward and westward for the NE. trades. In summer and fall they need not go as far south for steady trades as they do in winter and spring."

Maury then gives a copy of a letter addressed to him by Captain *Ranlett*, of the *Surprise*, observing at the same time—and with reason—that this ship made her westerly passage on a parallel too near to the "calms of Cancer;" that is, too far north. We will quote a portion of Captain Ranlett's letter, though it should be understood that the same route is not to be followed:

"Last year I crossed from San Francisco to Shanghai in the *Surprise*, and had a good run of 38 days across. Your 'wind and current charts' were not then out, I think; at least I had not seen them; I, for want of some such directions as you give, took my own course, and kept far to the north of the Sandwich islands, and had a tolerably good run all the way, with much fine weather, while the *Mystery* and some others went farther south in the old track, and had much wet and squally weather, and longer passages generally. This voyage I left nearly one month later; and, although I have your charts of the Pacific, I kept north of all the tracks given, and have had very light winds all the way across; in fact, my sails have flapped against the masts all the way. I sailed, after leaving San Francisco, 5,580 miles by log, without taking in a skysail or a royal studding-sail, the wind veering and hauling from ESE. to ENE. generally; weather fine as one could wish, but too hot to work in the sun much of the time. I passed between North and Sulphur islands, two of the Volcano group. The north and south islands of this group may be run for at any time, being high and bold; but the middle one is low to the eastward, and cannot be seen far in the night; the high hill is on the western side. I passed between Ousima and Kakirooma without difficulty, and saw no dangers, except a high lot of rocks about five miles from the NW. point of Kakirooma, and what appeared to be an island off the SW. end of Ousima. The seas break heavily all around this heap of rocks, and in places between it and the mainland of Kakirooma, and, although there is a wide passage between that and the shore, I would not attempt it unless surveyed.

"I arrived on the 22d October, after a passage of 55 days. Found the *Golden Gate* in before me; she sailed 10 days after me, and had a strong NE. wind all the way across, excepting a few days. The *Swordfish* was 42 days from San Francisco to Hong Kong; she sailed two days before me. *I presume she went well south.* I beat all the passages across last year, but this year was beaten by all, though I took the same track."

Our own advice, therefore, is: To reach the NE. trades as soon as possible, (vide § 67, and the following.) Then run to the west, on a parallel, not too near the northern limit of the trade-winds, thus avoiding the "horse latitudes." In order to do this choose some parallel between 20° and 15° N., keeping near 20° N., between June and October; and, on the contrary, in the neighborhood of 15° between December and March. At all seasons pass south of the Sandwich islands and N. of the Marshall group.

From May to October. From May to October, when the SW. monsoon prevails, keep to the northward of the Carolines islands, and leave the Marianas on the northern hand. If deemed advisable Guam island can be sighted. The island is low on the north coast, but quite mountainous and steep on the south side. Afterward steer for the strait of San Bernardino, with the NE. winds. These will gradually change to variable winds with squalls, and then to the SW. monsoon as the vessel approaches the Philippines. Enter the China sea by the passage between Mindoro and Luzon, and end the voyage as stated in § 161. Keep a lookout for leeway while approaching the coast of China.

From October to April. From October to April, when the NE. monsoon prevails, run north of the Mariana group, sighting, if convenient, the Farallons de Pajaros. These are volcanic rocks, the highest of which has an altitude of about 800 feet. Some captains pass between the Grigan islands and Assumption; the former are a mass of mountainous rocks, the most elevated peak being about 1,300 feet high. Assumption is a volcanic cone; its diameter at the base is about one mile, and its height about 2,600 feet. It will generally be best to pass to the northward of all the Mariana group, as we have before stated; quicker passages are generally made in this way.

The entrance to the China sea is by the Bashee islands.

Note on the passage via the strait of San Bernardino, on the voyage to Hong Kong, from June to October.—Before entering the strait of San Bernardino, it is recommended, *when the wind is from S. or E.*, to sight cape Espiritu-Santo, which can be seen for forty miles. Leave the island of San Bernardino on either hand, according to circumstances, then pass to starboard of Capul, and to port of the islets off the SE. point of Luzon. Look out for currents which often run here very violently. Sail around Ticao from the E. and N., and pass between Burias and Masbate, then between Marinduque and the islet of Banlon. Thence run between Mindoro and Luzon, and pass successively to the south of Green island and Maricaban; skirt point Santiago, and double from the north, and at a good distance, Ambil, Lubang, and Cabra islands. From this position a ship can run up, off the W. coast of Luzon, on a N. by W. course, as far as cape Bolinao. For further information vide § 161.

§ 102. ROUTE FROM SAN FRANCISCO TO THE SANDWICH ISLANDS.—After leaving San Francisco run for the trades, with the prevalent north-westerly winds common on the coast. (Vide §§ 24, 67, and the subsequent ones.)

From June to December clear the coast as soon as possible, steering WSW. for instance. The calms to the eastward of 128° and 133° W. will thus be avoided. The trades always blow south of the tropic of Cancer. Near the Sandwich islands the trades may possibly haul to E., or even to SE. From June to December.

From October to January, particularly, the wind may blow from SE. in the neighborhood of the tropic. From October to January.

The land should be approached from ENE., when all the possible winds will be fair. When making a landfall remember that the currents often run at the rate of 20 miles per day, and that calms and baffling winds are common to leeward of the islands.

§ 103. ROUTE FROM SAN FRANCISCO TO TAHITI.—The portion of this route from San Francisco to the trades is the same as that described in the preceding paragraphs. Afterward head, nearly, as if bound to Valparaiso, (vide § 71.) Cross the equator between 118° and 123° W., from May to October, and between 113° and 118° W., from October to May.

Beyond the equator run for the high lands of Fatu-Hiva, then through the Paumota group toward Tahiti. As this island is surrounded by a belt of reefs, be very watchful on nearing it. Artemise bank is especially dangerous. The following extract from *Wilkes* indicates the best passages through the Paumotas:

"When making passage from the north to Tahiti, my advice is to sight Rangiroa, (*Uliegen, Dean, or Nairsa,*) and to pass between this island and Tikahau, (*Krusenstern;*) this passage would appear to be the best and safest. If obliged to make the Paumotas more to the east, (*which will be the case when coming from Fatu-Hiva,*) you can sight Ahii, (*Peacock;*) the position of the W. point of this island is 14° 35′ S. and 146° 27′ W.; in this case the passage between Rangiroa (*Nairsa*) and Arutua (*Rurick*) can be taken. You can pass still farther to the eastward, between Raraka and Katiu, (*Sacken;*) before reaching these islands you can sight Taiaro, (*King,*) and afterward sail through the passage between Raraka, Katiu, and Tahanea. The wind often allows vessels to keep to the east of this last island.

"We would also recommend another more westerly route. By this captains should first take the bearings of Napuka island, (*Disappointment,*) and then steer about SE., so as to make Augatau island, (*Araktcheff.*) Afterward double successively Amanu, (*Moller,*) Hao, (*The Harp,*) Nengo-Nengo, (*Henry,*) and Hereherétua, (*St. Paul.*)

"In all this advice, I have taken into consideration the prevalent winds, or at least those which one has any probability of meeting. It is very evident that the Paumotas can be crossed in all directions by navigators at home in the group and by those willing to run the risks incident to this locality. I only give, however, the surest routes, and those which I would use myself. By these routes, several islands, where shelter can be found, are passed, and advantage may be taken of the resources of the group."

For further information concerning that part of the passage comprised between Fatu-Hiva and Tahiti, vide § 129.

CHAPTER IV.

ROUTES FROM EUROPE TO AUSTRALIA, NEW CALEDONIA, AND TAHITI, AND RETURN ROUTES.

§ 104. ROUTE FROM EUROPE TO AUSTRALIA.—This passage can be made by two different routes: first, by the cape of Good Hope, or the *southerly route;* second, by Suez.

1st. *The southerly route.**—Vessels leaving Europe should cross the equator west of 25° W. from November to May, and east of this meridian from May to November. In the SE. trades they should run a little free, and always attempt to cross the meridian of the cape of Good Hope south of 40° S. Much depends upon the parallel on which the passage is to be made; the length of the voyage especially is dependent on the captain's choice in this matter.

Yet the views of different authorities are widely divergent: Thus the instructions of the British Admiralty advise 39° S. as the proper parallel. In the extracts from these instructions, to be found farther on, useful observations are given on the passage to Sydney after leaving Tasmania. Horsburgh favors the same parallel, and expresses a very marked preference for the passage through Bass strait. In the present paragraph will be found quotations relating to the proper route to be taken when the passage through this strait is used. If Maury be consulted, it will be noted that he is strongly in favor of the extreme southerly route, between the latitudes of 45° and 55° S.; the choice of the parallel depends on the season, according to him, as well as on the winds, the qualities of the ship, etc., etc.

The voyage to Australia only being made by good, staunch, well-fitted-out ships, we think that the general instructions should particularly relate to the route to be followed under ordinary conditions. Our opinion is that merchant-vessels

* The Atlantic portion of this route is also treated of in the author's work on the "Navigation of the Atlantic Ocean."

should follow Maury's route, but without going too far to the southward. During the southern winter, and especially from November to February, they should keep on a parallel *south of* 43°; they can descend as far as 50° S., or even beyond that parallel. Quite rapid passages have been made, *in summer*, near 50° S. The weather is often better, the sea smoother than it is between 42° and 46° S., while the chance of meeting icebergs is the same.

During the southern winter, that is from May to September, the length of the nights (dangerous on account of icebergs) and the extreme cold will generally prevent vessels from going so far south. *If the route be chosen between* 42° *and* 45° *S., we think a good passage ought to be the result at this season.*

This advice accords with Maury's, and that of many officers who have made the passage.

The logs of the Dutch merchantmen bound from the English channel to Java, prove indisputably that *the ships which kept N. of* 40° *S. met twice as many gales as those which kept S. of the same parallel.*

This observation, being the result obtained from the mean of a great many voyages, seems to destroy the reasoning of certain navigators who think that by following the English Admiralty route they will have less bad weather.

This latter is also a longer route than the more southerly one, where the degrees of longitude are shorter, and it would also seem that navigators who choose it, willingly place their vessels in a zone where the chances of storms are twice as numerous.

It also appears probable that there exists an intermediate belt, situated between 40° and 43° S., or thereabouts, where the winds are unsteady, very variable, and accompanied by rainy and damp weather. This belt, in which some captains believe the scurvy is apt to be developed, probably changes its latitude, following the declination of the sun; it is found, for instance, between 39° and 41° S. from May to August; and between 41° and 43° S. from November to February.

If it be decided not to follow the route south of 42° or 43° S., we think that the parallel of 39° is to be preferred to that of 40° or 41° S.

Ships following the Admiralty route, on 39° S., will have

warmer weather; but will generally make longer passages. It will, however, be shortened by going through Bass strait. Sailing-ships should, however, avoid passing through this strait during the months of January, February, and March, when the winds come out from ahead or the east.

Sailing-vessels preferring the southerly route, between 43° and 52° S., more or less to the south according to the season, should not generally pass through Bass strait. They should rather round Tasmania to the eastward and make Sydney, as stated farther on at the end of our extract from the Admiralty instructions.

Auxiliary steamers following our advice and taking the southerly route, can shorten the last part of the voyage a little by running through Bass strait.

A ship will of course be exposed to *floating ice* if she follows the southerly route; but the chances of meeting it are small on the parallel 39° S.

We consider the Admiralty route as disadvantageous to the interests of ship-owners; we hold to our opinion, already stated, that the southerly route is the best; and that it presents no especial dangers to stout and well-equipped vessels. Finally, we reiterate that the chances of meeting storms are less on the southerly route than on that of the 39th parallel.

There now only remains to find out if there be a greater chance of meeting icebergs on 43° S. than on 45° or 50°. Now, nothing proves (vide § 43, observation No. IV) that vessels sailing on the various parallels between 43° and 52° S. should expect to see fewer icebergs at one edge of the belt comprised between these two parallels than at the other.

Icebergs are more numerous in high latitudes, (from 60° to 50° S.;) but they drift rapidly to the north and quickly leave these localities. On the contrary, after they have been carried beyond 50° S. they are set toward NE., then toward E. by the current, and remain in warm latitudes until they melt. Still it is best, especially in the southern winter, not to go farther S. than the 51st parallel.

Captain Guérin, commanding the St. Paul, gives the following account of his meeting with icebergs between 46° and 47° S. and 4° and 12° E.:

"On the 1st January, 1867, lat. 45° 51′ S., long. 4° 28′ E., sighted an iceberg to port, distant 6 miles; sighted another

2 hours afterward. Next day strong breeze from NW., with thick fog; position at noon, 46° 31′ S. and 9° 08′ E. At 8 p. m., an immense mountain of ice loomed up in the fog right under our bow; succeeded in avoiding it. At 10.15 p. m. we passed close by a small detached piece of floating ice. This was the last, though we kept the courses clewed up all the night.

At daylight we were completely surrounded by icebergs. There were seven large ones, and numerous little bergs about them. At noon on the 3d January, lat. 46° 56′ S., long. 11° 53′ E., took an ENE. course, in order to run out of these high latitudes; *the advantages of the southerly route being hardly apparent*, at least in January. We passed our last ice-island on the night of the 3d and 4th.

"*Nothing announced the approach of* these icebergs; the thermometer showed no sudden change in their vicinity. At 46° S. the temperature of the air was 41°, and the water 36½°. It was almost useless to keep a lookout at night, on account of the thick fog; and the greatest danger we had to fear was from the numerous little icebergs that showed for only a few feet above the sea. The principal bergs were from 160 to 330 feet high. They should, by rights, drift faster than the small ones; consequently, to avoid them, I think it is best to cross the groups on the side of the largest berg. The only peculiarities I noticed in this region were the thick fogs, the absence of birds, an unusually smooth sea, and some old pieces of wreck."

We extract the following table from the *Annales Hydrographiques*, (vol. 18.) The reader should also refer to § 43.

In the month of—	Latitude S.	Longitude.	Name of vessel.	Remarks.
	° ′	° ′		
January....	At 44 00	2 40 W	The White Swallow.	Met floating ice.
	45 33	10 40 W	....do	Met three icebergs, one of them 100 feet long and 150 feet high.
	43 40	16 40 W	The Malay........	Met twenty-six icebergs of different sizes.
	42 18	12 08 W	....do	Met thirty-four icebergs of different sizes.
June........	45 00	48 43 E	The Panama......	Met two large icebergs.
July........	49 32	8 30 E	The Beverly	Met an ice-field and icebergs.
	48 35	11 00 E	....do	Met forty icebergs.
	47 15	16 03 E	....do	Met thirty icebergs.
	46 50	20 23 E	....do	Met two icebergs.
	47 30	19 25 E	....do	Met one iceberg.
	48 08	37 45 E	...do	Do.
	48 44	43 00 E	....do	Do.
September..	46 00	12 00 W	The Gertrude.....	Met one large iceberg.
	47 30	10 55 W	....do	Met two small icebergs.
	49 24	1 39 W	....do	Met two icebergs.
	48 15	45 00 E	....do	Met one very large iceberg.
	52 10	37 07 E	The Great Britain.	Met two large icebergs.
October.....	50 09	29 26 E	The Marion.......	Met one large iceberg.
	40 39	46 10 E	....do	Do.
	53 12	21 23 E	The Auckland	Met a field of ice and two icebergs.
	51 10	26 20 E	....do	Met one iceberg.
November ..	52 26	19 42 E	The Oriental......	Met one large iceberg.
	52 20	27 47 E	....do	Met three large icebergs.
	51 03	32 21 E	....do	Met four large icebergs.
	51 20	37 06 E	....do	Met one large iceberg.
	53 51	86 40 E	....do	Met two large icebergs.
	43 38	8 15 E	The Flying Dutchman.	Met four large icebergs.
December...	52 57	95 20 E	The Oriental......	Met one large iceberg.
	43 33	18 10 E	The Kongleader ..	Met two large icebergs.
	48 45	37 24 E	The Malay........	Met one iceberg, well worn.

Having given these general considerations, we will now quote some of the most important passages from the *Admiralty* instructions, *Horsburgh*, *Maury*, and *Desnoyers*. We will afterward give a few extracts from the logs of ships which have made this passage. After a perusal of these various authorities each captain can form his own opinion, which, we candidly admit, may be very different from our own. At whatever conclusion the navigator may arrive, he will, however, find several authors for and several against his route. Our preceding advice seems to us correct for merchant-vessels, and is based on all past experience; the accumulation of twenty years' knowledge proving the correctness of Maury's theory, that the southerly route is the best.

The Admiralty Instructions.—The English Admiralty give the following instructions on the "Route to Australia:"

"Should the southern route to Sydney be preferred, a ship running down her longitude in 39° S., as recommended, and having arrived at 130° E., should get to the southward of Tasmania on the meridian of 145° E., before making the land, in order to avoid falling in with its rocky western coast in the night, from any error in the reckoning, or being caught on a lee-shore by a SW. gale. After rounding the South cape, a ship, not bound into Hobart-town, should give a berth of at least 20 or 30 miles to cape Pillar and the east coast of Tasmania, by which she will escape the baffling winds and calms which frequently perplex a ship in-shore while a steady breeze is blowing in the offing. This is more particularly desirable in summer, when easterly winds prevail, and a current is said to be experienced on the SE. coast running to the N. by E. at the rate of three-quarters of a knot at 20 to 60 miles off shore, while inshore it is running in the opposite direction with nearly double that velocity. From 30 miles to eastward of cape Pillar, the distance to cape Howe is 350 miles; and the course N. by E. After just sighting the cape stand to the northward for Port Jackson, remembering that there will probably be a contrary current running down the coast at a distance of from 20 to 60 miles from the land." (Vide § 29.)

The choice of a parallel.

Horsburgh's Instructions.—"Since the discovery of Bass strait, the passage through it is generally preferred to that around Van Diemen's Land, as it is equally as safe and greatly shortens the distance.

"A ship, having passed the island of St. Paul, and intending to pass through Bass strait, may get into lat. 39° or 39° 15′ S., then steer east on this parallel. As she advances, the variation will rapidly decrease; in about long. 132° E. there will be 1° E.; and having advanced 1° or 2° more to the eastward, she will begin to have more easterly variation; at King island, the west entrance to Bass strait, it is 8¼ East in 1863. Although the opinion appears to be held by many that the passage is best made by keeping a *high* southern latitude, the propriety of adopting the parallel of 39° S., as above recommended, has been already confirmed by the experience of Captain Erskine, of H. M. S. Havannah, who performed the voyage from the cape of Good Hope to Sydney in the very short period of 34 days, leaving the cape on the 3d July and arriving at Sydney on the 7th August. Captain Tudor, R. N., when in command of the East India steam-vessel Pluto, however, met with baffling winds on the northern parallel, and was obliged to run to the southward for the westerly wind.*

"Bass strait is the navigable route for all vessels passing round south of Australia, and is consequently much frequented.† King island divides its western entrance into two channels, that north of the island being far the safer of the two, and therefore the one constantly adopted. Route through Bass strait.

"Bass strait should be approached with caution by vessels coming from the westward, if not certain of their latitude, which ought to be correctly ascertained before they reach 143° E.; and the strait ought not to be entered in the night, unless the land has been previously seen or both latitude and longitude be known by observation. The parallel of 39° or 39° 15′ S., according as the wind may incline, is the best track for passing between King island and cape Otway; and a sight of either, or, preferably, of both, will point out the true situation.

"Westward of the north end of King island, at 30 miles distance, there are soundings from 65 to 70 fathoms, sand

* It is worthy of remark that Horsburgh, while advocating the route by the 39th parallel, should quote one bad and one good route.

† It is frequented by vessels which have chosen 3° S. for their easting—a route we do not advise.

which will indicate the proximity of this strait in thick weather. The only danger to be apprehended here is the Harbinger reefs, two patches situated nearly 6 miles to the WNW. of the north end of King island; but they are so far separated from it, and from each other, as to leave passages between them, in case of necessity, where the shoalest water found by the Cumberland schooner was 9 fathoms.

"Having passed the north end of King island, on which is a fixed light, visible 24 miles, a course should be made good from it, E. by S., for Sir Roger Curtis island, and part of this distance may be run in the night with a good look-out; the soundings in this track to the eastern part of the strait are regular, from 35 to 48 fathoms, fine sand and shells. The best track is on the south side of Sir Roger Curtis isles, and on either side of Kent groups, keeping near the southernmost island of the group. If the south channel is chosen, to avoid the Endeavor rock, then steer ENE., (E. by N. true,) if nearly before the wind, or on either side of this course as the wind may incline, taking care not to approach the northern Long-Beach, formed between Wilson promontory and cape Howe, which becomes a concave lee-shore with a SE. gale. This makes the channel south of Kent groups preferable, at times, to those north of them; but with a steady NW. wind and settled weather, either of the channels south of Redondo might be pursued occasionally. Then a course steered well to the eastward, to give a berth to the Long-Beach, and cape Howe may be rounded at any reasonable distance.

Anchoring-places.

"The most convenient places for anchoring in the strait with easterly winds are: under the NW. end of King island; on the east side of New Year isles, which anchorage is now called Franklin road; Port Phillip; Hunter isles, between Three-Hummock and Barren (or Hunter) islands, taking care not to anchor too close to the weather shore, lest the wind change suddenly; and on the west side of Wilson promontory, *in a case of necessity only*, for this place is dangerous if the wind change suddenly to SW., as a deep bay is formed between the promontory and cape Liptrap. There is also anchorage in the channel called Murray pass, formed by the islands of Kent group; and there is one anchorage in East cove, on the eastern side of the pass, and another farther to the northward, in West cove, (Erith island,) on its western side. Strong squalls may be expected off the land and eddy tides near the shore."

Maury's Instructions.—"Vessels that are bound to Australia, after crossing the line in 30° W., can generally reach 30° S. between 30° and 20° W. The great-circle distance thence to Melbourne is, *if it could be followed*, about 6,700 miles; but it crosses the barriers of perpetual ice, which forbid the passage through the antarctic regions. But if a vessel do not go south of 55°, she cannot accomplish the distance to Melbourne from the parallel and meridian of 30° in less than 7,300.

"The majority of vessels bound around the cape of Good Hope cross the meridian of 20° W. between the parallels of 30° and 35° S. Here they generally aim to make a course a little to the south of east. But the great-circle route to Australia would, were it practicable, require them to pass the parallel of 70° S. before crossing this meridian of 20° W. That route is the nearest which, being practicable, deviates the least from the great circle. From the moment that vessels leave the calm belt of Capricorn, between 20° and 30° W., this route is tangential to the parallel of the highest degree of latitude that they intend to make.

"Admitting that the tropical calm belt is left near 30° S., the distance via the 'composite' routes—for the parallels of 45°, 50°, and 55° taken as 'vertices,' and from the meridians of 30° and 20° W.—is as follows:

"1st. From 30° S. and 30° W. to 45° S. in 20° E.; thence east to 120° E.; and thence by tangent to Melbourne, is 8,000 miles.

"2d. From 30° S. and 30° W., to 50° S. in 30° E.; thence to 100° E.; and thence to Melbourne, 7,700.

"3d. From 30° S. and 20° W., by tangent, to 45° S. in 30° E.; and thence, as upon the parallel of 45° from 30° W., to Melbourne, 7,600.

"4th. From 30° S. and 20° W., by tangent, to 50° S. in 40° E.; and thence to Melbourne, as before, from 30° W., 7,300.

"5th. From 30° S. and 25° W., by tangent, to the parallel of 55° in 40° E.; and thence along this parallel to 90° E.; and thence, by tangent, to Melbourne, the distance is 7,300 miles.

"These tangential curves are arcs of great circles, and the navigator who will not take the trouble to get out these curves, so that he may follow them to and from the parallel, or 'vertex,' upon which he proposes to 'run down his longitude,' but prefers the rhumb-line course, must make up his mind to the loss to be incurred, for even in the cases above quoted, he will lose by the rhumb-line course from a few hours' to a day's sail, according to circumstances.

"At any rate, when he comes to view the route to Australia as here described, he will perceive that the route to the cape of Good Hope turns off from it about the parallel of 30° S., and that therefore Australian-bound vessels need not hug the trades as close as the cape-bound vessels do. Here, then, as you clear the belt of the SE. trade-winds there is a fork in the road—the vessel bound by the beaten track to the cape or Calcutta going to the east; but she whose destination is Australia should stand on to the southward, not thinking of hauling up to the eastward until she clears the calms of Capricorn and finds herself well within the region of the trade-like westerly winds of the southern hemisphere.

"She may then begin to edge away and haul up gradually to the eastward, crossing 10° W. between 40° and 50° S., according to the season, and reaching her extreme southern parallel in our winter months, near the meridian of 20° E. Upon this parallel (say from 45° to 52° S.) she should run along her 'vertex' till she cross the meridian of 100° or 120° E.,* when she may begin gradually to edge up for her port, but still keeping to the right of the rhumb line on her chart that leads to it. Hence, it will be perceived that Australian-bound vessels have nothing to do with the cape of Good Hope; *nor should they wish to go within six hundred or eight hundred miles of it.*

"The best crossing-place of 25° or 30° S., that the SE. trades will generally allow for the Australian route, is about 30° W., a few degrees more or less. The distance from it to Melbourne is about 6,500 miles, the arc of the great circle crossing the prime meridian between the parallels of 70° and 75° S., the meridian of 55° E. between the parallels of 80° and 82° S. Here it reaches its greatest southern declination, and begins then to incline northwardly. Australian-

*The author translates these meridians as 90° and 100° E., (merid. Paris.) Maury, however, gives them as 98° and 118° E., (merid. Paris.) —Translator.

bound vessels are advised after crossing the equator near the meridian of 30° W., *say between 27° and 32° as the case may be*, to run down through the SE. trades with topmast studding-sail set,* if they have sea-room, aiming to cross 25° or 30° S. at about 28° or 30° W., and so on; shaping their course, after they get the winds steadily from the westward, more and more to the eastward until they cross the meridian of 20° E. near 45° S., and afterward reach if possible 55° S. near 40° E. The *Nightingale* ran as far south as 57° S., and made the quickest time of all the passages we possess. Therefore, there will be a great advantage in keeping to the S. of E., as much as the ice, etc., will allow, without attempting to return north until the meridians of 90° or 92° E. are reached. The best plan is to make the extreme southern latitude between the meridians of 52° and 82° E., afterward bearing away more and more to the north and east as Van Diemen's land is neared.

"Such is the best route to Australia, the highest degree of south latitude which it may be prudent to take *depending mainly on the season of the year, the ice, the winds, the state of the ship, and the well-being of the passengers and crew.* If the winds are not good and strong, bear south to look for them. In our summer, one will not have to go so far south to look for these winds as he will in our winter. The shortest passages, therefore, will probably be made *in the southern spring and early summer*, when daylight, the winds, the state of the weather, and all, except ice, are most favorable for reaching high southern latitudes.

"I have endeavored to impress navigators with a sense of the mistake they commit in considering the cape of Good Hope as on the way-side of their best route to Australia. It is not only a long way out of the best and most direct route for them, but the winds also to the north of the 40th parallel of south latitude are much less favorable for Australia than they are to the south of this parallel. '*Sailing Directions' issued by the British Admiralty, I am aware, recommend the cape of Good Hope route, and the parallel of 39° S. as the best upon which to run down easting to Australia.*

"It is in the fall and winter months (not in summer, as

* From this it is evident that Maury does not advise that you should always cross—even the equator—very far to the westward, as some authors have accused him of doing.

the 'Sailing Directions' state) when the sea is most free from icebergs, for every one knows that icebergs are often seen in the North Atlantic in June, and not unfrequently in July. December and January* are the worst months for ice along the Australian route. By March† well-nigh all that the summer heat could set adrift has been borne north and melted; the southern winter is the time when the icebergs are held fast, for then they are forming for the heat of the next spring and summer to break out and set adrift.

"In recommending this new route, and a route which differs so widely from the favorite route of the Admiralty, I should remark that I do it, not because it is an approach to the great circle route, nor because it has anything to do with the *composite track*, but because the winds and the sea and the distance are all such as to make this route the quickest. I say the sea, because I suppose there is no more danger from icebergs, if a proper lookout be kept, than there is on the voyage between New York and Liverpool. I do not even see them mentioned in the voyage to Australia, except by three ships, namely: the *Malay*, on the 21st December, 1853, in lat. 48° 25′ S. and long. 35° 24′ E.; the *Oriental*, on the 11th December, 1853, in lat. 46° 25′ S. and long. 125° E.; the *Auckland*, on the 25th October, 1853, in lat. 53° 12′ S. and long. 21° 23′ E.

"Furthermore, *Horsburgh* says that H. B. M. *Guardian* fell in with one, in lat. 44° 15′ S. and long. 44° 30′ E., on the 24th December, 1789; also that the French ship *Harmonie* met one in April, at lat. 35° 50′ S. and long. 18° E. In short it can be said that icebergs are very rare in these localities.

"There seem to be two spots in the Pacific where icebergs are frequent: one near the meridian of the cape of Good Hope; the other in the neighborhood of the longitude of Australia. I therefore would advise navigators not to cross the prime meridian to the south of 45° S.; then to run for their extreme southern parallel—55° S. for example—near the meridian 40° E.

"In further proof that the route recommended in the 'Sailing Directions' of the Admiralty is too far to the north I have prepared tables, and so far as the facts de-

*From November to April according to Maury; edition of 1859.—*Translator.*

†May; *Maury* edition, 1859.—*Translator.*

duced from these tables go, they justify the assertion that for every degree you go south of the Admiralty route, you gain three days on the average, until you reach 45° and 46° S., for the averages of the tables are not below these parallels. I believe it will turn out that the best streak of wind, on the long run, is to be found between 45° and 50° S. It seems to be almost as steady between these parallels from the westward as it is anywhere from the east between the trade-wind parallels of 15° and 20°."

Captain Sallot Desnoyer's Instructions.—"There seems to be a choice between two routes in making this passage: First, the Admiralty route, near 39° or 40° S.; second, the route near 45° S., more or less to the southward of that parallel, according to the season and the winds. During the southern winter, when the nights are long and the weather cold, it is natural for one to hesitate before taking the southern passage, notwithstanding the fact that icebergs are less numerous than they are in summer. At this season the Admiralty route can be followed on the parallel of 42° S. I do not think there is any danger in running well to the southward in summer, for if icebergs should chance to be encountered, the long days will be of much assistance in avoiding them. Of course, the wind and weather will at all times be taken into consideration; and if settled winds are not found on the parallel chosen, the best thing to do is to run south, remembering that the farther south vessels go, the shorter their passage."

*The observations of various captains on the choice of a parallel.**—We will first quote the following extracts from volume 29 *Annales Hydrographiques*, relating to this very disputed question. Lieutenant Galeche, of the French navy, expresses himself in these terms:

"Maury advises vessels leaving the cape of Good Hope to run for the meridian of 32° E. near 45° or 50° S. latitude, thus approaching the great circle—which curves into still higher latitudes—as nearly as possible. This is without doubt the most direct route, being 300 or 400 miles shorter than the Admiralty route. By it the passage is shortened two days. But Maury himself admits that the voyage is boisterous and stormy, and should only be undertaken by stout ships. He speaks of the ships that have

* The extracts in § 105 will also be interesting in this connection.

made magnificent passages from Europe to Australia by descending to 55° S.; but makes no mention of those which have been lost, or have sustained extraordinary damage to their masts and rigging. Therefore I think that a man-of-war, bound to her station in China or Oceanica, and wishing to arrive there in ship-shape condition, *should not go lower than 45° S. in summer, and 40° S. in winter.* Bad enough weather will be found even on these parallels, but in spite of it a good passage may be made. Going from Réunion to Bass strait, in the *Marceau*, I ran down my longitude on the 40th parallel; and from 67° E. on 36° S. I had no easterly winds (September and October) until I reached Bass strait."

Captain Veillet, of the Hoogly, says: "I scudded along the parallel of 50° S. for two weeks, with a violent west wind after me all the time; but I would not follow this route in a weak vessel."

Captain Fernaud, of the *Réunion*, who made voyages to the East Indies for 15 years, makes the following remarks:

"The most convenient place to cross the equator, after leaving Europe, is between 23° and 28° W. After passing the Tristan islands, head for the parallel of 42° or 45° S., upon which run down the longitude. Here the sea will be found very much smoother and the weather better than on the parallels of 36° and 40° S. I have always experienced fine weather, variable wind, sometimes even exceedingly variable, but no gales; while between 36° and 40° the sea is very rough, the gales very violent, and the danger of damage to the ship great. After the heavy squalls there are generally two or three days of calm, when the ship will not make more than two or three miles per hour and labors greatly. The heavy squalls do not seem to be found lower than 40° S.; the wind begins from NE., jumps violently around to NW., and shifts gradually to W., then to SW., finally to S., where it dies away."

Captain Prouhet makes the following remarks, (*Ann. Hydr.*, vol. 31:)

"The bad weather we experienced in our passages toward Van Diemen's land gave us an entirely different idea of the 'brave west winds' promised by Maury. They undoubtedly blow, in the zone of variable winds, especially in high latitudes; but they are very variable, and suddenly shift, at times, to all points of the compass. One moment we would be sailing along with a moderate breeze, the next a squall would be upon us, strong enough to carry every mast by the board. We were continually shifting or taking in sail. The sea was also entirely different from what Maury would lead one to expect. Instead of the long swell that was to set us a little ahead of our dead-reckoning every day, we very often experienced a heavy cross and broken sea, the waves breaking in all directions, and making the vessel labor violently. Nevertheless the route to the east—advocated by this superior American officer—is a good one; but navigators must not fall into the illusion that they are going to have good weather; for I found that all the descriptions fell short of the reality. The speed by this route is more rapid, and as the crew cannot be drilled they will not object to the rough weather."

The following is an extract from the report of Captain Binet, of the *Isis*, on the passage from Réunion to Sydney, (vide Ann. Hydr., vol. 28 :)

"Left Saint Denis on the 20th February; lost the SE. trades on the 27th, lat. 38° S., long. 54° E.; and without experiencing a moment's calm, found a light breeze from NE., which soon freshened and shifted to NW.

"From that time till the 20th March, when we sighted cape Otway, (lat. 42° S.,) we made 200 miles per day, with fresh winds from NNW. to SSW. The only change in the weather was that the wind lessened in intensity, now and then, for a few moments. We experienced a few calms and light airs near land in Bass strait."

We would call special attention to the following quotation from the report of Captain Jouan, relating to the passage of the sailing-transport Bonite, from the cape of Good Hope to Sydney, (Ann. Hydr., vol. 27 :)

"The routes to Australia, as prescribed by Maury and the British Admiralty, are essentially different. The Admiralty directs that the 39th parallel is to be followed in the Indian ocean as far as the meridian of 107° or 112° E., whether bound through Bass strait, or to the south of Tasmania. Maury, on the other hand, advises a ship to gain as high a latitude as possible. The Admiralty route, if not very rapid, offered more inducements for a transport like the Bonite, as her people were ill clothed, etc. We left the cape on the 2d

May, and on the 24th were in sight of St. Paul, and south of the island. The length of the nights, and the want of a detailed chart of Bass strait, made me decide to go to the southward of Tasmania.

"I found *better* and *more settled weather on the parallel of* 46° *S.* than nearer the equator. Between the cape and the S. of Tasmania the currents were generally to the east, sometimes to the north. Along the east coast of Australia, from cape Howe to Port Jackson, the winds, during the first two weeks of June, were gentle, and from WNW. to WSW., freshening at night; the sky was clear, and the stars very brilliant. The westerly squalls, common here at this season, are apt to set vessels from the land. We also encountered several thunder-storms, during which the wind shifted.

"The barometer on this coast rises with the S. wind, and falls with the N. Near the land, and when the wind is from WNW., the glass varies from $29^{in}.76$ to $29^{in}.92$, as the wind hauls to N. or S. The rain-storms seem to have no appreciable influence on the barometer. On the evening of the 25th June we were in sight of Port Jackson light.

"The next morning there was a dead calm, and we were towed to our anchorage at Farm cove; at noon the thermometer stood at 59° or 60°. From cape Howe to Sydney the currents set me to the northward, when I was near land; and to the southward, when I was about 60 miles off the coast.

"While at Sydney I met several captains who were in the habit of making the voyage from London to New South Wales. All of them informed me that *in winter nothing was to be gained by going through Bass's strait.* They differed as to the best route through the Indian ocean; some recommending the route along the 39th parallel; others maintaining that ships should go higher, and on the parallels of the Prince Edward islands, Kerguelen land, and the Macdonald group, (52° 40′ S.)

"In these localities the wind is steadier, but the weather *stormy.* Icebergs make this route dangerous in summer. *Good passages* have been made *by both routes.* All the captains said that vessels should never keep between 40° and 42° S., as the weather is never settled, and calms and head-winds are common in that locality."

2d. *The Suez route.*—The reader should first refer to the observations on this route, which he will find in the "Navigation of the South Atlantic ocean."

Reference should also be had to § 148 of this volume.

From April to October a sailing-vessel or an auxiliary steamer can, if need be, *sail* down the Red sea without great difficulty, for, at this season, the winds prevail from the north. Inversely the passage from Aden to Suez will be impracticable during these months, and rendered too long for commercial purposes. During the rest of the year the N. winds prevail only in the northern two-thirds of the Red sea. In the southern third they blow from S. and SSE., often quite stiffly. From April to October.

Therefore, from October to April it is almost impossible for a sailing-ship to go either up or down the Red sea. From October to April.

Sailing-vessels wishing to pass through Torres strait should aim to arrive at the Malay peninsula between December and February, so as to take advantage of the NW. monsoon.

It is even then extremely difficult for a sailing-ship, or for an auxiliary steamer wishing to economize coal, to go through both the Suez canal and Torres strait. The only possible way of doing this is to descend the Red sea during the first of October, so as to arrive at the sea of Arafura in November or the beginning of December. In this manner the strong SE. winds of the Red sea will be avoided, as these do not set in until November. In the sea of Arafura the NW. monsoon will be found; but it should be stated that this is the worst season in the Indian ocean, especially to northward of the line, for the change in the monsoons takes place at this time.

In short, sailing-vessels, or auxiliary steamers wishing to economize coal, should not go through Torres strait.

The best season to go down the Red sea is between April and October, for then the SW. monsoon will be found N. of the equator; the easting can then be run down, and the line crossed in a good position to haul up on the port tack in the SE. trades. Bear away to the southward and eastward near 30° S., and pass south of Australia.

§ 105. ROUTE FROM EUROPE TO NEW CALEDONIA.—Navigators intending to make this passage should refer to the instructions given in § 104, as the route discussed in that paragraph is nearly the same as the one under discussion.

We have already stated that sailing-ships should pass to the southward of Tasmania. We will now point out the best crossings thence to Noumea:

As will be seen, the general rule for the last part of this voyage is, not to edge away too quickly to north, and to pass, or at least reach, the meridian of New Caledonia before clearing the parallel of 35° S. Auxiliary steamers can naturally make a more direct route.

The following crossings are to be preferred between Tasmania and Noumea:

In January, cross 40° S. between 165° and 166° E.; 35° S. and 30° S. between 168° and 169° E.

In February, cross 43° S. at 155° E.; 40° S. at 163° E.; 35° S. and 30° S. at 168° E.

In March, 40° S. at 165° E.; 35° S. and 30° S. between 170° and 171° E.

In April, 40° S. between 156° and 158° E.; 30° S. between 163° and 165° E., and thence directly for Noumea.

In May, 40° S. between 158° and 159° E.; 35° S. at 163° E., and 30° S. at 166° E.

In June, 40° S. between 157° and 158° E.; 35° S. at 165°, and 30° S. at 167° E.

In July, 35° S. between 164° and 166° E.; and 30° S. between 167° and 168° E.

In August, 40° S. between 154° and 156° E.; 35° S. between 161° and 162° E.; and 30° S. between 165° and 166° E.

In September, 35° S. between 165° and 166° E.; and 30° S. between 167° and 168° E.

In October, 40° S. at 160° E.; 35° S. between 165° and 166° E.; and 30° S. between 167° and 168° E.

In November, 40° S. between 160° and 162° E.; 35° S. between 166° and 168° E.; and thence for Noumea.

In December, 40° S. between 162° and 163° E.; 35° S. at 167° E.; and thence steer for Noumea.

For instructions relating to the last part of the voyage, and to the landfall, consult § 111. We will now give a few extracts relating to this passage, and we would especially draw the reader's attention to the voyages of the Sibylle and Garonne.

Passage of the Sibylle, Captain Brossalet, (*Ann. Hydr.*, 1871.) —"I kept a clean full on the port tack while running through the SE. trades of the Atlantic. Between 8° and 17° S. the trade-winds were quite squally.

"At 26° S. they commenced to haul to the east, and even NE.; and I was enabled to come up a little to the eastward, and crossed 2° E. at 41° S. Arrived at this meridian on the 19th October. Experienced variable winds from ENE. to NNE., which, at times, compelled me to take two reefs in the top-sails. These winds were, for three days, interrupted by light airs from NNW. to W. and SSW. At 41° S. and 2° E. I bore for the 45th parallel at 20° E.

"I expected to run down my longitude to Tasmania on this parallel, but just before I reached Marion and Crozet islands the wind—which had been from N. for several days—hauled to E.; so that instead of giving the islands a wide berth I had to run to leeward of them. I therefore had to run down my longitude on 47° and 48° S., and arrived at 150° E. on the 19th November. The wind, during this run, was continually varying from NNE. to NW. and W. and even to S., but the prevalent direction was N.

"It seemed to have a rotary motion, and after blowing strongly from N., it hauled to W., dying away a few hours afterward, while shifting to SW.; after an interval of calms and light airs it would again come out from NNE.

"I might have found the winds more to the westward in higher latitudes, but I doubt if I would have made a quicker passage, nor would the health of the crew have been as good. If I had intended to take a south route, I should have had to make up my mind early in the voyage to double the Macdonald islands (lat. 53°) to the southward. As the wind in this locality is squally, and often hauls to SW., and then to SE., I should have probably lost a great deal of time. Besides, Kerguelen land—situated in nearly the same meridian as the Macdonald islands, and extending from 50° 50′ to 52° 20′ E.—makes the situation less desirable.

"I did not deem it prudent to run between these two groups, as the winds are variable and the chances of obtaining observations rare.

"I incline to the opinion that the northerly route is the

better one. During the two days prior to my reaching this position the winds were inclined to haul to NNE. and NE., and I was uncertain whether I could double the islands or not, but I was enabled to do so without making a tack, thus economizing time and avoiding much anxiety. Even if you have to go about, there will be plenty of sea-room.

"The current averaged 1 knot per hour for 25 days, and set the ship to the eastward. This current seems to follow the variations of the wind, inclining to the N. or S., as the breeze hauls in either of those directions. During the 25 days the frigate made 5,539 miles, and, though the wind was not very fresh nor always fair, we averaged 221.5 miles per day. We nearly always carried whole top-sails; in the worst weather we furled the upper foretop-sail, and put two reefs in the main-sail. Once I was nearly lying-to under the two lower top-sails; but, even then, made over 5 knots per hour; the breeze grew light at the end of a few hours, and, the sea not having had time to rise, I again made sail.

"I encountered, at 108° E., a single iceberg, about 400 feet high and 1,300 feet long; its shape very irregular. When I sighted it the weather was fine, clear in the zenith, and slightly foggy in the horizon. It was distant about 4 miles when I made it out, and I lost sight of it when it was a little over 4½ miles off. The thermometer showed no signs of its proximity, even when it was within 1½ miles of us. The southern aurora was also very magnificent one night while we were in these latitudes.

"Thinking that I would have better weather, I passed midway between Australia and New Zealand, the result justifying my anticipations. I ran from 48° S. to Norfolk island in less than seven days; and, taking my departure from that island, I anchored in Noumea on the 28th November; 94 days from Toulon and 76 from St. Vincent."

Passage of the transport Garonne, (auxiliary steamer,) Captain Rallier.—"Arrived at Dakar, Senegambia, on the 27th August, 1872, and left that port on the night of the 2d and 3d September. Crossed the line between 16° and 17° W.; then, contrary to the usual custom—the Garonne making very good way close-hauled—I hugged the trades, and passed 450 miles to windward of Trinity island. I reached

Tristan da Concha 19 days after my departure, and on the very day when I should have anchored at St. Catherine had I not kept close-hauled. By steering this course I had, therefore, gained 1,800 miles on my route, and was from 15 to 20 days ahead of the time I would have made had I touched on the South American coast.

"In the southern part of the equatorial and Capricorn calm belts I met stubborn S. winds, which sensibly diminished our supply of coal; contrary to my original intention, I was therefore obliged to make the greater part of the passage under sail, using the engine only in cases of the most urgent necessity.

"I ran well south from Tristan da Concha, crossing 45° S. at 2° E., and 50° about the meridian of the cape of Good Hope; here I was delayed for 6 days by calms and light baffling airs; the fog was also thick and 'glacial,' which I attributed to the meeting of antarctic and Indian ocean currents.

"Thence I ran in an oblique course to the Macdonald isles, which I intended to make the vertex of our route. I wished to sight these islands, as (according to my judgment) they are located near the best route between the two capes, and should be used by vessels to rectify their positions. It also seemed to me best to give Kerguelen land a wide berth, as its exact position is doubtful and the winds in that locality are unfavorable. Unfortunately, I passed the Macdonald isles in the night, and could not obtain all the information I had wished. I was able, however, to fix the exact position of the islets off the western part of the group, as our chronometers proved to be only two minutes out when we reached Noumea. The principal peak of the island stands out in bold relief; we made it at a distance of 45 miles at sunset, and judged its elevation to be about 7,000 feet.

"After clearing the Macdonald group I found a fresh wind from N. to NNE., which drove me down to 54° S.; this was the only violent wind we experienced south of the 50th parallel, and it lasted only a short time. During the passage of 21 days in these latitudes the royals were only furled 10 times, and 76 hours in all. The sea was remarkably smooth; the wind moderate, and the only heavy swell from NW. The wind which caused this swell did not reach us. It probably came from a locality 600 miles distant, as we did not strike the corresponding wind until we reached 45° S.

"The temperature did not fall below 25°; this was not very cold, but was very much felt by the crew, as the weather changed so rapidly that they had not time to become acclimated. The nights were especially severe, and I had to serve out an extra allowance of grog. Still, few were taken sick, the air being dry and healthy.

"The danger from icebergs can hardly be exaggerated, especially during the foggy nights and the long snow-squalls. We met our first one at 51° 47′ S. and 35° 08′ E.; it was 270 feet high, and made such an impression on the crew that I knew I could count on a 'good lookout;' we also passed through a group of them on the 16th, 17th, and 18th October, between the parallels of 52° 23′ and 52° 06′ S. and the meridians of 92° and 104° E. At noon on the 18th, lat. 52° 06′, long. 105° 17′, we thought we saw land to the south. I ran for it, but at the end of an hour saw nothing to fully confirm my belief. These latitudes are so little frequented that land may possibly be there; but as signs of the scurvy were beginning to make their appearance among the crew, I could not delay any longer. If this land exists its position is 52° 30′ S., and 105° 20′ E.

"I bore away gradually to the north on the 14th of October, and recrossed 50° S. on the 20th, after making 100 degrees in longitude to the southward of this parallel, or an average of 5 degrees per day.

"To the north of 50° S. the weather completely changed, rain took the place of snow, the sea became heavy, and the wind violent, especially in the squalls, when only low sail could be carried. The wind was usually fair, though very variable, obliging us to be continually bracing and hauling. In short, this zone was worse than the more southern one, with the one exception that the weather was warmer. This higher temperature of the atmosphere acted favorably on our passengers, though our efforts to make a quick passage had to be renewed to avoid the tendency to scurvy caused by the dampness.

"We doubled South cape, Tasmania, on the 26th October, 26 days from the meridian of the cape of Good Hope, and an average of 200 miles per day.

"East of Tasmania I found one day's calm, and two of

head winds. The breeze then sprung up from the east, and we arrived at Noumea on the 4th November, 63 days from Dakar, with an average speed of 7.8 knots."

The opinions given in the following quotations differ materially from the preceding:

Passage from Europe to New Caledonia by the sailing-frigate Alceste, Captain Brosset, (*Ann. Hydr.*, 1871.)—"I ran out of the SE. Atlantic trades at 20° S. and entered the region of the 'variables' without meeting any calms. I headed to double the cape of Good Hope at 38° S.; thence, bearing to the S., I made for about 41° S.

"Made my easting between 40° and 42° S. At the 102nd meridian I again ran to the southward so as to double Tasmania at 46° 30′ S.; this mean route, between that of the British Admiralty and the one indefinitely laid down by Maury, appeared to me the best for the winter season. On this long passage I found very fresh and settled winds, and no too violent squalls; nor was the temperature very low even in the heart of this bad season.

"In this passage the various meteorological observations agreed exactly with those taken on my preceding voyage.

"The rotation of the wind is invariably the same; it comes out from NE., shifts to N. and NW., fresh and squally; then veers suddenly to SW. with clearing weather. The wind then oscillates between SW. and N. for a longer or shorter period, according to the latitude; it then hauls to S., dying away at SE., and comes out once more from NE., after a few hours' calm. The farther south you go the longer it takes the wind to complete this entire revolution; in other words, the wind is steadier between N., W., and S. in high latitudes; from which Maury's instructions 'to run well to the south when Australian bound.'

"These rotations of the wind occurred very frequently when the Alceste was N. of 38° S., but S. of that parallel the wind never got to the E. of S."

Passage of the Alceste from Réunion to Noumea.—"As I had to make this passage in the middle of winter I did not think I was justified in taking Maury's route, especially as the ship was crowded with passengers and prisoners in confinement. On the other hand, as the route of the British Admiralty seemed a little too far north, I took, with all due deference to Maury's rules, an intermediate route; and left

Réunion, with the intention of running down my longitude between the parallels of 41° and 42° S.

"I left Réunion on the 16th July. I ran out of the trades at 27° S. and 53° E., and the variable winds which then sprang up allowed me to run on the arc of a great circle passing 90 miles S. of Van Diemen's land. I then reached about 41° 30′ S. and 70 E. I then headed E. until I crossed 110 E. Thence, bearing to the south, I doubled Van Diemen's land (long. 145° E.) at 46° S.

"During the whole length of this 'composite' route the wind was nearly always from N. to SW.; shifting by the W. point of the compass, it generally blew very strong, causing us sometimes to scud under the fore-sail. Maury's charts show conclusively the prevalence of the west winds in these latitudes.

"I think I can deduce from my personal observations, that the farther one goes to the south, the longer it takes the wind to make a complete revolution of the compass.

"Thus I experienced, in the Alceste, between the parallels of 42° and 46° S., a wind that blew for 10 days from SW. to N. (shifting by W.) before it went completely around the compass; while in lower latitudes (between 33° and 40°) the time required for its complete revolution was never more than 4 or 5 days. Still farther to the northward, between 30° and 33°, it is not a rare thing to see the complete rotation peformed in 24 or 36 hours. Finally, in the vicinity of the trades, the wind runs around the compass even in half a day; whence the calms and baffling airs near the tropic of Capricorn.

"From all the data we have on the subject, we can conclude that there is probably some parallel, still undetermined, where the wind never blows from the eastern semicircle of the compass."

Passage of the Saint-Michel, Fradin, master, (vide Ann. Hydr., vol. 23.)—"We doubled Pernambuco during the night of the 8th of August, and experienced fine weather and light winds until the 15th. We were then at 33° 33′ S. and 29° 15′ W.; steered south for the parallels of 43° and 45° S., on which we intended to make our longitude. On the 5th September we were in lat. 43° 18′ S., long. 9° 40′ E. Thence to cape Otway experienced moderate winds from

NW. to SW., and quite favorable currents. In my opinion, the passage should be made *between the parallels of* 48° *and* 50° S., for here the wind is more settled, the current stronger, and the sea smoother. After running to the east between 43° and 45° S., we arrived at the boundary of changeable winds; every fourth or fifth day we had a few hours of calm; then the south winds, shifting to SE., became squally; they afterward hauled to the E., NE., N., and from NW. to SW., from which point they blew for a long time. These sudden changes caused a heavy sea, which made the ship labor greatly, and deadened her headway. At daylight on the 12th October, we were out of Bass strait. From the 12th to the 16th October we had fine weather with light winds from NE. to NNE.; the 17th, wind from SSW. to S.; making good headway; on the forenoon of the 19th we passed close to the E. of Lord Howe islands, and about 9 miles from the Ball pyramid. Weather fair and wind from S. to SE. On the evening of the 22nd we were in sight of 'mount d'Or.' We anchored at Noumea on the 23rd."

We will finish with an extract from the observations of Captain Rion-Kerangal, of the *Isis*, on the passage from Réunion to New Caledonia:

"The westerly winds never failed us in high latitudes; they were as settled as the SE. trades of the Atlantic, and strong enough to enable a smart sailer to make 240 miles per day. These winds usually prevail from 35° to 55° S.

"From 35° to 40° S. the NW. wind is dominant; from 43° to 55° the wind from W. and SW. prevails.

"On the Admiralty route the winds are more moderate and variable than they are farther south; this passage is also damp and rainy; and the scurvy is liable to break out in these latitudes. On the 45th parallel we find, it is true, a very strong wind, but it is steady, the sea being constantly from the west, the sky clear, and there is no dampness when the winds are from the south. It is colder on this route, but on the other hand the health of the crew will be better.

The zone between 40° and 43° S. is to be avoided, as the wind is variable and the sea irregular and very heavy; "solar" winds are also very frequent in this locality, and the barometric changes very great.

"The Sibylle left Réunion 24 hours after the Isis, and arrived at Noumea 5 days before her, although their sailing qualities were about equal.

"The Sibylle made a portion of her longitude on the parallel of 36° 30′, and afterward bore south for Maury's route, at 45° S., crossing 40° and 43° S. at nearly right angles.

"The Isis ran down her longitude on 39° 30′ S., the route given by Horsburgh. Near Bass strait the NE. winds forced her to the south of Tasmania, but the *Sibylle* was ahead of her, having doubled Van Diemen's Land.

"The route between 40° and 43° S. is not followed much, *captains being careful to avoid this locality.* The winds in this zone will be found farther to the north in the southern winter than they are in summer.

"Maury's route is better than the Admiralty's for well-equipped vessels.

"I found a vessel from Bordeaux at Noumea, which by following Maury's route had made the passage in 105 days."

§ 106. ROUTE FROM EUROPE TO TAHITI.—The track from Europe to Tahiti by the cape of Good Hope follows the parallel of 50° S., or the belt comprised between 45° and 50° S. The passage is 45 degrees longer in longitude than that to New Caledonia.

Ships bound to New Caledonia commence to bear north at 137° or 142° E.; while those destined for Tahiti should run to the east until they cross 173° or 168° W., a difference in longitude of 50°. Tahiti is situated in 17° 30′ S., and Noumea in about 22° S., a difference of 5° in latitude. The difference to Tahiti is therefore, in round numbers, 55°, and the passage from 18 to 20 days longer. The whole voyage rom France to Tahiti can be accomplished in from 110 to 130 days *at most;* while by the cape Horn route, touching at Valparaiso, it will take *at least* from 120 to 130 days to reach Tahiti.

1st. *The cape Horn route.*—Quick passages have been made by this route, but instances of them are rare. In the "Navigation of the Atlantic" instructions are given on the passage from Europe to and around cape Horn. The reader should also refer to § 44 of this volume. The last part of the voyage will be found in §§ 93 and 94.

2d. *The cape of Good Hope route.*—Probably the passage from Europe to Tahiti via the south of Australia will be generally quicker than the one by cape Horn. The instructions given in §§ 104 and 105 should be followed until the ship is south of New Zealand, after which the crossings should be approximatively as follows:

In *January*, cross 45° S. between 174° and 176° W.; 40° S. between 161° and 163° W.; 35° S. between 154° and 156° W.; and 25° S. between 147° and 149° W.

In *February*, 45° S. between 173° and 175° W.; 40° S. between 154° and 156° W.; 35° S. and 25° S. between 147° and 149° W.

In *March*, 45° S. between 169° and 171° W.; 40° S. between 155° and 157° W.; 35° S. at about 153° W., and thence for Tahiti.

In *April*, 45° S. between 177° and 179° E.; 40° S. between 164° and 166° W.; 30° S. between 153° and 155° W.

In *May*, 45° S. between 174° and 176° W.; 35° S. between 163° and 165° W., and then head for Tahiti.

In *June* and *July*, 45° S. between 174° and 176° W.; and 35° S. between 157° and 159° W.

In *August*, 45° S. between 174° and 176° W., and 35° S. between 154° and 156° W.

In *September*, 45° S. between 174° and 176° W.; 35° S. between 151° and 153° W.; 30° S. between 148° and 150° W., and head for port.

In *October*, 45° S. between 174° and 176° W.; 35° S. between 158° and 160° W., and 30° S. between 152° and 154° W.

In *November*, 45° S. between 177° and 179° E.; 40° S. between 165° and 167° W.; 35° S. between 156° and 158° W.; 30° S. between 152° and 153° W., and 25° S. between 150° and 151° W.

In *December*, 45° S. between 174° and 176° W.; 40° S. between 157° and 159° W.; 35° S. and 30° S. between 146° and 148° W.

Ships generally pass west of Rapa island, and some distance to the eastward of Vavitao. For instructions concerning the end of the voyage, vide § 140.

§ 107. Route from Australia to Europe.—There are also two ways of making this passage, viz: the cape Horn route and the westerly route by way of the cape of Good Hope or Suez.

1st. *The cape Horn route.*—The cape Horn route is the best and quickest for sailing-vessels at all seasons of the year.

It passes right through the middle of New Zealand, therefore the islands can be doubled either to N. or S.; or after leaving Sydney vessels can take either Cook or Foveaux strait. We think that it is *always* best to pass to the southward of New Zealand and between "*The Snares*" and *Auckland*.

If the weather shows every prospect of remaining fine and the wind comes out ahead, there is no special reason for not taking Cook strait in the good season, especially if on board an auxiliary steamer.*

The passage through Cook strait.

A ship deciding to take this strait will find shelter at port Gore or Guard bay in case of strong SE. winds; if not up to these ports when the wind begins, she can anchor at Port Hardy or in the harbor of Croisilles on the west side of D'Urville island. The latter anchorage is the better, for a heavy swell, caused by the tide, sets through the inlet at Port Hardy. On approaching Cook strait with the *N. W. wind* mount Egmont will be sighted, situated in 39° 18′ S. and 174° 05′ E. It is a regular cone and very high, the diameter of its base being about 30 miles, and its altitude about 8,500 feet; its summit is always covered with snow. On the contrary, when nearing the strait with the *S W. wind*, endeavor to pass by cape Farewell, situated in 40° 30′ S. and 174° 42′ E.; but be careful to avoid the dangerous bank which extends to the east of the cape for 17 miles. This can be done by heading for Burnett or Knuckle mountain, a remarkable peak on the western coast of Massacre bay. Knuckle hill is about 2,000 feet high, has two round summits, the northern one being the highest, and lies 9 miles SW. from cape Farewell. The light-house on Bush-end point can be advantageously used as a leading mark to clear Farewell spit.

The light is white with a red sector, revolves every minute, and is visible 17 miles. It is built on a wooden scaffolding with white and red bands, and is about 120 feet above the water; position of light, lat. 40° 33′ S., long. 173° 01′ 45″ E. The red light shows between S. 27° E. and E. 29° S., or in the direction of the spit, and is hidden by the sand-dunes from S. to E. 26° S. The vessel should be at least 4 miles from the N. edge of the *red* light when she opens it.

* Vide §§ 8 and 111 on the uncertainty of the weather in these latitudes.

Only auxiliary-steam vessels and ships bound to Otago should use Foveaux strait; it is 15 miles broad, except west of Ruapuke island, where it is less than 10. The variable wind and weather make navigation through this strait dangerous for a sailing-vessel, and a sharp lookout should at all times be kept for SW. and NW. squalls. The passage through Foveaux strait.

The current along the SW. coast of Middle or Tavai-Pounamou island sets south. *Vessels striking the W. wind out of Melbourne* can take Bass strait, and pass south of Kent islands, heading about E. by N. in order to double Wright rock and Endeavor reef; thence pass south of New Zealand between "The Snares" and Auckland island.

If after leaving Melbourne an E. or NE. wind is found, run SW. and between cape Otway and King island. After doubling the north point of this island steer about S. by E. and give the western coast of Tasmania a wide berth, for sea-room will be necessary in case of one of the SW. squalls frequent in these parts; thence steer for the passage north of Auckland.

After passing New Zealand make for the parallel on which it may be decided to run down the longitude. The choice of a parallel.

There are fewer parallels from which to choose on this route than on the one from Europe to Australia. Cape Horn being in 56° S., the easting will have to be made in quite high latitudes, in order to double the cape with plenty of sea-room. Ships passing either north or south of New Zealand will then find themselves in respectively 45° or 49° south latitude; even then they will be south of the Admiralty route of 39° S. From this it results that this passage is necessarily made in high latitudes, and that the voyage is *rarely a long one*, for the winds are here very stiff from the west.

Without fixing on any particular parallel, we will only indicate the route which we consider the best.

We think that a ship should at first steer to the east between 48° and 53° S., more or less to south or north, according to the season.

Farther south icebergs and cold weather will be encountered. By referring to § 43 it will be seen, however, that floating ice only passes beyond the 50th parallel from October to April, and that it may not be met south of 50° even at this season. There is less danger of meeting ice from April to October. *Therefore the extreme southern limit for this route to the east should be 50° S. from October to April, and 52° or 53° S. from April to October.*

If willing to run every risk for the sake of a *quick passage*, make for 57° or 58° S., and run to the eastward on that extreme parallel; *but we cannot advise* such a high latitude.

In short, 50° S. during the southern summer, and 52° S. during the southern winter, would seem to be the extreme limits. During the long days of December and January a ship might run a little higher; for instance, after crossing 98° W., bear to the south for the parallel of the Diego-Ramirez islands, near 76° W. Both winds and currents will of course always be favorable for doubling cape Horn.

The voyage through the Atlantic.

Instructions relative to the voyage from cape Horn to Europe will be found in the "Navigation of the Atlantic ocean."

Never hurry to the north, and of course do not pass west of the Falkland islands. We would also state that sailing-vessels bound to Europe should never, unless under very extraordinary circumstances, cross 30° S. west of the 28th meridian.

In January, February, and March cross 30° S. between 14° and 16° W.

In April, May, and June, 30° S. between 19° and 21° W.

In July, August, and September, sailing-vessels should cross 30° S. between 8° and 10° W., and auxiliary steamers 30° S. near 27° W.

In October, November, and December, 30° S. between 7° and 9° W.

In general, the farther 30° and 40° S. are crossed to the eastward the quicker the passage.

Supply-ports.

In the author's work on the Atlantic (pages 172 and 173) a few details were given in regard to Port Stanley.

We take from *Ann. Hydr., vol.* 30, the following extract relating to the Falkland or Malouines islands:

"Coal and provisions can be obtained at these islands. Rear-Admial Hastings states that he procured from "Dean and Son" 408 tons of coal in 16 hours. This firm also had all kinds of ship-stores. According to the admiral, every ship bound around the cape should revictual at the Falklands."

Another account of these islands (*Ann. Hydr.*, 1872) gives an entirely different idea of the islands:

"The pilotage is not all that could be desired. I would not advise a ship to touch at Port Stanley when she can reach La Plata or Rio. The smallest repairs take 2 or 3 months, and a captain wishing to careen his ship may be forced to wait weeks for a still enough day in which to do it."

The following passage is from the report of Captain Launay, commanding the *Virginie*, which vessel touched at Port Stanley in March, 1873:

"In a commercial point of view this port is important, as 30 or 40 ships in distress usually touch there every year; some of these are condemned on account of the high price of labor. Provisions, &c., are quite moderate. Soft coal costs (coal was very high in 1873) $18 per ton; tallow, 10 cents per pound; fresh beef, 10 cents per pound; strong mutton, 10 cents per pound. If provisions are brought off in the ship's boats the prices are lower. Canned fruits and vegetables are abundant."

The following circular has been issued by *The Falkland Islands Company:*

"Good water from the government reservoir is worth 2*s.* 6*d.* per ton when taken off by the ship's boats, and 12*s.* 6*d.* delivered alongside. The price is 16*s.* when less than 5 tons are taken. If delivered outside, it brings 20*s.* Ballast can be found ashore, or delivered for 7*s.* per ton. Fresh beef and mutton are worth 4*d.* per pound; vegetables sell from 1*d.* to 2½*d.* A supply of coal is always on hand."

We will conclude by repeating that *Saint Helena* is *the best* supply port for sailing-vessels, but that Port Stanley may be useful in case of necessity.

The following instructions from Horsburgh, Fitz-Roy, and Maury may be useful:

Horsburgh's instructions.—"Ships from Port Jackson or Tasmania, bound to Europe in the summer months, and perhaps at all seasons, may expect to make a quicker passage around cape Horn than by any other route, for the prevalence of westerly winds in high southern latitudes is favorable for that passage.

"Ships pursuing the route from Port Jackson round cape Horn have in general made favorable passages round cape Horn; but, as stormy weather and high seas may be expected at times in high southern latitudes, this route ought not to be chosen in a leaky or crazy ship, and those who pursue it ought particularly to keep a good lookout for ice-islands, both to the westward and eastward of cape Horn.

"Icebergs are most constantly found between 133° and 110° W. during the winter season, probably drifted from a large extent of undiscovered land to the southward."

Fitz-Roy's instructions.—"In crossing the Pacific, toward the east, in southern latitudes, a ship should not go beyond 50° S. till near cape Horn, as there is usually much ice southward of that parallel, especially in the eastern part of the South Pacific, and occasionally it is met with some degrees farther north in autumn (February, March, and April,) after long continuance of westerly gales.

"A few hundred miles may be saved in distance out of about twelve thousand by going into very high southern latitudes, but at the risk of encountering ice, and with the certainty of a very cold, disagreeable climate. This applies equally to Australian passages by the cape of Good Hope, where great-circle sailing has been carried too far by some ships.

"In the long, dark nights of an antarctic winter, when the moon is not nearly full, *ice* (especially the low, less visible floes which are not many feet above the surface of the water) is especially to be guarded against by the most vigilant lookout, and by keeping under manageable sail, in readiness to alter the course instantly if danger is suddenly reported.

"In the summer of the southern seas there is so little night that ships may run with security, provided that (even in broad daylight) a good lookout ahead is invariably maintained under all circumstances. Foggy weather is comparatively rare, unless very far south.

"The distance on the great circle between the SW. cape of Van Diemen's land and cape Horn is 5,100 miles. The average length of the voyage (from 17 logs) from Melbourne to cape Horn is 35 days; the distance sailed, about 5,500 miles."

Maury's instructions.—"The same 'brave west winds,"

which take vessels so rapidly from the meridian of the cape of Good Hope eastwardly along the parallels of 50° to 60° toward Australia, will also bring them over eastwardly along the same parallels toward cape Horn.

"The investigations which have been carried on at this office, concerning the winds of that part of the ocean, forbid me to recommend the Admiralty route to any homeward-bound European or American vessel, under any circumstances whatever; always assuming that these directions are intended for ships that are seaworthy, properly fitted out, and sound. The average passage to Europe by this Admiralty route is 120 days. Ships may occasionally find the easterly winds as low down south as the directions of the Admiralty suggest; but it is the exception, not the rule, so to find them. In proof of this I refer to the pilot-charts of that part of the ocean, and shall quote other authorities.

"Returning by way of cape Horn homeward, the best route is to get south of the parallel of 40° as soon as you can, and then shape the course direct for cape Horn, recollecting that the farther you keep south of the middle of the straight line on your chart from Van Diemen's land to cape Horn, the nearer you are to the great-circle route, and the shorter the distance—the difference by the great circle and by the straight course on the charts being upward of 1,000 miles.

"In the passage from Australia to cape Horn, by keeping between the parallels of 40° and 60° all the way, you will, I am of the opinion, feel more or less the warmth and set of a current that passes south of Australia from the Indian ocean. Whether the boisterous weather to which a warm current in such latitudes would give rise will compensate for the advantages to be gained in other respects, must be left for experience to determine. For my own part, I do not suppose this current to be as strongly marked as is our Gulf Stream in the Atlantic; though the passage from the capes of the Delaware to Liverpool may be considered as affording us the means of judging pretty accurately as to this passage from Australia—the chief difference being, I suppose, in the climate and the gales.

"The climate in the Pacific, along this route, will be found not quite so mild as is that along the European route in the Atlantic. But the gales in the Atlantic are probably more frequent and violent than they are in the South Pacific—at any rate, I suppose that such will be found to be the case until you reach the regions of cape Horn.

"The Australian routes present occasional opportunities for fine runs. In the South Pacific ocean, below the parallel of 35° or 40° S. and away from the influence of the land —as along this route, especially from New Zealand to cape Horn—the westerly winds blow almost with the regularity of the trades; and a fast vessel, taking a westerly gale as she clears the New Zealand islands, may, now and then, run along with it pretty nearly to cape Horn.

"These winds are already beginning to be known so well to the Australian traders, that it is usual for them, I am told, when bound home by this route, to strike top-gallant masts before leaving port. It is a voyage that tries ship and crew; but of all the voyages in the world, that part of it between the offings of Australia and cape Horn is perhaps the most speedy for canvas. There it may outrun steam."

2d. *The westerly route.*—The Australia Directory advises ships bound from Australia to Europe or Hindostan, from September to April, to proceed to the westward, passing south of Australia and by Torres strait during the southern winter. Maury strongly opposes these directions, and *with reason*, as may be seen by reference to the extract in the first part of this paragraph.

As, however, the opinion of the British Admiralty has much weight, we would refer the reader to § 172, where will be found a quotation from the Directory.

Still, the cape Horn route is, in our opinion, the best on which to make the return passage to Europe.

Vessels bound to Singapore, Cochin-China, or Réunion, can round Australia either to N. or S. Instructions relating to this passage will be found in §§ 171 and 172.

Auxiliary steamers will find this west route the most rapid, especially as they can make the return passage from Australia through the Suez canal.

A ship starting from Sydney, *between May and August*, can count on the SE. monsoon to set her through Torres strait, (vide § 171.) SE. winds will also be experienced while crossing the Indian ocean near the parallels of 12° or 15° S. Cross the equator a little to the north of the Seychelles, and run for Aden, under low steam, if necessary. Before October a ship will probably have strong head winds all the way up the Red sea.

The worst time of the year to leave Sydney is between July and January. The NW. monsoon will at this season prevent the passage through Torres strait, and as the route north of New Guinea will be too long, the cape Horn voyage will almost invariably be preferable. In January, February, and March, there is no reason why vessels should not run south of Australia, especially if their port of departure be Melbourne or Port Adelaide, (vide § 172.) After doubling cape Leeuwin run to the west with the SE. trades between 22° and 17° S.; but do not cross 15° S. before reaching 62° and 64° E. Thence bear north, when the SE. trades will be replaced by the NW. monsoon, or rather by the variable winds from NW., SW., and SE., which generally come under the head of the NW. monsoon.

The NE. monsoon prevails at some distance north of the line, except toward the end of March and the beginning of April, when the change in the monsoons takes place.

At this time vessels will have to steam for Aden. From October to April the winds are from S. and SE., in the southern part of the Red sea, and as far north as the 17th parallel.

The Suez route is therefore not favorable for sailing-vessels, and only for auxiliary steamers, *from the beginning of May to the end of July*, when they can pass through Torres strait. Of course, we do not bring the mail steamers into these considerations.

The cape of Good Hope route seems hardly admissible for auxiliary steamers that can take the Suez canal, nor would we recommend it for sailing-vessels.* It is always a tedious passage, particularly when doubling the cape, in June, July, or August. During these months the weather off the cape of "Tempests" is alternately stormy and calm. Cyclones also are frequent from December to March, in the neighborhood of Réunion. By all means take the cape Horn route.

§ 108. ROUTE FROM NEW CALEDONIA TO EUROPE.—1st. *The cape Horn route.*—The reader should refer to the first part of § 107, where he will find useful information.

* Vide p. 103, "Navigation South Atlantic."

The greatest difficulty on this route for a sailing-ship will be doubling the N. and NE. capes of North island, New Zealand. We will, therefore, give the following instructions, which, if not absolutely indispensable, are at least very necessary for passing north of New Zealand. The passage through Cook or Foveaux strait is not advisable on account of the exceptionably bad weather which prevails near the west coast of New Zealand, (vide § 8.) Ships often have to lie-to in this locality with a lee-shore close aboard. Presuming that the ship is not pooped, (as a certain frigate was,) she will be forced to run through Cook strait, with a gale of wind after her from NW., a bad horizon, no land in sight, and with the danger of being "broached-to," etc. In fact, the climate to the west of New Zealand is about the worst in the world, and if she should chance to fall in with a good spell of weather, it is not likely to last more than 12 hours. Even an auxiliary steamer will find it to her advantage to double cape Otou to the northward.

Therefore, after leaving Noumea, run to the southward and eastward for New Zealand; if in an auxiliary steamer do not hesitate to use even one-third of the whole allowance of coal; if in a sailing-vessel remember that the length of the passage to Europe depends greatly on the time spent on this, the first, part of the voyage. From Noumea to 30° S. the wind will be from SE. but very variable. After leaving New Zealand use every means to make to the east and south; if, therefore, the wind should come out from S. and SSE., take the starboard tack; if it should haul ahead, do not keep off to the north of east, but go about immediately. South of 30° S. the wind will become still more variable, when take advantage of it by either beating or scudding, as the case may be. Here steam will outrun sail. The worst months for doubling New Zealand are those between the beginning of September and the end of February.

After leaving cape Otou, or North cape, it will be comparatively easy to double East cape and to pass east of Chatham islands. After crossing 48° or 49° S. between 158° or 163° W., run down the easting to northward of 50° S., from October to April; and north of 52° S. between April and October. Beyond 98° or 93° W., commence to bear away around cape Horn. We once more repeat that when detained for an unusually long time between Noumea and New Zealand, the whole first half of the voyage to Europe will be prolonged.

But, beyond cape Horn, it will be the captain's own fault if he strike north too quickly and thus lengthen the passage; in other words, the passage will be a long one if 30° S. be crossed to the westward of 28° W. Navigators should particularly remember this and cross 30° S. somewhere between 7° and 21° W., according to the season and the instructions in § 107, (vide also § 14 "Navigation of South Atlantic.")

In the "Navigation of the Atlantic ocean" it is stated that the line should be crossed at the following points: in July and August, at 21° W.; in September, at 22° W.; in June, October, and November, at 23° W.; in January, at 25° W.; during the remainder of the year at 24° W.

After crossing the NE. trades rap-full the easting for Gibraltar, or the channel will have to be made well to the northward, as the west winds keep to high latitudes in this part of the Atlantic.

There are many instances of quick runs from New Caledonia to Europe.

The Garonne, Captain Rallier, left Noumea on the 6th December, 1872, and anchored at Brest on the 7th March, 1873, 90 days at sea. The Orne left New Caledonia on the 8th June, 1873, put in at St. Helena on the 9th August, and arrived at Brest on the 11th September, 94 days at sea, and 1 day at St. Helena.

2nd. *The westerly route.*—It is generally conceded that the cape Horn route is the only one for sailing-ships. Small auxiliary steamers, fearing the rough seas of high latitudes, can take the westerly route, especially if they are bound to one of the Mediterranean seaports, and pass through Torres strait and the Suez canal. In this case, they should leave Noumea between the beginning of May and the last of July, or at the furthest not later than the first days of August.

The SE. monsoon will carry them through Torres strait without difficulty, (vide §§ 139 and 171.) Thence the track lies through the Indian ocean, between the parallels of 12° and 15° S., and crosses the line N. of the Seychelles. Finish the passage as stated in § 107.

There is nothing to be gained by taking this route at any other season, for a ship would have to pass either south of

Australia in January, February, or March, or north of New Guinea from October to January. In the latter case a large detour through Gillolo or Dampier straits will have to be made, (vide §§ 139 and 171.)

Sailing-vessels should therefore never take the west route, and small auxiliary steamers only when they can leave Noumea between May and August, or when bound to one of the Mediterranean ports. Large auxiliary steam-vessels, frigates and corvettes for instance, can make a quick trip by this route, through Torres strait, and thence to Suez; always provided they start between May and August. At all other seasons they should take the cape Horn route, particularly if bound to one of the Atlantic ports of the French coast, and wishing to save coal.

The following account of a west passage is from the log of the dispatch-ship Guichen, (screw,) Captain Perrier:

"Left Noumea on the 1st July, 1871; passed through Torres strait by the Raine island channel, on the 11th and 12th July. Arrived at Kupang (Timor) on the 19th, and left on the same day. Ran through the Indian ocean between the parallels of 12° and 15° S., with steady winds from SSE. to E. Anchored at Mahe (Seychelles) on the 12th August; coal-bunkers still full. Left on the 18th August, and reached cape Guardafui under sail; then got up steam. Arrived at Aden on the 28th August, and left the same day. Found a stiff breeze from NNW. and rough sea, in the southern part of the Red sea.

"Head winds forced us to anchor, on the 2d September, at Djeddah to take in coal. Arrived at Suez on the 9th September, and at Port Saïd on the 11th. Seventy-two days from Noumea to Port Saïd."*

§ 109. ROUTE FROM TAHITI TO EUROPE.—After leaving Tahiti keep the sails full and make as much to the southward and eastward as the variable direction of the trades will permit.

The prevalent west winds do not generally come north of 35° S. After finding them the ship should follow as nearly as possible a great-circle route to cape Horn. Thus, after

* The reader will note that this passage was made during the favorable season, and that the only difficulty experienced was the head winds in the Red sea. The Guichen arrived at Rochefort on the 12th October —103 days from Noumea.

crossing 40° S. near 140° W., head her for 50° S. in the neighborhood of 123° W.; and 55° S. near 103° W.

Probably the passage will be nearly as quick, and the chances of meeting ice less, especially from October to April, via the following route. After reaching 40° S. steer nearly SE., if the winds allow. When near the *highest latitude*—50° S. from *October* to *April*, or 52° S. from *April* to *October*—run down the easting, keeping a little to the northward of this extreme latitude-limit; for south of it the ice is dangerous. Do not commence to bear south until beyond 98° W., or rather, near 93° W.; when steer to cross the parallel of Diego Ramirez islands near 76° W.; then double cape Horn.

It is impossible to indicate the crossings which should be made between Tahiti and the parallels of 35° or 40° S.; for they will be different according to the season and the direction of the wind. The wind, south of Tahiti, may be said to be generally from E. and SE., yet even here it often blows from all points of the compass. The observations of two or three consecutive months go to prove that the direction of the wind is variable for almost every day.

The only advice therefore that can be given is to make south as soon as it can be done without hugging the wind; the west wind will be found near 35° or 40° S. The passage from cape Horn to Europe is the same as that in § 107.*

Passage of the sailing-frigate Alceste, Captain Brosset, (*Ann. Hydr.*, 1871.)—"I left Tahiti on the 18th November, or, in other words, at the beginning of the winter months, for the groups S. of the line. At this season the trades are often interrupted by variable winds and long calms, and the Alceste took 14 days to make the 660 miles to the parallel of 30° S.

"At 30° S. I ran out of the belt of calms and variable winds; and at 39° S. met the west winds. These blew freshly from N., W., and S., and as they never got to the eastward of the latter point, I made rapid way. At 53° S. and 88° W. I met a violent westerly gale. It sprung up from SW. at 1 a. m. on the 23d. The aneroid barometer indicated its approach by a slight rise. The wind shifted according to the law for the southern hemisphere; that is, in a direction against the hands of a watch."

* The Sibylle made the passage from Tahiti to Toulon in 107 days.

CHAPTER V.

ROUTES FROM THE PORTS OF AUSTRALIA OR ASIA TO THE EAST.

From Port Adelaide.

§ 110. ROUTE FROM AUSTRALIA TO THE WESTERN COAST OF AMERICA.—A ship starting from Port Adelaide should proceed south of Tasmania. Off Melbourne the winds are generally westerly, and enable vessels to run along the south coast of Sir Roger Curtis islands, and thence, either north or south of Kent group.

If the wind come out from the eastward, as it is likely to do in January, February, and March, pass west of King island, and well to the west of Tasmania. For the passage through Bass strait vide § 104. The reader should also refer to § 107.

From Sydney.

If the point of departure be Sydney, the beginning of the voyage should be in accordance with the instructions in § 107.

In all cases, after a vessel is once to the eastward of New Zealand, she should head—as near as the wind will allow—for her *extreme* parallel, or rather a little to the *north* of it; this extreme latitude will be 50° S. from October to April, and 52° S. during the rest of the year. Thence make the easting, but do not get to the south of these parallels if it can be avoided.

Bound to Valparaiso.

If the point of destination be Valparaiso, do not make any northing until near 103° W. Cross 48° or 47° S. near 98° W.; and approach the land to the southward of your port. Look out for a norther near the coast of Chile, especially during the bad season, from the end of May to September, (vide §§ 29, 69 to 71, and 110.)

Bound to Callao.

Vessels bound to Callao can begin to bear to the northward after crossing the meridian of 123° or 120° W. They should strike the SE. trades near 90° W., and then steer for their destination; being careful to keep it bearing to the northward of NE., to avoid the necessity of hugging the wind in case it haul ahead.

A ship, about to make passage to Panama, can follow a more westerly route than the above. She should sight cape San Francisco, and finish the voyage as stated in §§ 48 and 54. Bound to Panama.

If going to Mexico, she should follow the same route, entering the SE. trades near 30° S. and 93° W., and crossing the line, either to east or west of the Galapagos islands, according to the season. For instructions relative to the last part of the voyage, vide § 49. Bound to Mexico.

When bound to San Francisco, Maury says to cross 45° and 40° S. betwee 150° and 140° W.; and the equator between 130° and 120° W. Information concerning the last part of this passage will be found in § 50. Bound to San Francisco.

Maury's instructions on the passage from Australia to California.—"In coming out of the Victoria ports go south of Van Diemen's land, or through Bass strait, as you have the winds and find it expedient.

"Being south of Van Diemen's land makes it convenient to pass south of New Zealand, if the wind be fair, as in the majority of cases it will be. Having passed south of New Zealand steer for the parallel of 40° or 45° S., between the meridians of 150° and 140° W.; thence for the equator between 120° and 130° W., crossing by a north course both the horse latitudes of the southern hemisphere and the equatorial doldrums; then run through the NE. trades as best you may, keeping a "rap-full," and running up into the variables beyond the horse-latitude calms of the northern hemisphere, if need be, to complete your easting, and make your port.

"If the winds be not fair for passing south of New Zealand, try Cook strait in preference to passing to the north of New Ulster.

"If you pass through Cook strait, then stick her well to the eastward, and take the eastern passage. On this passage you should run down your easting pretty well before you get far enough north to be bothered by the baffling winds of the horse latitudes south. If these come as low down as 38° or 40° S., stand north the moment you feel them till you get the SE. trades; then cross these and the NE. trades, both as obliquely to the eastward as they will permit, with fore-topmast studding-sail set.

"On this passage you will have, finally, to run down

your easting when you get in the variables beyond the NE. trades, and, of course, you will aim to reach the parallel of 38° or 40° N., or even a higher one north, to do this. How far you will go north depends somewhat upon the distance you may be west of California when you lose the NE. trades. If you be only a degree or two from the land you will steer straight for your port without caring to get to the northward of it; but if you be ten or twenty degrees to the west of it, or even farther, then, of course, the distance to be run makes it an object to turn out of your way and go north in search of good winds.

"The most difficult and uncertain parts of this passage will be in the time required to cross the three belts of calms, and to clear the winter fogs of California. But for these the eastern passage, from Victoria to California, would be one of the most certain passages in the world.

"The distance from Victoria to California cannot be accomplished under canvas, by the eastern route, much short of 8,700 miles. But driving captains, with clipper-ships under them, may expect to average, one trip with another, along this route, not far from 200 miles per day; for I feel assured there is no part of the ocean in which the winds generally admit of more heavy dragging and constant driving than they will in the extra-tropical regions of the South Pacific; say on the polar side of 43° S."

§ 111. ROUTE FROM AUSTRALIA TO NEW CALEDONIA.—After leaving Sydney run to the eastward, without bearing north, until about 450 miles from the Australian coast. Then commence to make the northing and gain the exceedingly variable SE. winds, which are here termed trades.

The length of this passage, under sail, will be very irregular; perhaps only 8 days; perhaps over three weeks. From June to September the voyage will be quite easy, and vessels can run straight away north. During the rest of the year the east winds come farther south; it will then be advisable to run farther to the eastward on the parallel of Sydney. The following crossings will give an approximative idea of the mean route:

In *January*, vessels may follow the parallel of Sydney to 160° E.; cross 30° S. near 167° E., and then easily bear to the northward.

In *February*, cross 155° E. between 33° and 35° S.; 165° E. between 31° and 33° S., and 30° S. between 166° and 167° E.

In *March*, 155° and 160° E. between 33° and 35° S., and the parallel of 29° S. between 166° and 168° E.

In *April*, 155° and 160° E. between 33° and 35° S., and 30 S. between 163° and 165° E.

In *May*, 160° E. near 33° S., and the parallel of 30° S. between 165° and 167° E.

From *June* to *September*, vessels will be able to pass close to the east of Ball pyramid, and to cross 30° S. in the neighborhood of 162° E., whence the passage will be easy.

In *September* and *October* they can reach 30° S. without difficulty between 162° and 164° E., and thence for Noumea.

In *November* and *December* they can follow the parallel of Sydney until they are near 162° E., and then bear away in order to cross 30° S. well to eastward of 165° E., especially in December.

The light-house on Amédée islet has greatly facilitated the approach to Noumea. One very important injunction is still necessary, which is to make the light between NNE. and E. by N. The approach to Noumea.

If the light be sighted, bearing NE. for instance, from deck on a clear day, the reefs will be at least 15 or 20 miles distant. A ship endeavoring to make the light-house in any other bearings than the ones mentioned above, may be on top of the rocks even before she sights it. Therefore place the vessel to the SW. of the light when at least 60 miles off and lay a NE. course.

The tower is a good mark in daytime. Ships generally run in by one of the Bulari passes, between To and Toombo reefs. There is a small shoal between the two passes, on which the sea always breaks. The south pass is the narrowest, being about 1,400 yards wide; it lies between the shoal and Toombo reef, and being to windward, is the better one of the two. The other pass lies between To reef and the shoal; it is about 1,800 yards wide, and can be taken in case of a head wind from westward. There are no dangers in either channel. Dumbea pass can be used if the ship fetch to leeward of the Bulari passages.*

* These instructions on the approach to Noumea are from the work of *M. Bouquet de la Grye.*

In § 6 the reader will find indications on the wind-system of this locality. Reference should also be had to § 138 for an account of the gale experienced by the Morceau near Sydney. The following accounts, furnished by Captain Jouan, of the Bonite, contain much useful information. The voyages of the Bonite were made prior to the erection of the light-house at Noumea, in 1865:

"The distance in a straight line from Sydney to Noumea is about 1,020 miles. Half way lies a group of reefs, between 29° and 32° S.; these reefs are terminated by two highlands: Howe island, (2,600 feet,) and Ball pyramid, a gigantic leaning obelisk, which, from a distance, has the appearance of a ship under sail. The position of these rocks is not accurately put down on the charts, and the existence of some of them is doubtful. Vessels generally pass to the southward of them on the voyage from *Sydney* to *Noumea;* and to the northward, on the *return voyage.* I think that it is preferable to pass south of them when you leave Sydney in the winter, as the west wind is then fresh on the extra-tropical coast of Australia, and carries a ship rapidly toward the trades. On the other hand, after leaving Noumea, I would pass north of them, as vessels generally make the coast of Australia near Moreton bay, where they strike the NW. wind, and, if compelled to beat, have the current in their favor. In no case would I attempt to shorten the passage by getting into the neighborhood of Eliza, Seringapatam, Elizabeth, etc., reefs. The weather is rarely settled in this locality, the group appearing to be situated near the limit of both the New Caledonian and Australian winds; the currents about it are also strong and irregular.

"The portion of the ocean lying between New Caledonia, Australia, and New Zealand—from 152° to 177° E., and from 20° to 40° S.—is about the worst place in the world for variable and bad weather. The W. and NW. winds blow a gale, on the west coast of New Zealand, nearly all the year round. In winter, from June to September, the wind on the east coast of Australia is very stiff from WNW. to WSW. These winds meet the regular trades, south of New Caledonia, at this season, causing calms, and sometimes violent squalls. In summer, from May to October, the wind is ordinarily from NE., and the weather fine on the Australian coast; sometimes a *hot wind* blows from

NW.; the weather is then close and oppressive, but at the end of two or three days the wind jumps around to south with heat lightning. Keep a good lookout for this shift of wind, as it is very rapid and may dismast your ship. While this south wind blows the sea is usually rough, and a voyage is rarely made through this portion of the ocean without experiencing a gale. During the southern winter the wind prevails from W. to SW. for quite a distance off the coast of Australia. At this season quick passages are made from Sydney to Noumea; inversely the return voyage is long. Sometimes vessels are a month going from Noumea to Sydney.

"A three years' experience on the W. coast of New Caledonia has convinced me that the winds there are prevalent from the westward during one-third of the year. After the fresh ESE. winds have blown, in gusts, with a few drops of rain, for 2 or 3 days, (barometer $29^{in}.80$ to $29^{in}.92$,) the wind hauls to E. with an overcast sky; then to NE. with steady rain, and to N., NW., and W.; the weather does not become fine until the wind has reached SW. and SSW., and then it very soon comes around again to SE. and ESE. With the NW. wind the weather is uncertain; sometimes close, calm, and extremely hot, with violent squalls at intervals.

"On her first voyage the Bonite left Sydney on the 11th July, 1860, and steered south, passing 45 miles from Ball pyramid. From the 13th to the 21st variable winds from WSW. to S., weather very fine, barometer between $30^{in}.16$ and $29^{in}.96$. On the 21st, 120 miles SSW. from Noumea, wind SE. to ESE., usual direction at this season. The next morning, about 9 o'clock—after a squally night—saw the reefs distinctly. The mountains being enveloped in a heavy fog were not sighted till long afterward. This coast is often hidden by clouds, and it is necessary to be careful while steering for Bulari or Dumbea passes, for if once set to leeward it is difficult to beat up against the NW. current.

"In a second voyage, the Bonite left Port Jackson on the 22d December, 1860, with a good breeze from S., (bar. $29^{in}.84$.) This was followed by a rain-squall from the same point, during which the barometer stood at $29^{in}.68$. The wind then came out strong from SSE. with a heavy swell; I steered to pass to the northward of Elizabeth,

Middleton, etc., reefs. At noon on the 25th, position of ship, 29° 28′ S., 158° 27′ E. On the 28th, being 180 miles SW. of Noumea, experienced several thunder-storms; wind variable and light from N.; thermometer 79°. On January 3d sighted land to leeward of Dumbea pass. The Bonite could not make anything against the current, though all her sail was set. At nightfall weather looked very ugly; barometer falling rapidly; lay-to on the starboard tack. After sunset the wind sprung up furiously, raining in torrents, barometer 29in.37; force of the wind equal to that of the heavy squalls of India. This gale ended on the 6th January, after driving the ship 180 miles from Noumea. After two days of respite had another lighter gale. We did not reach our anchorage until the 11th January.

"Third voyage. The Bonite left Sydney on the 26th March, 1862. Found a light breeze from ENE. to NNE., and good weather outside. I tried for some time to make to the southward, but a squall from SSE. made me give up the idea. At sundown on the 6th April, sighted New Caledonia, very far off. During the night the wind shifted to NW., W., and WSW. Ran for land under all plain sail; very stiff breeze, weather overcast. At 3.30 p. m. made the reefs; no land in sight all day. A short time after they showed up plainly, and we found that we were in a kind of gulf, formed by the reef S. of Bulari pass. Not being able to double these reefs, took the south tack, in order to give them a wide berth; sky overcast in the west; a bad appearance. Violent squalls and very heavy sea during the night. The weather being more moderate and the wind from ESE., on the 10th, we succeeded in entering. Fifteen days at sea.

"In a fourth voyage, made in 11 days, (August and September,) the Bonite took the southerly route. The weather was nearly always fine, except on the 26th, 150 miles from Noumea, where we had a very strong breeze and heavy sea from E. For several days the barometer stood between 30in.20 and 30in.23."

The following quotation is from the log of the sailing-frigate Isis, Binet, (vol. 28, *Ann. Hydr.*:)

"Left Sydney (in April) with a light breeze from SSE. Next day a light breeze and calm. Afterward experienced a day's NW. to SW. winds. The SE. wind then began to prevail, blowing light and gentle from ESE. to E.

§ 112. Route from Australia to New Zealand.— Vessels bound to Otago harbor should pass through Foveaux strait, (vide § 107.) Those bound to Port Nicholson should take Cook strait, (vide § 107.) Those destined for Auckland or the bay of Islands, should double cape Otou (north) and pass N. of New Zealand.

Leaving Sydney *from the beginning of September to the end of April*, and bound to Auckland or the bay of Islands, first head SE. and then keep S. of the 35th parallel until the ship is in the neighborhood of 170° E., when commence to bear to the NE. By steering in this manner a vessel will meet fewer head winds, as they blow from the east quite frequently north of 35° S. during the southern summer. When bound out of Sydney, from May to the end of August, a more direct route can be taken.

Vessels leaving Melbourne or Port Adelaide should begin the voyage by passing through Bass' strait, or going S. of Tasmania, as described in § 107.

§ 113. Route from Australia to Tahiti and the Sandwich islands.—A ship *leaving Sydney*, from May to August, can pass north of New Zealand, thence south of the Kermadec islands, and then make her easting south of 30° S., crossing that parallel between the meridians of 158° and 153° W. She can generally run between Vavitao and Tabuai.

1st. To Tahiti, from May to August.

Cook strait is to be preferred at this season, (vide § 107 for the passage through this strait.) After leaving the strait run down the easting, doubling Chatham islands to the northward, and crossing 30° S. near the meridian of Rapa island. During the *first season* a ship can generally pass to the westward of Moorea, and then go about for Papiete. During the other season make Tahiti from the eastward, and pass between Anaa and Maitea islands, as stated in § 93.

From the beginning of September to the end of April.

Vessels starting from Melbourne or Port Adelaide can either go south of Australia or through Bass strait, according to circumstances, (vide §§ 107 and 110.) If they take the latter route, Foveaux strait can be used, (vide § 107,) though we think that the passage between the Snares and Auckland islands will be preferable in either case. The parallel of 40° S. should always be crossed between 158° and 153° W.; and that of 30° S. between

153° and 143° W. Cross both 40° and 30° S. farther to the west in June and July than in December and January.

For instructions concerning the first part of this passage vide § 110.

2d. To the Sandwich islands. From Sydney go through Cook strait, (vide § 107.) From Melbourne and Port Adelaide pass between the Snares and Auckland. Always cross 40° S. near 143° W. Thence pass near, and a little to the westward of Pitcairn and Oeno islands. Cross the line in the neighborhood of 133° W., and finish the voyage as described in § 95.

1st. During the SE. and SW. monsoons. § 114. ROUTE FROM SINGAPORE TO THE MOLUCCA ISLANDS.—A ship bound from Singapore to the Moluccas, from *May to September*, will find the SE. monsoon south of the equator, and the SW. monsoon north of it. She should, at this season, run south of the Anambas and Great Natuna, thence between Charlotte and Louisa reefs, being careful to give Friendship and north Luconia shoals a good berth. The north points of Balambangan and Banguey can be passed close aboard and the exit made through Balabac strait; thence pass through the Sulu group, double the N. point of Celebes, and make south through Molucca passage.

Balabac strait. Balabac strait has three channels: the two near Balabac are narrow and rarely used; the southern one near Banguey is the best. Approached from the west, the high mountain of Kinibalu in Borneo is a good mark; its position is lat. 6° 5′ N., long. 116° 40′ E. There is also a conical peak on the NW. coast of Banguey island close to the shore; it can be seen for 40 or 50 miles; and is situated in lat. 7° 19′ N., long. 117° 06′ E. If short of water run south of Balambangan and anchor about a mile and a half from shore, when the peak on Banguey bears NNE. Good water will be found in a small river bearing E. A launch can cross the bar, but the crew should be armed.

Approaching the southern (and best) channel of Balabac strait from the west, guard against being set too far to leeward, during the season of the SW. monsoon. Fifty or sixty fathoms of water will be found about 35 miles off the west coast of Balambangan; and the north coast can be approached to within 5 miles. Steering E. by N., 5 or 6 miles off the north coast of Banguey, a ship will sight the Mangsee islands, bearing ENE. She should keep on the Banguey side of the channel, and thus avoid the reef which extends for 9 miles west of the Mangsee group; the position of this reef may be recognized by the green color of the water over it. The soundings will show a coral bottom at 7, 14, and 17 fathoms; 3 or 4 miles from the coast of Banguey the depth is about 6 fathoms. When bound east, and the Mangsee islands bear NNE., a small sand-bank, surrounded by reefs, will be sighted, bearing ESE.; pass at least 3 miles to north of this, where over 8 fathoms of water will be found.

Do not head to the southward of ESE. ½ E. after leaving Balabac strait, until the soundings show from 7 to 8 fathoms; then stand for the south coast of Cagayan Sulu; thence steer so as to pass close to the south coast of Pangutaran, and through the channel between this island and Obian.

No soundings will be found on approaching this channel *from the W.*, and vessels should keep well to the S. during the SW. monsoon, as the current near the W. coast of these islands sets rapidly to N. On nearing the south point of Pangutaran island, keep it bearing to the N. of E., and run close to the island until through the channel, and thus allow for any drift.

Sulu anchorage is 33 miles ESE. of the southern extremity of Pangutaran.

In all cases double the W. point of Sulu by leaving it to the northward, and pass south of Pata and the adjacent islets. On the contrary, run to the northward of Tapul, Taluk, and Kabinguan islands. Tapul is high; the two others low. The ship will be on soundings nearly the whole time in this neighborhood, and can anchor if the currents should prove violent, or beat up during the night in clear weather.

Once clear of the Sulu islands, head for the north point of Celebes and Molucca passage.

The SE. monsoon prevails in this passage prior to the month of October, and the NW. monsoon in November. In the first case the track, after leaving the north point of Celebes, passes *near* and to southward of the Tifore islands, thus striking the SE. monsoon well to the east. After the month of October steer from the north point of Celebes straight for Buru and Amboina.

2d. During the NE. and NW. monsoons. On the passage from Singapore to the Moluccas, from *October* to *May*, the NE. monsoons will be struck north, and the NW. monsoons south, of the equator. At this season the passage should be made through Carimata strait.

After leaving Pedra-Branca, run for 27 or 30 miles on an E. by S. course, if the wind permit; then 24 or 27 miles ESE., in order to get well clear of the Geldria and Frederick banks; thence steer to leave Saint-Barbe on the north, and run for Soruetou. If the wind come out from W. or SW., Biliton can be doubled to the northward; thence pass south of Ontario reef and around the Montaran islands.

The best route through Carimata strait, especially in overcast weather, when coming from the NW., is to pass *to the E.* of Ontario reef. Vessels should keep from 30 to 45 miles from Soruetou and steer SSE., when the W. point bears at least 3° to west of N., and thus give Ontario reef a wide berth. After keeping the point on this bearing until well clear of the reef, steer so as to pass about 20 or 30 miles from Borneo. The depths along this coast will vary from 14 to 17 fathoms, increasing sometimes near the banks. South of Rendezvous* the depth of water is from 19 to 21 fathoms until within 30 or 33 miles SW. of Pulo-Mankap, nor should you approach any closer than this distance with a large vessel. If uncertain of the longitude, it would be well to sight Borneo, if possible, and then head SW., after doubling Rendezvous island, thus clearing Mankap banks.

In clear weather vessels can also go to the westward of Ontario reef. It will then be necessary to double Soruetou at a distance of 15 or 18 miles, and to head SE. by S., or SSE. when that island bears NE. If it be not advisable to round the bank, the west point of Soruetou should be brought to bear to the east of N. 8° E. Steer in this way until 25 miles south of the island, when the west point will be on the horizon, and will bear N. by E. By keeping on the same course a vessel will sight Montaran islands, and pass 15 or 18 miles to eastward of them; thence steer SS. by S., and leave Cirencester and Discovery banks on the west, and the reefs SSW. of Pulo-Mankap, on the east. Alongside of these banks do not run into less than 15 or 16 fathoms, nor into more than 20 fathoms near the banks on the west coast. The soundings are very irregular.

Sometimes called Kumpal island, (lat. 2° 44′ 30″ S.)

After running out of Carimata strait, head so as to pass about 10 miles south of Great Solombo; thence make your easting of 150 miles between the parallels of 5° 36′, and 5° 50′, S. A ship can run over Laars shoal (5° 43′ S.) if *sure of her latitude*; if not, she should run farther south, and not double Brill (*Taka-romata*) shoal in the night-time, as the currents are very strong and irregular. If running east on 5° 43′ S., the Tonyn islands will be sighted from the masthead in clear weather; they are situated near Brill shoal. Continue to the eastward after doubling the Tonyn islands, and sighting Tanakeke, pass it at a distance of 12 or 15 miles.

When the wind is *SW. during the day* and the weather *clear*, head for Salayer strait, passing south of Mansfield bank. If Bonthein mountain be visible, steer straight for Middle island, when the peak bears between N. by W. ½ W. and N. by W.; and then pass between this island and the one to the south.

During the *night, the wind being uncertain and the weather overcast*, the best plan is to pass inside of Mansfield bank and along the coast of Celebes. The soundings here extend from 6 to 9 miles off the coast, and a ship can anchor if necessary. On this route keep from 4 to 6 miles off the Celebes coast, until Bonthein mountain bears N. by W. ½ W. After doubling Mansfield bank to the north, keep 12 or 15 miles off the coast. As soon as Middle island is sighted, keep well out to sea until it bears E. by N., thus avoiding Amboina bank; thence steer between Middle and South islands.

Once clear of Salayer strait, Amboina and the Banda islands are easily reached by passing successively to south of Hegadis, Groenwout, and Binonko islands. But vessels *bound through Pitts passage*, from October to May, should run around the S. point of Buton, and skirt along the shore until they reach the E. point, passing to the west of the Wangi-Wangi islands. They should then bear north for Weywongy island, and thence run for the S. point of Xulla-Bessi; the currents in this locality set to the south and are very strong. If drifted to leeward of the N. point of Buru, they should pass to the southward and eastward of this island, between Buru and Manipa.

1st. During the NE. and NW. monsoons.

§ 115. ROUTE FROM SINGAPORE TO TORRES STRAIT.—The NE. and NW. monsoons blow from October to April; but sailing-vessels should not leave Singapore for Torres strait until November, and then invariably pass through Carimata strait, as stated in § 114; thence they will have a choice of two routes through the Java sea: The first, *the southerly* route, especially adapted to sailing-vessels; the other, the northerly route, more advantageous for steamers and auxiliary-steam vessels.

The southerly route.

The southerly route through Java sea.—After leaving Carimata strait, steer for Baean island, and pass it to the northward and eastward. According to Captain MacKenzie, a ship near Bawean *at night* should run for Giliang (or Pondi) island, a little to the eastward of Madura, and afterward cross Sapœdie strait between Giliang and Sapœdie.* If, on the other hand, she should lose sight of Bawean *before nightfall*, she can run for Kangeang island, and pass it to the south and between it and the islet of Urk; thence the route is north of Lombok, and the course SE. for 8° S. Keep on this parallel, as it is free from all dangers, until north of Ombay strait, when run between Ombay and Kambing, or between Kambing and Babi, and afterward between Wetta and Timor, and between Kissa and the NE. point of Timor; this latter passage is 18 miles from the Timor coast. Once clear of the Java sea, run straight through the sea of Arafura for Torres strait. If deemed advisable the high and wooded islands of Karimon-Java can be passed 10 or 12 miles to the northward, and after running south of Hastings rock, the route already given may be rejoined in passing between Kangeang and Urk.

The best and safest route is that which passes south of the chain of islands lying east of Lombok. To follow it a ship should conform to the preceding instructions relating to the manner of reaching the passage between Kangeang and Urk, or to those referring to the passage between Pondi and Giliang, and afterward run for Lombok or Allas strait.

Horsburgh states that the northerly currents in Lombok strait will lengthen the voyage, and that Allas is the best strait for vessels coming from Carimata during the NW. monsoon.

Babi strait is never advisable. After leaving Allas or

* *Sometimes called Galiœn, or Respondi island.*

Lombok straits run about SE. by E. for 150 or 200 miles, then 200 more E. by S., changing the course if necessary to pass south of Damo island. This island is situated to the southward of Rotti. Coal may be obtained at Kupang, but the anchorage there is not very safe during the west monsoon. Steer east from Damo, and thus run a few miles north of Echo bank, over which there are hardly 12 fathoms of water. After a run of about 500 miles from Damo island, cape Croker will be sighted to the southward. After passing about twenty miles from capes Croker and Wessel, head for the Wallis islands if bound through Endeavor strait, or for Booby island if making for Prince of Wales channel.

The northerly route through the Java sea.—Some authors recommend the route through Salayer strait during the season of the NW. monsoon, (vide § 114, 2d part.) This is a good route for auxiliary-steam-vessels and steamers, but we think that the passage through the south of the Java sea will be a better one for sailing-vessels. A ship taking the route through Salayer strait should steer east until the peak of Kambyna bears NW., then run SE. by E. for Ombay strait. She can pass between Ombay and Kambing, or between Kambing and Babi; thence between Kissa and the NE. point of Timor, and run for Torres strait. The northerly route.

At this season, (*from May to September*,) only steamers can make the passage from Singapore to Torres strait. They should take the route through the south of the Java sea, as given above. They should keep to northward of the islands, on the parallel of 8° S., until they are N. of Ombay strait; and then pass between Ombay and Kambing, or between Kambing and Babi, and thence as already stated. Coal can be obtained at Kupang, where the anchorage is good at this season. Kupang can be approached from the north through Ombay strait, or more directly through Alloo or Pantar straits. 2d. During the SE. and SW. monsoons.

§ 116. ROUTE FROM SINGAPORE TO THE WESTERN COAST OF AMERICA.—At this season (*from October to May*,) the passage may be made in two ways, either by the easterly routes, or by taking the route south of Australia. 1st. During the NE. and NW. monsoons.

The easterly routes are two: first, that by the N. of New Guinea, described in § 175; second, that by Torres strait, (vide §§ 115 and 175.) For information concerning the route south of Australia, vide § 176.

The route south of Australia. In a good vessel, the passage south of Australia will probably be quicker and safer. If this route be decided upon, follow the directions given in § 176, and keep a clean full through the SE. trades of the Indian ocean; thence make south for the steady west winds; these once found head SE. for the parallel on which you intend to run down the easting; this parallel should be between 46° and 50° S. at this season. The track is close to the northward of Auckland island, and east of New Zealand is the same as that given in § 107 or § 110. Information concerning floating ice may be found in §§ 43, 104, and 107.

The easterly routes. Navigators deciding to take one of the easterly routes, and preferring the one by the N. of New Guinea, should conform to the instructions in § 175, and cross 10° S. near 172° E., and thence keep west of the Fiji group and east of New Zealand.

Between 45° and 48° S. the west winds prevail, but run still farther south, say to about 50° S., if bound to any of the ports of South America or around Cape Horn.

The Torres strait route. Vessels choosing the route through Torres strait should first head according to the instructions given in § 115, then run through Prince of Wales channel and Bligh passage, and along the southern coasts of New Guinea and the Louisiade group. The winds in this locality are generally from W. to NW.; therefore southing and even easting can be made with facility.

Vessels running north of New Zealand will have a quicker passage; but it will be difficult to double cape Otou (North) at this season unless advantage be taken of every chance for making to eastward as well as to southward. If it be found impossible to double cape Otou run through Cook strait, (vide §§ 8 and 107.) For information concerning the latter part of the voyage, vide § 110.

2d. During the SE. and SW. monsoons. At this season (*from May to October*,) ships bound to cape Horn and the ports of South America will find it greatly to their advantage to go south of Australia, (vide § 176,) but they will encounter the severe weather of the southern winter, and should be well fitted out in every way.

The prevalent SE. winds will render the passage difficult as far as the straits of Sunda, but once in the Indian ocean a vessel can run across the trades rap-full, and probably find the west winds at 30° or 32° S., and thence bear away to the SE. and run down to 50° or even 52° S. Pass either N. or S. of the Auckland islands, and keep to the eastward on the same parallel, then make north again as stated in § 110. For information in regard to floating ice, vide §§ 43, 104, and 107.

But for Californian or Mexican-bound ships the northerly route is preferable. They should, on leaving Singapore, cross the China sea, as stated in §§ 153 and 156, and enter the Pacific near the Bashee group, if sailers, or through the strait of Formosa, if steamers; 35° N. should be reached as soon as possible. Vessels taking the strait of Formosa will be less liable to meet typhoons, or if they should happen to fall in with one, there are plenty of ports of refuge. (Vide §§ 20, 157, and 159.)

By running through the Bashees, a ship will be the sooner out of the dangers of the China sea, and can take advantage of the full force of the Black current, or Kuro Siwo, (vide § 37.) Still, typhoons are common in this neighborhood, especially in July and August, and even until November. The region of west winds once reached, the voyage should be finished according to the instructions given in §§ 119 and 120. Most of the easting can be run down between the parallels of 40° and 45° N., where both wind and current are generally favorable.

We have stated that vessels bound to South America should take the southerly route; but it should be understood that they can also take the northerly route. After meeting the west winds they should run down their easting north of the parallel of 40° N., commencing to bear south after passing 158° W., and the NE. trades should be left near 118° W. and between 10° N. and the equator. The last part of the track lies south of the line, as stated from §§ 68 to 72.

§ 117. Route from Saigon to the western coast of America.—This route is nearly identical with the preceding.

Thus, if it be decided to go south of Australia, during the season of the NE. monsoon, (from October to May,) run out of the China sea according to the instructions given in § 151; thence according to §§ 176 and 116. If one of the

During the NE. monsoon.

easterly routes be decided upon, run through Carimata strait, as stated in § 151, and then follow the advice given in §§ 115, 116, and 175.

During the SW. monsoon. During the SW. monsoon (*from May to October*) staunch auxiliary steamers bound to South America will find it greatly to their advantage to go south of Australia. They can fill up with coal at Singapore and make the first part of the voyage as described in § 170. They should then follow the route given in §§ 116 and 176. But auxiliary steamers making passage to the coast of North America, and all sailing-vessels bound to any American port on the west coast, should take the northerly route at this season, and pass through Formosa channel or the Bashee group, as stated in § 116.

1st. During the NE. monsoon. § 118. ROUTE FROM CHINA TO VALPARAISO, CALLAO, AND PANAMA.—Starting from Hong Kong at this season, (*from October to May*,) the best and shortest route is that south of Australia. After sailing down the China sea, as stated in § 151, follow the instructions given in §§ 116 and 176. If, however, the port of departure be Shanghai or Yokohama, the northerly route will be the best. The ship will probably strike the west winds between 35° and 40° N., and should finish the voyage as described in the present section, under the head of SW. monsoon.

2d. During the SW. monsoon. The only practicable route at this season (*from May to October*) is the northerly one. After leaving Hong Kong steer for the Bashees in a sailing-vessel, or for Formosa channel in an auxiliary steamer. Run down the easting near the parallel of 45° N., where both the winds and currents are generally very favorable, (vide § 116.) The NE. trades should be struck between 148° and 143° W.; and the SE. trade-winds near 5° N. and between 118° and 123° W. From here shape the course as stated in §§ 68 and 72.

Maury gives a more westerly route, advising ships to run into the NE. trades near 152° E.; to cross the equator in the neighborhood of 172° E., and thence to pass either east or west of New Zealand. The last part of the voyage is easily accomplished.

But we think this route hardly advisable for *sailing-vessels*, as several groups of islands are in the way; it is also difficult to double New Zealand, and there are many dangers in the route. An auxiliary steamer can, of course, steam around the northern end of New Zealand, (vide § 108.)

Below is found the main portion of a letter addressed by Maury, in 1854, to a Boston merchant, relating to the proper route from Hong Kong to Valparaiso:

"To reach Valparaiso from Hong Kong you have to make nearly 180° of longitude, and the question is, in which hemisphere will you run down this easting? If in the northern, you will have, for the sake of the winds, to run to the north of your place of departure; and, if in the southern, you will, for the same reason, have to run to the south of your port. But the 'brave west winds' of the southern hemisphere will decide the question for us.

"This point being settled, the question is, will you run down for those winds by passing to the east or the west of Australia? Clearly not to the west, if you take your departure from Shanghai or Japan. From Hong Kong there is room for difference of opinion, and I have not observations enough on the winds and currents of those seas to enable me to decide. The shortest distance from Canton, west of Australia, is about 500 miles less than it is east of New Zealand, and 800 miles less than it is by the south side of that island and east of Australia; and the route east contemplates your going as far as the variables of the northern hemisphere, say between the parallels of 30° and 35° N., in order to get far enough east to clear Australia. The question of going west of Australia is debatable only during the strength of the NE. monsoons, or from October to March, inclusive. During the rest of the year east of New Zealand is the only route.

"I recommend the western route only in the NE. monsoons and when they do not admit of a good offing for the eastern route. In December the Flying Cloud made the run from Hong Kong to Java-Head in 7 days. When the winds are fair for such runs as that, the western route is the passage; and the question as to routes, like the route north or south of Ireland, from Liverpool to New York, ought to be decided at the moment of coming out of port, and finding how the wind is.

"Before I go further in discussing routes I will state the shortest practicable distance by the several routes from Hong Kong to Valparaiso:

	Miles.
From Hong Kong via strait of Sunda and south of Australia	11,400
From Hong Kong via 33° N. and 150° E., to 0° and 163° E., and south of New Zealand.............	12,200
From Hong Kong via 33° N. and 157° E., to 0° and 170° E., and south of New Zealand.............	11,900
From Shanghai via 33° N. and 157° E., to 0° and 170° E., and south of New Zealand.............	11,100
From Shanghai via 33° N. and 150° E., to 0° and 163° E., and south of New Zealand.............	11,500
From Japan via 33° N. and 150° E., to 0° and 163° E., and south of New Zealand..................	10,900
From Japan via 33° N. and 157° E., to 0° and 170° E., and south of New Zealand	10,400

"So you observe that the route east of Australia and south of New Zealand is the longest, and the route west from Hong Kong is 500 miles shorter than the route which passes east of New Zealand, and this is the route which, I think, experience will probably prove to be the best in the long run; certainly from Shanghai and Japan it is the best.

"I give the preference to the east side of New Zealand, because better winds are found along that route, and which will probably more than make up for the difference of distance from Hong Kong.

"I take it that a vessel steering from 30° or 35° N. in the Pacific, and entering the NE. trades in April, will be able to make, with a good 'rap full,' a course between SE. and SSE. to the line, and, after crossing the line and entering the SE. trades, she will be able to make a course through them with not more than one point westing. From the equator, and between 170° and 175°, (west of New Zealand,) is plain sailing;* therefore, if, after turning to the southward and eastward from 30° N., or whatever be the parallel attained, the winds will, without pinching, allow you to cross the line between 170° and 175° E., do so, and then stand as straight as the wind will allow you for the 'brave west winds' of the extra-tropical south, shaping your course for 50° S. about the meridian of 140° W., taking care not to recross the parallel of 45° to the west of 90° W. If it be found practicable to accomplish this route, the distance will

* The author thinks that ships should always pass east of New Zealand.

be about 11,000 miles. I am particular in stating these distances, because intelligent navigators, in case they be pinched, will have no difficulty in determining which side of New Zealand to pass. Of course it will be understood that there is no virtue in the parallel of 30° N.; I only indicate that as the lowest parallel upon which, in the month of April, good westerly winds prevail. Now, with all these preliminaries before us, the instructions are, after getting an offing from Hong Kong, make the best of your way to the meridian of 150° E., without making any southing; and the nearest way to get there, that is by great circle, is to reach, say the parallel of 30° N., long. 137° E. So you observe that it is not much out of the way to run up to 30° or even 35° N., for the sake of better winds. With a smart ship a good navigator on this route can reach the line in 25 days; in April it may be done in 18, and perhaps sooner in other months; it will take him thence 15 days to cross the SE. trades and get into the 'brave west winds' of the South Pacific. Suppose he gets them in 48°, long. 180°, he will be into Valparaiso in 25 days more.

"So, tell your captain that you expect him to make the passage, if he succeed in getting clear of the Asiatic coast without delay, in about 70 days. He ought to average 175 miles a day. Caution him, after he gets south of the SE. trades, not to be deceived by the first spurt of westerly winds. He should reach 48° or 50° before bearing away for his easting."

§ 119. Route from China to Mexico and California.—Vessels always take the northerly route from Hong Kong or Shanghai to the Mexican or Californian coasts, running to the northward and eastward at first, as stated in §§ 160 and 163. Although the winds of the SW. monsoon are favorable, typhoons are to be feared, especially from July to November. After the month of November, that is during the strength of the NE. monsoon, the difficulty of beating to northward will be counterbalanced by the fact that the weather is generally less inclement.

The region of westerly winds once reached, the route follows the arc of a great circle as nearly as possible; it runs up as high as 48° N. from April to October, and keeps between 40° and 45° N. from November to March. Here both wind and current are favorable.

Maury says: "All vessels from China or Japan will first make for the variables, which they will find strong and good from the westward, between 35° and 40° N. in winter and spring; between 40° and 45° in summer and fall.

"Vessels from China may follow pretty closely the great circle route, which crosses the meridian of 180° in about 50° N. The route in the Pacific is free from icebergs, and is not more foggy than that in the Atlantic. As to the relative fury and frequency of the gales, I cannot speak."

The reader should also refer to the following paragraph for further instructions:

§ 120. ROUTE FROM YOKOHAMA TO SAN FRANCISCO.—This is an easy passage. The westerly winds will be found near 40° N., and the current will be favorable for the ship almost as soon as she is out of sight of land.

Vessels should follow a great circle route as nearly as possible, though it is not advisable to pass 48° N. in summer, (from April to September,) nor 44° and 45° N. in winter, (from October to March.) The only drawback to the voyage is the fogs on the Californian coast, which are especially common from April to November. The land-fall should generally be made to the northward of the destination. Reference should also be had to §§ 119 and 121. The following extract is from the U. S. Coast Survey Report for 1867:

"A vessel making the great circle track to the eastward would have the great Japan stream in her favor to about 43° N. and 156° E., or about 1,440 miles; then the cold Behring sea current and the end of the Alaska current to 47° N. and 157° W., or 1,980 miles; finally to San Francisco, about 1,860 miles, passing through the great bend of the Japan stream, where so many indications of land have been recorded, and where the weather is almost invariably thick and bad in summer and cold and boisterous in winter. On this track the summer winds would generally be favorable, and, with good weather, it would be altogether the desirable route; but with thick, foggy weather for nearly the whole of this distance, undetermined velocity and direction of the currents, (except in general terms,) great variability of climate to passengers and cargo, and extra hazard and risk to life and ship, some great positive advantage over all these must exist to warrant the adoption of it.

"The commercial advantages of the steam route to China, through the warmer and more equable latitudes, must always outweigh any merely theoretical and shorter but more hazardous route. A study of the currents, winds, and weather, on the lower latitude route, will lead to the conclusion that is being solved practically. From the south end of Japan to San Francisco, a course very little north of a direct line on a Mercator projection carries a vessel across the great Japan stream, in part through the axis of the main branch flowing eastward, across the northern part of Flicureus whirlpool, and across the California stream, with favorable or light winds the greater part of the distance."

CHAPTER VI.

ROUTES FROM THE PORTS OF OCEANIA.

§ 121. ROUTE FROM THE SANDWICH ISLANDS TO SAN FRANCISCO.—A ship starting from the Sandwich islands should run north on the starboard tack for the westerly winds and then make her easting, approaching her port from the northward. From July to September she may have to run as high as 44° or 45° N., for the west wind, though they may be struck during the rest of the year, especially in winter, even before the parallel of San Francisco is reached. Fogs are very frequent on the coast of California throughout the year, and especially from April to November.

Captain Wood gives the following instructions:

"The passages from the Sandwich islands to any part of the NW. coast of America are made by standing to the northward till the westerly winds are reached, when the run into the coast is easily made, taking care, however, if bound to a port to the southward of you, not to bear up till well in with the land, when northwesterly winds will be found to carry you down to the southward.

"On this coast, as a general rule, the land should be made to the northward of the port you are bound to, as in almost all cases the wind and current prevail from the northward from Vancouver island to cape Corrientes of Mexico.

"Though lying between the parallels of 19° and 23° N., the Sandwich islands are often visited during the winter months with strong breezes and gales from S. and SW., but for the rest of the year the trade-wind blows pretty steadily."

Maury's observations on this route.—"From the 'Islands' to San Francisco, the course is to the northward; so steer with a rap-full, and, as the winds will let you, lay up till they are found to be fair. The navigator, as a rule, will always have to go to the northward of San Francisco to be sure of good winds, which are frequently found near the parallel of 38°; but sometimes, as from July to September, inclusive, as far as 45° N.

"The islands, such as the Society and Sandwich, that stand far away from any large extent of land, have a very singular but marked effect upon the wind. They interfere with the trades very often, and turn them back; for westerly and equatorial winds are common at both these groups, in their winter time. Some hydrographers have taken those westerly winds of the Society islands to be an extension of the monsoons of the Indian ocean. Not so; they are local, and do not extend a great way either from the Sandwich or Society islands.

"These winds at the Sandwich islands often come from the south as well as the west; and on such occasions they afford vessels bound for any of the Pacific ports of North America a fine opportunity of running to the northward, clearing the NE. trades, and getting the westerly winds of the variables beyond."

The following passages were made by Captain Paty, who has been running constantly on this route ever since 1837:

"*From Honolulu to San Francisco.*

Clipper-brig Zoe, October and November, 1853... 14 days.
Clipper-brig Zoe, January, 1854.................. 13 days.
Clipper-schooner Restless, April, 1854........... 13 days.
Clipper-schooner Restless, May and June, 1854... 16 days.
Clipper-schooner Restless, July and August, 1854. 21 days.
Clipper Francis Palmer, February, 1855......... 11 days.

"This last is the shortest ever made. The St. Mary's, Captain Bailey, made the next quickest passage, leaving Honolulu at the same time.

"The track up (from Honolulu) requires the most skill in navigating; the track down is pretty generally understood. The average of passages from Honolulu to San Francisco is, in length, to the passage down, as 6 to 5. Therefore, *ten* days down is no better than *twelve* days up, and *vice versa*."

§ 122. ROUTE FROM THE SANDWICH ISLANDS TO PANAMA.—The first part of the route is the same as that described for San Francisco in § 121. Vessels should not bear south until they have passed beyond the meridian of 138° W.; this will bring them into the NE. trades near 133° W. The voyage should be finished as described in § 68.

§ 123. ROUTE FROM THE SANDWICH ISLANDS TO VALPARAISO AND CALLAO.—After leaving the Sandwich islands keep a clean full, on the port tack, through the NE. trades, and run out of them in the neighborhood of 148° W. A vessel will strike the SE. trades somewhere between 10° N. and the equator, (near 10° between June and September, and near the line between December and March;) after crossing them a trifle free and passing to the west of the Paumotas, the west winds will be found either at 30°, 35°, 40°, or even 45° S., according to the season, as these winds reach lower latitudes when the sun is in the southern hemisphere; thence the run to the eastward is easily made. Do not stand to the northward, or for the SE. trades, until the port of destination bears to the north of NE. The landfall should be made to the southward of your port, except at Valparaiso, during the bad season. (Vide §§ 21, 69 to 71, and 110.)

James Wood's instructions.—"In making a passage from the Sandwich islands to the coast of Chile or Peru, the best way is to stand across the trade as near the wind as the topmast studding-sail will stand. This, as the direction of the wind is in general from ENE. to E., will enable you to make Tahiti, and pass the Society islands by one of the clear channels to westward of them. It is of little use trying to fetch to the eastward of these, as not only do you lose much time by hugging the wind too close, but also the strong current, which sets to the westward from 20 to 40 miles per day, is pretty sure to drift you that much to leeward; and even if this were not the case, so difficult, tedious, and dangerous is the navigation among the archipelago of low coral islands which lie to the eastward, that, unless you can weather the Marquesas altogether, it is better even to bear up than to entangle yourself in such a labyrinth. After passing the Society islands stand to the southward, till on or about the 30th parallel, when the westerly winds will be found. These will carry you into the coast, care being taken not to bear up when within the influence of the southerly winds till near enough to the land to insure keeping them down to your port."

§ 124. ROUTE FROM THE SANDWICH ISLANDS TO EUROPE.—The first part of the passage is identical with that described in § 123. The west winds will generally be found near 35° or 40° S.; cross these parallels near 158° W., or more to the eastward if the trades allow. Thence run to the southward and eastward as much as possible, so as to put your ship between 47° and 50° S. from October to April, and between 48° and 52° S. from April to October. Then run east until near 98°, or rather 93° W., and cross the parallel of the Diego-Ramirez islands in the neighborhood of 76° W. This route is quite free from ice.

It should be understood, however, that a ship in *great haste* need not make such a detour as that above stated; but steer so as to reach the parallel of cape Horn at about 98° W. But the danger from meeting icebergs will be great on this route, and will necessitate a sharp lookout, (vide § 43.)

The voyage will end as described in § 107.

§ 125. ROUTE FROM THE SANDWICH ISLANDS TO NEW CALEDONIA AND AUSTRALIA.—Vessels leaving the Sandwich islands should shape their course so as to run out of the trades near 168° W.; and cross the line between 168° and 173° W. The track lies to leeward of the following low islands: Swallow, (with a lagoon in the center;) McKean, (coral island, about 20 feet above water;) Gardner, (an atoll, visible 15 miles, a few trees upon it;) Mitchell islands, lat. 9° 27′ S., long. 179° 54′ E., (covered with cocoa-nut trees, and visible about 10 miles;) it then runs west of Meek shoal; a good lookout is necessary near this coral bank, as the water does not always break over it. The South Pacific Directory for 1871 gives its position as 10° 40′ S. and 179° 08′ E. Its longitude according to the French charts is 178° 30′ E. Rotumah should be sighted and passed to the westward. There are two small islets about 2 miles north of this island; one of them is very low. The route also lies west of Hammond reef, the position of which is doubtful. Vessels bound to New Caledonia should sight Maré island and run through Havannah passage with the prevalent E. and SE. winds.

If the wind should come out from W. or even from NW. or NE., east of the pass, it will be better to give the isle of Pines and the Great reef a wide berth, and enter by Dumbea pass, as stated in § 111.

If bound to Australia, pass west of Mathew island; thence

steer about WSW., and clear the great New Caledonian reef well to the southward. For the end of the passage, vide § 138.

As the Pacific is not thoroughly explored between the equator and the south of New Caledonia, a sharp lookout must be kept in that portion of the route. The winds in this locality are generally from E. and SE. Occasionally heavy squalls blow near the New Hebrides and New Caledonia, (vide §§ 4 and 26.).

§ 126. Route from the Sandwich islands to China.—Make your westing between 20° and 15° N.; keeping farther to the *north in summer* than in winter. Information concerning the latter part of the voyage is given in § 101.

§ 127. Route from the Sandwich islands to Tahiti.—This passage presents certain difficulties, and should only be undertaken by vessels that can lie very close to the wind.

Mr. Biddlecombe, Master H. M. S. Actæon, says: "On leaving the Sandwich islands, you should stand south till in the latitude of the southern part of Hawaii, when you should haul your wind to cross the line, if possible, in the longitude of Tahiti, as the SE. trade breaks you off when you first make it, and then you do not fetch it within several degrees. It is tedious to get to the eastward in the latitude of Tahiti, owing to the strong westerly current; therefore, you should lose no chance of preserving your easting."

The most favorable time to make this passage is from the end of March to the 15th of June. But the most difficult season in which to reach the equator on the meridian of Tahiti is during the months of July, August, and September, when the SE. wind prevails, south of 10° N. During October, November, and December the direction of the wind will be a little more favorable; and in January, February, and March the wind will keep well to the northward and eastward until you reach the line. But navigators should always bear in mind that even in April, May, and June it will be difficult to fetch Tahiti without going about, unless with an exceedingly fine sailer.

If the ship be not weatherly, or is a very small auxiliary steamer, the better plan will be to run up beyond 30° N., on the starboard tack, and make the easting on that parallel; even half the allowance of coal may be usefully expended on this part of the passage. Sailing-vessels will be delayed by calms and baffling airs unless they run well north, especially from July to November. They should, therefore, recross 30° N. at 153° or 150° W., if they can possibly do so, and then stand across the trades with the wind abeam, and try to reach 10° N. at 140° or 138° W., as soon as possible. If the wind then haul to the south, they will be far enough to windward to fetch the equator to eastward of Tahiti.

Once south of the line, they can take one of the passages through the north of the Paumotas, (vide Wilkes's instructions in § 103.)

§ 128. ROUTE FROM THE MARQUESAS TO THE SANDWICH ISLANDS.—This is not a difficult passage. Starting from the Marquesas, a vessel should cross the equator as far east of 143° W. as she can, with a clean full on the starboard tack. From the end of June till November the SE. trades usually blow as far north as 10° N. Later, and especially from February to June, the NE. winds reach nearly to the line; naturally this last season is the worst for a passage from the Marquesas to the Sandwich islands. The chances of meeting calms near the equator in this locality are small; and the frequency of baffling airs from the *west* is certainly overdrawn by some authors. Information concerning the last part of this voyage can be found in §§ 95 and 99.

Biddlecombe remarks as follows on this passage: "Cross the equator, if possible, to the eastward of 145° W., as you will then be enabled to steer for Hawaii, or a degree to the eastward of it, if you should fall in with the NE. trade early, although you seldom meet it till you are in 10° N. The variable winds are generally westerly,* and the current runs with the wind; but if you get easterly variables you may expect to be set a long way to the westward, as the currents run more strongly in that direction than in any other. You should, therefore, cross the line well to the eastward, to insure your fetching to windward of Hawaii. In passing Hawaii, do not go nearer than 40 miles to it, as vessels often get becalmed for many days together under the land."

§ 129. ROUTE FROM THE MARQUESAS TO TAHITI.—This is an easy voyage, with a fair wind all the way. It only de-

* I do not think that this assertion is often realized.

mands attention near the Paumota group and Tahiti, (vide §§ 103, and 93, 94, and 98.)

Captain Richard Foy makes the following remarks on this passage, (*Ann. Hydr.*, v. 29:)

"The voyage from Nuka-Hiva to Tahiti is generally accomplished in five or six days; course about SW. ½ S.; this brings you a little to windward of the central islands of Taiara and Raraka, to starboard of Faaite island, and then to windward of Faarava.* Or, if deemed advisable, run between Toau, or Elizabeth island, and Aura, and then near Greig (Niau) island, sighting the lofty peaks Orohena, of Tahiti, rising about 7,360 feet above the horizon."

§ 130. Route from Tahiti to San Francisco.—Keep a clean full and cross the equator, if possible, on the starboard tack between 152° and 148° W.; or, in other words, as far to the east as the variable ESE. trades will allow. The best season is from the end of June till November, when the SE. trades reach nearly to 10° N. From February to June the NE. trades come well south toward the line. Few calms and usually a westerly current will be found. Run through the NE. trades on the starboard tack.

From June to November. After crossing 10° N. well to the eastward, the starboard tack will fetch a ship to windward of the Sandwich islands.

From November to June. Advantage must be taken of every favorable shift in the wind that will set the ship to the northward and eastward, remember, however, that it is never well to lie too close.

The SE. wind inclines to the eastward near the equator, and to NE. to northward of the line. In order to make to windward of Hawaii, cross 10° N. near 148° W. If this cannot be done, run off and pass well to leeward of the Sandwich islands, thus avoiding the calms nearer the group. For information concerning the beginning of the voyage, vide § 135; for the last part, vide § 121.

§ 131. Route from Tahiti to the Gambier islands, Tubuai, Valparaiso, Callao and Panama.—Ships bound to Valparaiso, Callao, and Panama should run for the region of west winds, as stated in § 109. These winds will usually be found beyond 35° S. With these winds vessels make to the eastward between the parallels of 35° and 45° S., and, if their port of destination be Valparaiso, approach the land to the southward, except during the season

* *Faarava* should probably read *Anaa.*—Translator.

of northers, (vide §§ 29, 69 to 71, and 110.) Vessels bound to the intermediate ports and Callao should not again enter the SE. trades until their port bears to north of NE.; they should also make their land-fall to southward of their port, (vide §§ 69 to 71, 110, and 123.) Vessels bound to Payta or Panama should not run into the trades until cape Blanco bears north of NE.; while those for Panama should sight cape San Francisco, and end the voyage as stated in §§ 48 and 54.

The following remarks by Captain Foy may be of interest:

"Every vessel bound from Tahiti to Gambier or Valparaiso, should double the north point of Moorea at a distance of 8 or 9 miles and pass to westward of that island. This precaution is necessary, as the mountains of Tahiti intercept the prevailing winds. Time will therefore be saved by keeping well off shore; still if the wind be from N. or NW., she may take the channel between Tahiti and Moorea, except from June to November, when the sea is heavy and the currents strong. This passage is 10 or 11 miles broad.

"Once clear of the group, steer about SSE. or even a little to the eastward of that point, as the current sets to west. Cross the meridian of Tubuai near 20° S., and stand on the same course to 22° S., thus clearing the doubtful low islands situated a little to the W. of the meridian of Tubuai and beyond 22° S. Vessels bound to Tubuai should then head a little west of S.; after running about 64 miles on this course, they ought to make the island off the lee (starboard) cat-head. At a distance, Tubuai resembles a "ham."

Passage of the Sibylle, Captain Brossolet, (Ann. Hydr., 1871.)—"I ran from Tahiti to 27° S. in four days. Wind moderate and from SE. to ESE. The trades then died away and were succeeded by calms and rain-squalls, lasting for several days; headed S. to clear this locality. Struck the W. wind on the 9th February, near 36° S.; attempted to steer a great-circle route for Valparaiso, but the wind not holding, I did not reach 40° S. till the 16th. Here the wind set in steady from WNW. and I ran at the rate of from 7 to 10 knots, until I reached Valparaiso, on the 7th March, 1869; 35 days at sea. One day during the passage the wind grew so strong that I feared I would have to lie to. Near the Chilean coast it died away calm.

"Though I passed within 3 or 4 miles of the supposed

position of Tabor* island, in broad daylight, I saw nothing of it."

Passage of the Alceste, Captain Brosset.—"Maury does not mention the passage from the Society islands to Valparaiso. The route, however, is quite simple.

"You should run through the SE. trades rap-full, and when you reach the region of west winds make a great-circle route for Valparaiso. Maury's pilot-charts prove this to be the best route.

"I left Tahiti on the 31st October. Experienced calms and light variable airs instead of the usual stiff trades. Did not clear the Tubuai group until the 6th November. On the 7th ran into a NE. wind, at 30° S., and steered a great-circle route; the wind shifted from NE. to N. and NW. on the 9th; lat. 34° S., long. 147° W. The great-circle route brought me to 40° S. During the voyage the wind followed the usual laws, and was as a general thing moderate. I arrived at Valparaiso on the 30th November after a passage of 30 days."

1st. Route to New Caledonia.

§ 132. ROUTE FROM TAHITI TO NEW CALEDONIA, NEW ZEALAND, AND AUSTRALIA.—As a general rule vessels should run N. of the Tongas from July to October, and S. from November to June. By making the following crossings the voyage will be easy and the wind fair:

In *January*, cross 18° S. at 160° W.; 22° 30′ S. at 175° W., or S. of the Tongas; and make a straight wake for your port.

In *February*, run S. of the Samoas, cross 17° S. at 175° W., and 20° S. about 178° W., then steer westward.

In *March*, 20° S. at 154° W.; 22° S. at 160° W.; and 22° 30′ S. at 175° W., south of the Tongas.

In *April*, 20° S. at 152° 10′ W.; 22° S. at 155° W.; 23° S. at 160° W.; thence pass N. of Minerva reefs, and sight Fearn and Mathew islands.

In *May* and *June*, 15° S. at 152° W.; 13° S. at 155° W., and follow this parallel to 165° W.; thence run to southward of the Samoa group; cross 18° S. at 176° W., and pass west of the Tongas and south of the Fijis.

In *July*, 15° S. at 152° W.; 12° 30′ S. at 160° W.; 15° S. at 169° W., and run along this parallel to 180°; cross 18° S. at 174° E., and 22° S. at 170° E. This route carries the

*Should probably read Tabon.

ship south of the Samoas, and north and west of the Fijis.

In *August*, 15° S. at 151° 10′ W.; 14° S. at 160° W., and keep on this parallel to 165° W.; 17° S. at 175° W., and 20° S. at 178° W.; in other words run south of the Samoas, north and west of the Tongas, and south of the Fijis.

In *September*, 18° S. at 160° W.; 21° S. at 170° W.; 22° S. at 175° W., and thence to the westward. This route lies south of the Tonga group.

In *October*, 15° S. at 152° 10′ W.; and follow this parallel to about 175° 50′ E.; thus passing south of the Samoas and north of the Fijis. Thence cross 22° S. at 170° E. Onaseuse, or Hunter island, can be sighted; it is situated in lat. 15° 31′ S., long. 176° 19′ E., and is a cultivated and inhabited volcanic island.

In *November* and *December*, 20° S. at 152° 10′ W.; 22° S. at 160° W.; and thence make the westing south of the Tonga group.

After making these crossings, as nearly as circumstances will allow, run through Havannah pass, with the prevailing E. or SE. winds. But if to eastward of Havannah passage the wind come out from W., NW., or even NE., give the isle of Pines a wide berth and enter by Dumbea passage, as stated in § 111.

The following observations on this route are by Captain Foy, (*Ann. Hydr.*, vol. 29:)

"After leaving Papiete for Noumea, vessels run down their westing with the wind astern. They keep a good distance from Morea island, and leave the Borabora group on their starboard hand; thence they run through the Cook group. The principal island of this group is Aitutaki, and may be recognized by its remarkable hummock, which slopes to the southward. This island should be left well to port; still, if a shift of wind to west, or any other cause, should compel a vessel to run through the group, she can double Aitutaki to southward, keeping a good lookout for shoal water. Thence the route should be resumed, so as to profit by the trades. Tonga-Tabou may be recognized by the shape of Eooa, which is a round, large island with two summits; it is situated south of Tonga. Vessels may pass between these islands, or south of all. In the N. channel the currents set toward the eastern coast of the island; while well off shore and south of Tonga their direction is westerly. I sighted Fearn and Mathew islands, and left them to port; also Walpole to starboard. The first is about 400 feet high; the second, 340 feet; and the third, 210 feet at its northern extremity, and about 480 feet at its southern. The distance between the N. and S. summits of Walpole island is about 4,140 yards. A short distance to the west is the isle of Pines, a low sugar-loaf, covered with pines, and about 870 feet high.

"The length of the voyage is from 16 to 17 days, and the sailing distance about 2,475 miles."

Passage of the Bonite, Lieutenant Jouan commanding, (*Ann. Hydr.*, *vol.* 21.)—"Left Tahiti at daylight on the 13th September. Gentle breeze from ENE. until the 17th. Sighted Pylstaart island during the night of the 23d and 24th and lost it, bearing NE. by N., the next morning at 9 o'clock. Rain-squalls and variable airs. At 11 a. m. on the 29th sighted Fearn island; this is a barren rock and quite steep-to. The volcanic island, Mathew, was reported in sight during the afternoon. At sunset on the 30th made Walpole island, very low on the horizon. As the wind was light and variable during the night I did not reach Havannah strait till 4.30 p. m. on the 1st of October. It was then too late to attempt the passage, especially as the tide was running ebb, so I stood off and on during the night and ran through at daylight. Although it was nearly low water, the current was still very violent, with a heavy chop sea at the entrance to the strait. Notwithstanding a good breeze from the east, the Bonite only forged ahead about a knot per hour, until the tide turned; when I ran through Woodin channel, and anchored off Noumea at 5.15 p. m."

2d. Route to New Zealand.

From April to November.

From April to November pass north of the Tongas, and cross 17° S. at about 175° W. Thence, running to westward of the Tonga group, cross 20° S. near 178° W.

Make the westing on the parallel of 20° S., thus passing to southward of Batou-bara (Vatu-rera) island; when steer SW., and cross the Tropic near 178° or 177° E.; when the ship will probably meet the SW. or NW. winds, which will easily enable her to make the bay of Islands.

From November to April.

From November to April, after passing north of the

Cook group, run south of the supposed position of Nicholson reef,* and cross the tropic at 173° or 175° W. Run to westward of the Kermadec isles, and thence for the bay of Islands, with the wind abeam.

The foregoing routes apply only to vessels bound to the bay of Islands. If bound to port Nicholson or Otago, it will be better to follow the route given for the passage to New Caledonia, and cross the tropic south of the isle of Pines; thence, after striking the west winds, run for Cook's strait, (vide § 107,) if bound to Port Nicholson; or Favorite strait, if for Port Otago.

Captain Foy makes the following observations on this passage, (*Ann. Hydr.*, vol. 29:)

"The route from Tahiti is nearly identical with that to New Caledonia, keeping to the tropics until it reaches 168° W. A ship should pass 60 miles south of Nicholson reef, as it is long and dangerous; cross the tropic at 173 W., south of Tonga-Tabou; and head for the bay of Islands, running west of Raoul island, one of the Kermadec group. From December to May the wind is nearly always abeam, from NW. or SE. On approaching New Zealand a vessel should keep a little to windward of Kororarika, so as to be sure to fetch the anchorage.

"The bay of Islands is easily recognized by a large rock, with an opening, which forms a good landmark. New Zealand is visible, in clear weather, at a distance of 30 miles; and when first sighted looks like a vast plateau covered with low hummocks. The land will gradually make out to the eastward as you approach; it looks like a group of islands, and is surrounded by peculiarly-shaped rocks, with daylight showing through the large openings in many of them. At a distance these rocks have the appearance of a vessel under canvas, the white strata in them looking like sails."

Follow the track given in the first part of this paragraph, under "the route to New Caledonia." Pass south of the isle of Pines and the Great reef; thence run some distance 3d. Route to Australia.

* The position of Nicholson reef is doubtful; its existence even is uncertain. Beveridge shoal has recently been located by R. Ad. Roussin; it is 5 miles long N. and S., and 3 miles E. and W., and is just awa-h, with a lagoon in the center.

to the west before standing to the southward. For information concerning the last part of the voyage, vide § 138.

From March to August. § 133. ROUTE FROM TAHITI TO CHINA.—From March to August, head about NW. to 3° or 2° S., and follow one of these parallels to 163° or 168° W. Thence bear north for the NE. trades. Keep to northward of the Ratack and Ralick groups. From May to October, steer from the north of the Marshall group for the south of the Marianas, keeping to northward of the Carolines, and enter the China sea through the strait of San Bernardino, (vide § 101, note.) The SW. monsoon prevails at this season.

From August to March. From August to March, and especially after October, the route is a little more to the southward. It passes within sight of the Suwarrow islands, and runs north of the Samoas; thence steer NNW. and pass west of Duke of Clarence, Duke of York, and Gardner islands. Cross the line at 178° W. or 180°, and keep east of the Ratack chain. After doubling these islands to the N., pass between Grigan and Assumpcion, or rather north of the Marianas, (vide § 101,) and enter the China sea to northward of the Bashee group. The NE. monsoon prevails at this season.

§ 134. ROUTE FROM TAHITI TO THE MARQUESAS ISLANDS.—This passage, though quite difficult under canvas, can, with the assistance of steam, be rendered comparatively easy. Vessels could easily fetch Nuka-Hiva if the Paumota group were not in the way. As it is, they should be very careful if compelled to go about in the middle of the group. An auxiliary steamer should always enter through the strait between Fakarava (Faarava) and Faaite islands; keep to windward of Raraka and Taiara, and leave the group by running to windward of Tika, (or Tikei.) From this point it is generally possible to fetch Nuka-Hiva.

The following instructions are due to Captain Foy, (*Ann. Hydr., vol.* 29:) "The approach to the Marquesas being simple, the delicate part of the route is that through the Paumotas. The wind in the western part of the group is generally from NE., and in the eastern part from SE., though variable in both instances.

"After leaving Tahiti and doubling point Venus you should lie very close and take advantage of every shift in the wind, and if lucky, you will be able to run through the passage between Fakarava and Faaite the next day at daylight. If possible keep to windward of Raraka and Taiera islands, and leave the group by passing to windward of Tikei, if the wind incline to shift to SE. or ESE. But if the wind freshen from ENE. to NE., it will be difficult to make easting, as the current sets west, or very often south, so strongly that it will be difficult to make headway against it, in which case you will be compelled to cross the Paumotas between Vliegen (Nairsa) and Arutua islands, and if possible run out to windward of Manihi island.

"Once clear of the Paumotas, beat to eastward as far as possible, as northing can be made at any time, especially if the SE. trades be found north of the largest of the Napuka group. Make the land at Roa-Poa, which lies south of the principal island, and thence run for Tai-o-hae bay, where there is an anchorage in from 7 to 9 fathoms. The harbor is sheltered, except from SW. and S.

"This passage can generally be made in 15 or 16 days, and the weather, especially among the islands of the Paumota group, will be quite bad."

Lieutenant Parchappe makes the following observations on this voyage, (*vide Ann. Hydr., vol.* 12:)

"The Paumota islands are very low, of coral formation, and here and there covered with clumps of trees. Most of them have a lagoon in the center. They are rarely visible at a distance of over 12 miles.

"Formerly vessels bound from Tahiti to Nukahiva, used to run to southward and then to eastward of the Paumota group; since the position of the islands has been better determined, the route is generally through the archipelago."

§ 135. ROUTE FROM TAHITI TO THE SANDWICH ISLANDS.—The reader should first refer to § 130, where, speaking of the route from Tahiti to San Francisco, we stated that, from June to November, vessels could easily fetch to windward of the Sandwich islands if they crossed 10° N. at 148° W. During the rest of the year it will be difficult to cross the 10th parallel so far to the east, and almost impossible to make Hawaii.

Captain Beechey gives the following instructions: "From the time we passed Maitea we endeavored to get to the eastward and to cross the equator in about 150° W. longitude, so that when we met the NE. trade-wind we might be well to windward. There is otherwise some difficulty in rounding Hawaii, which should be done about 40 miles to the eastward to insure the breeze. The passage between the Society and Sandwich islands differs from a navigation between the same parallels in the Atlantic, in the former being exempt from the long calms, which sometimes prevail about the equator, and in the SE. trade being more easterly. The westerly current is much the same in both, and if not attended to in the Pacific will carry a ship so far to leeward that by the time she reaches the parallel of the Sandwich islands she will be a long way to westward, and have much difficulty in beating up to them."

§ 136. ROUTE FROM NEW CALEDONIA TO SAN FRANCISCO.—For instructions concerning the first part of the voyage, vide § 108, (cape Horn route.) Vessels should double New Zealand to northward, and bear away for 48° or 50° S., keeping on this latitude until they reach about 138° W. Thence the track is to northward and eastward, and enters the SE. trades near 128° W. Finish the voyage as stated in §§ 50 and 110.

§ 137. ROUTE FROM NEW CALEDONIA TO VALPARAISO, CALLAO, AND PANAMA.—§ 108 also contains instructions for this passage. After passing north of New Zealand vessels gradually make to the southward and eastward. From October to April they run down their easting between 45° and 50° S.; from April to October, between 47° and 52° S. The voyage is generally ended as stated in § 110.

§ 138. ROUTE FROM NEW CALEDONIA TO AUSTRALIA.—This is a voyage of fair winds. In order to meet them a ship should first steer for Sandy cape, and then round off to the southward so as to reach the parallel of Moreton island, near 157° E. The course is afterward S., with variable winds and a current generally favorable. The land should be made to northward of the port.

In § 111 the reader will find a long extract from the instructions of Captain Jouan, concerning the proper route from Sydney to Noumea. We give below an abstract from the log of the Bonite, Captain Jouan, for three voyages between Noumea and Sydney, (Ann. Hydr., vol. 26:)

First voyage. "Left the 23d November, 1860; cleared the reefs at 6 p. m.; wind SE.; breakers close to leeward; outside gentle wind from WSW. Becalmed for 2 days on the tropic;

afterward 2 days' wind from E. and ENE., which carried us close to Middleton island. There the weather became overcast and variable, wind generally from NW., falling light and calm about 2 p. m., and followed by unsettled weather in the evening, with a thunder-storm from southward. During the night of the 3d and 4th December, experienced a *brick-fielder*, the wind shifted suddenly to SSE., squalls violent; notwithstanding this shift of wind, the upper strata of clouds still drifted to NW.; barometer 29in.72; thermometer, 73°. The bad weather lasted only a short time; gentle breeze from NE. on approaching the coast. Anchored at Sydney on the 7th December, after a passage of 13 days.

Second voyage.

"Left Noumea on the 26th February, 1862, with a light breeze from N. Had to let go again inside the reefs, as the wind shifted to SW. Ran out on the 1st March with a NW. wind, but was hardly clear of the reefs when it again shifted to W. and SW. The weather became fine as we drew away from the land. But the barometer, generally high with south winds, fell to 29in.57; stars brilliant; sea phosphorescent. Dead calm and heavy SE. swell during the afternoon of the 3d; suddenly a violent squall from SE., and an extraordinarily heavy sea. Fine weather on the 5th; barometer rising gradually to 30in.12; light breeze from SSE. to SE., shifting to E. and NE. On the 10th, being 210 miles NE. of Sydney and 120 miles E. of Hawke head, the wind shifted to NNW., blowing very stiffly; bar. 29in.92; sky very clear to NNW., and slightly overcast to E. At daylight, the wind died away and after a day's tow we reached Farm Cove just in time to avoid a *brick-fielder*.

Third voyage.

"Cleared Dumbea pass at 9.30 a. m. on the 2d July, 1862, with a light breeze from NW. Heavy surf on the reefs. At 10 a. m. wind shifted to W. With the exception of 2 days' calms and light variable airs from SSE. to NNE., the whole passage of 24 days was one succession of squalls; during the short periods between the squalls the wind was light and always from the westward. Lightning on the horizon always showed the direction from which the wind was to be expected."

Captain Richard. Foy makes the following remarks upon this passage, (*Ann. Hydr.*, vol. 29:)

From December to May.

"The passage to Moreton bay, Newcastle, and Sydney can always be made with fair winds from December to May. The route is direct, but as the current sets to the southward, it is best to head for the land at least a degree to the north of your port. Once in sight of land of course your future action will depend upon the direction of the prevailing winds. The approach to the coast is rendered dangerous by the frequency of the heavy wind and rain squalls from ESE. to ENE. They rarely last for more than 2 or 3 days. It is then advisable to keep well off shore until the weather changes. When bound to Newcastle the land should be made near port Stevens."

The heavy squalls mentioned by Captain Foy are the same as those referred to in § 6. The following account, though long, we consider necessary to give our readers an idea of the bad weather in these parts:

Passage of the Marceau, Lieut. Galache, commanding.—"The Marceau left Noumea for Sydney on the 5th February, 1868; weather fine, breeze light from SSW. On the morning of the 7th the wind was light and varied from ENE. to NE. Same weather till the 13th. The weather became more dull as we approached the Australian coast, the barometer falling from $29^{in}.92$ and $29^{in}.96$ to $29^{in}.84$ and $29^{in}.80$; this is the normal fall, and in these localities betokens fine weather, calms, and light northerly airs. I sighted successively the paps called the Three Brothers, situated north of Camden haven, cape Hawke, and the land near Sugar-Loaf point.

"The foggy weather and heavy appearance of the sky betoken a rain-squall from NW. and a shift of wind to SE. or S., for the next evening, (*burster or brick fielder.*) At this season these signs never fail, but I expected to make Sydney by noon the following day, and the shift of wind rarely takes place before 3 p. m.

"I therefore kept about 15 or 20 miles from the coast, and made moderate headway with the light wind from N. and the current which set to the S. at a rate of about 1½ knots per hour. On the night of the 13th we sighted port Stevens light and headed for Broken bay, judging that the current would drift us to the south.

"The breeze died away, however; the sky became overcast, and a single violent flash of lightning illuminated the horizon from S. to E.; this flash was the only warning we had; the barometer was steady and everything seemed to indicate a squall from SE. About 2 a. m. the SE. horizon became overcast, and the wind came in hot and cold puffs, as it often does before a rain-storm. Furled the light sails, and clewed up the mainsail; had taken one reef the night before; braced up the yards and received squall with the sails full. It did not blow very hard, and a quarter of an hour afterward reset the topgallant sails and mainsail and headed for Sydney. Lost sight of port Stevens light to NNW., at $3^{h}45^{m}$, Sydney bearing NE. by N., distant about 50 miles.

"From 2 to 4 a. m. the wind did not freshen or change, but the sky looked heavy and threatening; barometer steady. After 4 o'clock the wind hauled to the southward, and grew rapidly stronger. I could not keep the ship on her course; but still hoped, with the assistance of the current, to reach Sydney or at least Broken bay, situated 15 miles to the northward. Ran in for the land, hauled on the port tack, but could not make it out on account of the heavy rain and overcast sky; dead reckoning put us about 25 miles off shore at 6 a. m. All the appearances betokened bad weather.

"I immediately made everything snug aloft and went about to the eastward, the wind having hauled to SSW.; made 5 or 6 knots per hour; very heavy sea running, occasioned by the S. wind acting against the current; made from 25 to 30 miles on a good course from 7 a. m. to noon.

"On the night of the 14th and 15th the weather was passable; but on the morning of the 15th the wind was very strong and blew violently during the whole day; the sea was frightful during the whole of the 14th and 15th; the wind remained steady from S., but after the 15th shifted to SE. Sounded on the evening of the 15th and the morning of the 16th, but found no bottom at 82 fathoms. If we had, I should have considered the Marceau in a very dangerous situation, as the depth of water in this locality is 100 fathoms at a distance of only about 30 miles off the coast. Besides, in my preceding voyages I had carefully noted the color of the water off Sydney; within 30 miles it is of a dark olive green, easily recognized, especially on the crest of the waves. Where we were it appeared green or light blue.

"Lighted the fires and steamed on the evening of the 16th, thus reducing our leeway from 8 to less than 5 points. Course NE., Sydney bearing nearly NNW.

"During the night of the 16th and 17th the gale was at its height, the wind striking the masts and rigging with terrific force, and nearly throwing the ship on her beam ends; the sea was frightful, and broke all over the ship, inundating the hold, and sweeping everything before it.

"As the wind blew furiously, first from SSW. and afterward from ESE., it caused a frightful cross-sea, which sometimes struck us to windward and again to leeward, loosening the gripes on the boats, and rushing down the hatches.

"During all this time, from the 14th till the morning of the 17th, the barometer only varied between $29^{in}.80$ to $29^{in}.88$.

"After this the wind hauled to the E., and I went about to the southward on the morning of the 17th. Contrary to my expectations, the wind shifted to ENE. in the afternoon, and I again spread the fires, which I had banked in the morning. Near midnight the wind moderated and hauled to NE. and NNE., the barometer commencing to fall slowly; at 6 a. m. on the 18th I headed W. on the starboard tack, the wind growing lighter. About 8 o'clock the weather cleared and the rain ceased. Our position at noon was 65 miles ESE. of Jervis bay, and 120 miles SSE. of Sydney.

"Although I was quite sure that the easterly squalls had sagged the ship to SSE., I hardly expected that, after a three days' gale from S., I should find the southerly current as strong as it was.

"Navigators acquainted with these localities state that, although the strong S. winds occasion a surface-current toward the N., when the east wind again sets in, the mass of water rushes back in a SSE. direction, with a speed proportional to the length of time it has been kept from its normal direction.

"We sighted point Jervis light on the evening of the 18th; weather calm; at 2 a. m. spread fires and reached port Jackson at 8 p. m. on the 19th."

§ 139. ROUTE FROM NEW CALEDONIA TO SINGAPORE CHINA, AND JAPAN.—Starting from Noumea, bound to Batavia and Singapore, at this season, (from April to October,) a ship should, as soon as the passes are cleared, head NW.,

1st. During the SE. and SW. monsoons.

and then run E. of Fairway reef. Thence, with the SE. trades fair, the route lies to northward; giving Bampton and Mellish reefs and Diana bank a wide berth. Run through Torres strait, and finish the voyage as stated in § 171.

Bound to Saigon or Hong Kong, between March and the end of July, take the same route, with the SE. monsoon after you. Cross the Java sea and enter the China sea through Carimata strait. Here the SW. monsoon prevails, and lasts till the end of October. For information concerning the end of the passage, vide § 173.

The following voyage was made in October and November, or the season of change in the monsoons, when calms and light baffling airs are frequent.

Passage of the Guichen, (screw,) *Captain Perrier, from Noumea to Batavia.*—"Left Noumea on the 16th October, 1869; headed straight for Raine island; the east winds being very light, had to get up steam. Ran through Raine island pass on the 26th, and anchored at Sommerset on the 27th. Fresh beef was obtained here, but no coal. Left for Torres strait and Timor on the 28th. Found a dead calm in the Arafura sea. Headed for the N. coast of Timor, so as to make the anchorage of Dula; arrived off this place on the 3d November; no coal here either; kept on without anchoring. Arrived at the Dutch coal depot of Kupang on the 4th. Distance run 2,700 miles, without filling up with a fresh supply of provisions, and with a very light wind through the Coral sea.

"Left Kupang on the 7th November, and anchored at Sourabaya on the 11th, after crossing Lombok and Madura straits. Left for Batavia on the 16th, and arrived at that port on the 18th. Total length of passage from Noumea 33 days; steamed nearly all the way."

During this season (from October to April) sailing-vessels, and auxiliary steamers wishing to save coal, should take the *northerly* route for Hong Kong, Saigon, and Singapore. Auxiliary steamers carrying enough coal to make the passage from Noumea to Kupang, a distance of 2,700 miles, can take the Torres-strait route. The anchorage at Kupang is tolerably good at this season. The route from Kupang to Batavia passes through Lombok and Madura straits; thence the passage is direct for Singapore. Vessels will be compelled to steam nearly the whole time. 2d. During the NW. and SE. monsoons.

But, as we have stated, the northerly route is preferable for sailing-vessels and auxiliary steamers. Once to eastward of Fairway reef, stand to the northward, and pass between the Solomon and Santa Cruz islands. From this point there is a choice between two routes.

1st route. *If bound to Hong Kong or Saigon*, make to the northward, and cross the line between 162° and 166° E. The track passes a little west of Kusaie or Ulan island, (Carolines;) and either E. or W. of the Providence isles, according to circumstances; keep north of the Marianas, or between Grigon and Assumption, and enter the China sea by the passage north of the Bashees. For further information, vide § 173.

There is but one route to *Yokohama*. Pass to westward of the Volcano group, and cross 30° N. a little west of the meridian of your port. Navigators should recollect that the Kuro Siwo sets to NE., (vide § 37.)

2d route. Vessels *bound to Singapore* can keep to the eastward of the Solomon islands, and then skirt them either to N. or S. They should leave New Iceland to the southward, and enter the Celebes sea by passing between the Serangani islands; reach the Sulu sea by Basilan strait and the China sea by Balabac strait. Thence to Singapore they should follow the return route to that given in § 154 for going from Singapore to Palawan passage. Useful instructions concerning the first part of this voyage will also be found in § 173.

Passage of the screw Transport Turn, Captain Martin.— "Left Noumea, bound to Saigon, on the 14th October. Steamed for 36 hours. Experienced ENE. and E. trades as far as 11° S. Here the wind died away; lighted two fires and steamed 1,500 miles to 9° N.; dead calm all the way. Ship to eastward of the Solomon islands on the 19th; sighted Ulan island on the 26th. Passed between Ulan and Providence islands and crossed the Mariana group.

Light breeze from NNE. on the 28th, latitude 9° N.; stopped the engine and made sail. Settled trades on the 29th. On the night of the 2d and 3d November ran between Rota and Agrigan with a fine breeze from ENE. On the 10th sighted the Balintang islands, and doubled Luzon to the northward between Balintang and Batan island. The trades held to this point; but instead of finding the NE. monsoon in the China sea, the wind grew light and shifted to E., SE., and SSE., with rain-squalls and calms. As the wind died away completely on the 15th, got up steam. Made cape Padaran on the 16th, and cape St. James on the 17th; anchored the same day at Saigon, after a passage of 34 days."

Passage of the Saint Michel, Captain Fradin from New Caledonia to Singapore, (Ann. Hydr., vol. 23.)—"Left Noumea on the 17th November, and were outside the coral reefs by noon; wind SE.; gentle breeze until the 20th, when we were in 15° 40′ S. and 161° 29′ E. As the wind showed signs of holding, I decided to run W. of the Solomon group. The wind died away on the 24th. Rainy weather and variable winds from S. and E., to N. on the 25th. Several of the islands of Solomon group in sight at daylight of the 25th; ran along the coast of Hammond island at a distance of about 20 miles; at noon, point Nepean bore about NNE. The same weather on the 26th, Dyston island about 20 miles off. The weather still rainy and the wind light on the 27th and 28th; ship still among the islands. At daylight on the 28th the largest of the Treasury islands bore NE. by N., distance 25 miles. From Noumea to the Solomons the current was favorable, and about 1 knot per hour; among the islands it was variable, but generally in our favor. Position at noon of the 28th, 7° 24′ S. and 155° 06′ E. Set a course to clear the rocks lying NW. of the Treasury islands, and ran up along the shore of Bougainville; wind light and variable. Our dead-reckoning put us at daylight on the 29th near the NW. islet; instead, we had been set far to WSW. Headed for Bougainville, and at 12.30 p. m. we were a mile and a half from the island, in 22 fathoms, good holding-ground. Position 6° 52′ S. and 154° 50′ E. Continued our route on the 29th, and at daylight of the 30th we were 7 or 8 miles off shore; fine weather and very light breeze from SE. to E. At 8 a. m., noticing a heavy ground-swell a short distance off, sent a boat to examine the locality; found a coral reef 825 feet long and 330 feet broad, with 3 to 6 fathoms of water over it; position of reef, 5° 58′.50″ S. and 154° 44′ 57″ E. Bouka island is in sight from this point, but well to NNW.; also several islets lying SE. of Bouka island. Kept along the coast of Bougainville at a distance of 6 or 7 miles. At $10^h 45^m$ two more coral rocks reported, one on each bow; wind being light, was compelled to pass between them; all clear at 11^h. The one to port extended from NNE. to SSW., and was about half the size of the reef met in the morning. The one to starboard was much larger than its neighbor and extended from NW. to SE.; found 4½ fathoms of water upon it; position 5° 52′ 26″ S. and 154° 36′ 24″ E., and about 7 or 8 miles off the coast of Bougainville. Soon after noon we sighted four islets 7 or 8 miles off Bouka. The breeze freshened a little, and we passed close to Sir Charles Hardy island. Gentle breeze all night from SE. to ENE. At daylight, on the 1st December, sighted what we supposed to be one island, but on coming up to it at noon it turned out to be four islands. At 4 p.m. the north point of the most southern island bore about E. Continued on a NNW. course and sighted St. John island, bearing NW. by W.; it has at a distance the appearance of two distinct islands. At 6^h the N. point of the most northern island bore about E. Position of this island, 4° 17′ S. and 153° 00′ E. Passed close to eastward of St. John island, and at noon on the 2d were at 3° 43′ S. and 153° 53′ E.; light winds from NW. to W.; sighted Oraison island, bearing WNW.; and Antoiny Kaan on the 4th; position at noon, 2° 45′ S, and 153° 30′ E. Light airs during the night of the 4th and 5th. Sir Charles Hardy islands, sometimes known as the Green islands, extend for about 23 miles in a SSE. and NNW. direction. The current is variable and the weather uncertain in this locality. Light NW. breeze on the 6th; more land in sight; position, 2° 12′ S. and 153° 32′ E. Hauled the starboard tacks on board on the 7th, with breeze from NW. to NE.; position, 1° 56′ S. and 153° 23′ E. Same weather on the 8th; position, 1° 41′ S. and 153° 27′ E. Light squalls from NE. and N. on the 9th; position 1° 03′ S., and 152° 54′ E.; swell, etc., showing us we were in shoal water, probably over Syra shoal. On the 10th, 11th, and 12th, light breezes and strong currents against us. Wind freshened a little from NE. on the 13th and 14th; position at noon on the 14th, 1° 37′ N. and 150° 28′ E. Breeze a little fresher on the 15th, 16th, and 17th; current setting W. and SW.; position on the 17th, 4° 45′ N. and 144° 21′ E. From the 17th to the 20th variable winds, with squalls, after which the wind generally came out from NE. to N.; position, 5° 06′ N. and 136° 50′ E. Position on the 21st, 5° 23′ N. and 134° 55′ E.; headed for the passage between St. Andrew's and Current islands. Position on the 22d, 5° 02′ N. and 132° 20′ E. Strong and very variable currents from the 17th to 25th. Ran between Palmas and Meangis islands, but did not see them as we made the passage during the night of the 24th and 25th. At daybreak on the 25th sighted Mindanao, Serangani, and Hummock islands; passed south of the two latter. At 9 a. m. E. point of Serangani bore N. 13° E., and the W. point of Hummock N. 34° W. From the 25th to the 30th light baffling airs and calms, and strong tidal currents, alternately favorable and contrary. On the evening of the 29th sighted Sibago islands, situated near the entrance to Basilan strait; gentle breezes from NE. to NNE. Entered the strait about midnight, with fresh breeze, made little headway about 2 a. m., as the flood was running strong against us; tide turned about dawn, when we soon cleared the strait. Strength of tide from 4 to 5 knots. Passed the town of Samboangan at 7 a. m. Position at noon on the 30th, 6° 52′ N. and 121° 48′ E. On the 31st at daylight we entered the Sulu sea, and were becalmed off the western shore of the group of islands situated to the westward of the strait. It is best to leave them to port. Fresh breeze from NNE. and NE. during the middle of the day; steered for Balabec strait. Lagayan Sulu was in sight at daylight of January 1st. Shaped our course to pass to N. of the principal island and between it and the islet closest to it; but as soon as I saw the pass, decided to run between the islet and the other small islands lying to the N.; were to the westward of this group at 9.30 a. m. My impression is, however, that either of the above-mentioned passages can be taken in *very* clear weather, but in ordinary weather it is best to run well north of them all. Position at noon 7° 09′ N. and 118° 02′ E.; gentle wind from NE. and NNE. The islands we had just left, the entrance to Balabac strait, and Banguey island were all in sight at sundown. Kept under low sail during the night, as we did not wish to run through the strait before daylight; sounded and found about 10 fathoms, bottom clay and mud; at 8.30 p. m., vessel touched, but was off again in 2 or 3 minutes, ran clear of the bank on the starboard tack, and anchored in 10 or 11 fathoms of water; reefs to southward and eastward, but the sea appeared clear between SE. and NE. Mangsee islands bore N. distant about 8 or 9 miles. Between our anchorage and Mangsee (Mangui) islands there was a reef of rocks and sand, covered at high water. The E.

point of this bank bore NNE., and the W. point NNW.; there is a passage between it and the most northern of the 5 islets; there is also a passage between the bank and the Mangsee islands. This last passage seemed to be the best, but we could not fetch it under sail, and passed between the W. point of the bank and the islet. At 9.30 we were in the center of the main channel, steering for Banguey island; Balambangan being a little on the starboard bow. Sighted all the dangers put down on the charts as situated SW. and W. of the Mangsee islands; left them on our starboard hand. The soundings showed from 13 to 22 fathoms. Therefore vessels should not attempt to *run through Balabac strait at night*, and should keep a man at the masthead if the passage is made during the day. I also think that the *best* pass is the one between the Mangsee islands and the reef of rocks and sand. Keep a sharp lookout if it be covered; pass close to port of the Mangsee islands, and head a little to starboard of the most northern islet; you will thus run to starboard of the bank, which can always be seen in the daytime. The dangers SW. and W. of the Mangsee islands should be left on your starboard hand. On the 2d January, about 10.30 a. m., we were clear of the strait and in the China sea. No observation at noon, as the weather was squally; fine weather during the afternoon; wind from NE. to ENE. Same weather on the 3d; position 6° 33′ N. and 113° 54′ E.; ship making 7 knots; steered to double Louisa reefs to the northward. They were in sight and only half a mile off to the S. at 5.30 p. m.; only a few rocks showed above the water. Continued fine weather on the 4th and 5th; currents setting to S. and SW. Some very large trees drifted past us. At noon on the 5th raised point Pulo-Laut and Solo-Kong bearing NW. by N. ½ N.; shaped the course to pass between Pulo-Laut and Great Natuna. At 5 p. m. sighted the SW. point of Pulo to the northward; at 6 p. m. Semoine islet bore W. and 'the Rock' S. by E.; the latter distant 14 miles, and Semoine 8 miles.

"Weather still fine. Passed close to N. of Anambas islands, on the 6th; position at noon 3° 25′ N. and 105° 49′ E.; at 4 p. m. the most western of the Anambas islands bore about S., distant 3 or 4 miles; thence we steered for Singapore strait; at 8 p. m. made out the 'Rock,' situated SW. of Mobur.

"At daylight on the 7th we were off the strait; and at 10 a. m. inside Pedra-Branca light. Anchored at Singapore in the evening after a passage of 51 days from Noumea."

§ 140. ROUTE FROM NEW CALEDONIA TO TAHITI.—This is a difficult route for sailing-vessels. For information concerning the first part of the voyage the reader should refer to § 108; we would especially call attention to the instructions for doubling New Zealand to the northward.

The meridian of cape Otou or that of the bay of Islands once passed, the difficult portion of the voyage is over. Vessels should stand to the eastward with the prevalent though variable west winds. No rule can be given as to where a ship will strike these winds, as they are liable to be found on very different parallels, even during the same season. However, it will generally be the best plan to make the following crossings:

In *January*, cross 35° S. between 178° E. and 178° W.; follow 36° or 37° S. from 178° W. to 163° W.; cross 35° S. at 155° W.; 30° S. at 151° W.; 25° and 20° S. near 149° W.

In *February*, follow 36° or 37° S., from 178° W. to 150° W.; cross 35° S. at 148° W.; thence make the northing to 25° S.; and steer for Tahiti.

In *March*, follow 35° or 36° S. to 158° W.; and thence steer so as to reach the tropic near 149° W. The SE. winds will often allow this route to be followed.

In *April*, the south winds will frequently enable vessels to run north of the Kermadec islands. They should afterward cross 30° S. at 168° W.; 31° S. at 163° W.; 30° S. again at 153° W.; and thence running E. of Tubuai for Tahiti.

In *May*, vessels usually pass N. of the Kermadec group, with the south winds, which then incline both to the eastward and westward.

There is no need of running into a higher latitude than 28° S.; cross 25° S. near 158° W.; pass between Rurutu and Tubuai; cross 20° S. near 149° W.; and thence for Tahiti.

In *June*, the same route. After running down the easting on 28° or 27° S., cross 26° S. near 158° W.; and thence head for Tahiti.

In *July*, vessels will usually be compelled to pass S. of the Kermadec group; and thence steer directly for 27° S.

between 154° and 156° W.; after crossing 25° S. near 151° W., they should run E. of Tubuai.

In *August*, the Kermadecs can easily be cleared to southward; thence the track reaches 30° S. between 153° and 155° W., and runs to eastward of Tubuai.

From *September* to *January*, pass to southward of the Kermadec group. In *September*, keep between 32° and 33° S. until you reach 153° W.; cross 25° S. at 149° or 150° W., and run east of Tubuai. In *October*, reach 34° S. as soon as possible and at about 180°; and keep between 34° and 35° S. until across 166° or 164° W.; cross 30° S. between 153° and 154° W.; sight Tubuai; and make 20° S. near 150° W. In *November*, the same route on 34° S. to about 160° W.; cross 30° S. between 152° and 153° W., and pass to windward of Tubuai. In *December*, a ship can reach 35° S. between 175° and 177° W., and thence keep between 36° and 37° S. until she is near 152° or 150° W.; she should cross 35° S. at 148° W., and 30° S. at 146° W., passing either E. or W. of Lancaster reef, according to circumstances; to eastward of Vavitao; and then heading for Tahiti.

As already stated in § 108, the worst months for doubling New Zealand are those between September and February. Voyages will usually be long at this season; still a good sailer can generally make the passage from New Caledonia to Tahiti in from 50 to 55 days.

We can also conclude from the foregoing facts that the NW. monsoon only prevails, in the western part of the South Pacific, to the northward of New Caledonia. On the contrary, south of 20° S. the SE. trades, though *variable*, exist during the entire year, their southern limit following the declination of the sun the same in these localities as elsewhere.

Abstract from the log of the sailing-transport Bonite, (Ann. Hydr., vol. 21.)—" Left Noumea on the 2d August; becalmed between the Signal island and Dumbea passage; breeze springing up from SSE. in the afternoon, we were outside the reef at 4 p. m. Very fine weather till the 5th, wind hauling from SE. to E. and NE. It shifted on the 6th to NW. and NNW., bringing very rainy weather. On the night of the 6th and 7th passed close to southward of Norfolk; light wind from SSW. to SSE. Captain Jouan wished to sight Raoul (Kermadec group) in order to correct the chronometers; but on the night of the 11th and 12th, after excessive lightning to eastward, the wind hauled to ESE. and headed him off to the northward. Position on the 13th, 27° S. and 176° W. As we could make neither northing nor easting, went about on the south tack; stood on till we reached 32° S.; the wind shifting successively from E. to NE. and NW., with rain and very heavy sea, then to W. and SW. On the 20th, the wind being steady from SW., headed NE. for Vavitao island; but the breeze soon shifted to the eastward and we were compelled to again run south of 30° S.; near 140° W. moderate winds from W., S., and SSE. On the 27th sighted Rapa to NNE. distant 42 miles. From Noumea to Rapa island noted a current of 3 miles per day setting to the westward. After leaving Rapa the winds were from E. to NE. Sighted Mehelia* on the 1st of September, and anchored at Tahiti on the 2d, after a passage of 31 days." Captain Jouan states that the Bonite was a slow sailer; and that he would have gained at least 4 days if he had not attempted to sight Raoul island. He thinks that he ought to have passed through, or even to southward of the Kermadec group, and followed 30° or 31° S. until he reached the meridian of Rapa, instead of attempting, as he did, to steer for Vavitao. Between 29° and 32° S. the rotation of the wind was in accordance with the law for this hemisphere, viz, from SSE. to E. and NE., fine weather; then to NNE. with stiff breeze blowing in gusts, overcast sky, close heavy weather, and falling barometer; then a very fresh NW. wind with rain; and last to WNW., W., and frequently a sudden squall from SW. The sea having risen quickly, then commenced to grow smoother; the barometer rose, and the weather was fine when the wind got back to S. and SSE. It generally takes the wind 5 or 6 days to make the complete rotation.

The following quotation from the *Ann. Hydr., vol.* 28, refers to the passage made by the *Isis*, Captain Binet, during the months of April and May:

"Applying the experience I had gained by my preceding voyages, I beat to the southward in search of the favorable winds; these I found near the parallel of 33° S. They varied from N. to WSW.; and I ran from the meridian of

* Mehelia should probably read Maitea.—*Translator.*

cape Otou to Tubuai island in 12 days. I headed for Tubuai after reaching 166° W., and ran from this island to Tahiti with light trades, varying from E. to SSE."

Captain Richard Foy remarks as follows on this passage, (*Ann Hydr., vol.* 29:)

"Setting sail from Noumea you will first have the wind abeam, and as soon as you have cleared the small pass, well abaft the beam. Take the SW. outer pass; leaving Prony ledge to port or to windward, and the Signal island to starboard. Once outside, brace up and make for the variables. Vessels generally keep between 30° and 36° S., but sometimes have to run as far south as 40° in search of the W. wind, when they again bear to the north and sight Kemin island. It is flat, wedge-shaped, and slopes to the W. Or the tropic can be crossed near 158° W. Approach Bora-bora or Moorea, and then go about for Tahiti. The average length of the passage is 42 days, the worst part being between 30° and 36° S."

We will close our remarks on this route with the following quotations, relating to the voyages of the Alceste and Sibylle:

First passage of the Alceste, Captain Brosset.—"The route from New Caledonia to Tahiti is as follows: first, through the SE. trades to 32° or 33° S., in search of the W. winds; it passes N. of New Zealand and S. of the Kermadec group; thence it follows the arc of a great circle, and again enters the trades near the meridian of Tahiti, and thence makes directly for that island.

"As the wind was not favorable at the commencement of our passage, we could not follow this route.

"Left Noumea on the 10th September. Calms and light airs from ESE., SE., and SW., during the first three days; only 180 miles to southward of my point of departure on the 13th. The wind then becoming settled from SSE. and S., I decided to haul up on the E. tack, and reached 175° 59′ E. and 26° 06′ S. On the 17th, the wind having headed me off and hauled to ESE., I went about. As I approached New Zealand the weather became squally and threatening. On the 20th the wind shifted to SE., SSE., and SW., and blew a gale; the barometer fell to $29^{in}.49$; the sea became enormous, and the wind having changed, I again went about on the starboard tack, but could not make anything to the eastward, as the heavy head sea, caused by the previous wind, greatly deadened the headway of the vessel. I was then in 32° S. and 174° E., south of the Kermadecs, which I had doubled in moderate weather. I bore up about midnight and headed to clear the N. point of Raoul island. During the bad weather, of which we have just spoken, the wind shifted in a direction corresponding to that of the hands of a watch; that is contrary to its usual manner in these latitudes. I kept between 30° and 31° S. till I reached 155° W.; fine weather, and moderate wind from SW. to SSW., and from S. to SSE. Stood to the northward after crossing 155° W. at 30° 30′ S.; steady wind from ESE. to E., all the way to Tahiti. The trade limit seemed to lie between 30° and 31° S.; though I made my easting between these parallels I experienced no calms or baffling airs. I proved several times that the least deviation to the northward of my route caused a shift of wind to SSE.; while if I headed, even a trifle, to the southward, the wind was inclined to haul to SSW. or even SW.

"Despite all detentions, I made the passage from Noumea to Tahiti in 25 days."

Second passage of the Alceste, Captain Brosset, (Ann. Hydr., 1871.)—"Left New Caledonia for Tahiti on the 7th October. Found the wind from E., and ran due south. Passed close to westward of Norfolk island. Lost the trades at 30° S., and after 24 hours of calms and light airs struck the west winds. Passed between New Zealand and the Kermadec islands; and ran down my longitude on 32° S.; wind on this parallel steady between N. and W. After crossing 155° W. bore to the northward. Met the trades at 31° S., and assisted by them reached Tubuai on my 19th day out. Calms, squalls, and variable winds from N., NE., and NW. during the remainder of the passage. Did not reach Papiete until the 2d November, after a passage of 27 days.

"When the trades haul to SSE., near New Caledonia, as they are likely to do, I think it is best to haul up on the starboard tack. If a vessel keeps on the port tack she will be set far to the westward and lose much time. In the preceding voyage I met the SSE. winds 150 miles S. of Noumea; and by hauling up on the starboard tack made a good passage."

Passage of the Sibylle, Captain Brossolet, (Ann. Hydr.,

1871.)—"Left New Caledonia on the 14th December and arrived at Papiete on the 15th January. Experienced frequent calms. The passage, however, would have been made in 26 days if I had not met a series of squalls and local head winds off the coast of Papiete. I learned at Papiete that the weather was exceptional. I therefore am of the opinion that vessels should head straight for Papiete, if the wind permit, after they reach the latitude of Norfolk island; and that the passage is usually made too far to the southward. Each time I went south of Norfolk island I met calms. Some authors advise vessels to keep to the northward immediately, in the rainy season, but this seems hardly advisable, as the dangers from reefs and gales, as well as the hot weather, counterbalance the slight chances of meeting west winds."

§ 141. ROUTE FROM NEW CALEDONIA TO NEW ZEALAND.—The length of this passage is quite variable, and the route differs according to the port of destination.

Bound to Auckland. Vessels bound to Auckland should set the course as if they were going to Tahiti, (vide § 140,) or should follow the instructions concerning the first part of the voyage to Europe, (vide § 108.)

By referring to § 108, it will be seen that from September to February advantage should be taken of every shift in the trades, in order to make as much to the southward and eastward as possible. It will probably be found difficult at this season to reach the meridian of the bay of Islands; but from March to July vessels can commence to make easting soon after leaving Noumea, and can run to the southward at any time.

Bound to Port Nicholson or Otago. Vessels bound to Port Nicholson or Otago should run south, rap-full, with the east winds; and when they have reached the prevalent west winds, still keep to the southward, being careful not to get too close to the west coast of New Zealand. They should take Cook's strait, (vide § 107,) if bound to Port Nicholson; and Foveaux strait, (vide § 107,) if bound to Otago.

We will complete our instructions on this route by giving an account of two voyages made by the Bonite, Captain Jouan, (*Ann. Hydr., vol.* 26 :)

"*First passage.*—Left Noumea on the 11th November, 1861. Encountered a series of winds from SSE. to ENE., blowing a gale, with rain, and a very heavy sea. Put in to the bay of Islands on the 2d of December. Experienced a series of heavy squalls in this short distance, which delayed me for some days. Although it was the middle of summer we had as bad weather as is usual during the winter months, except that it was warmer. We were 21 days from Noumea to the bay of Islands, and 6 days from the bay to Auckland. The barometer, during all this time, did not at all correspond to the changes in the weather; it varied from $29^{in}.72$ to $30^{in}.04$.

"*Second passage.*—Fine weather all the voyage. Left Noumea on the 16th November, 1862, with a stiff wind from E.; it soon grew light and shifted to NNE. and N. Anchored at Auckland on the morning of the 11th day; heavy dews near the coast of New Zealand; barometer steady between $29^{in}.96$ and $30^{in}.12$; temperature of the air between 75° and 68°; water the same, except near the shore, where it was about 2° lower. The currents set to SE., speed about 17 miles per day."

§ 142. ROUTE FROM NEW CALEDONIA, OR THE FIJIS, TO THE SANDWICH ISLANDS.—On leaving New Caledonia, a ship should follow the instructions given in § 108. It will not, however, be necessary from November to March to run lower than 36° or 38° S., and only to 33° or 35° S. from March to November. Pass either N. or S. of the Kermadec islands, according to the season, as stated in § 140, and do not bear to the N. until 135° or 134° W. is reached. The course should be laid so as to pass S. of Pitcairn island and thence east of the Paumotas and Marquesas; head to strike the NE. trades between 5° and 10° N. and 133° and 138° W. From this point the route is easy if you are careful to keep to windward of Hawaii, (vide §§ 128 and 135.) 1st. From New Caledonia.

If the point of departure be the Fiji or Tonga islands, take the southerly route just described. Run through the SE. trades on the port tack; the west winds will generally be found near 30° S. Strike the SE. trades again to eastward of the Paumotas. This is the shortest and surest route.* 2d. From the Fijis or Tongas.

Wilkes, however, advises a direct route for vessels leav-

* Vessels can also run through the SE. and NE. trades on the starboard tack and make their easting near 30° N. This is the preferable route for vessels leaving the Tongas between October and March, and always the best for ships from the Samoas.

ing the Fijis, or any point E. of the 180th meridian. His route does not, of course, apply to vessels whose point of departure is Noumea, nor would we advise navigators to follow it in any case.

As the vessels of the U. S. squadron parted company as soon as they were clear of the Fiji group and made the passage to the Sandwich islands by entirely different routes, much useful information may be drawn by a perusal of the logs of the several ships:

"After the squadron had cleared the reefs of Mali passage, (Vanua-Leou,) I made signal to the Porpoise to part company for the purpose of visiting the eastern part of the group. I afterward dispatched the Flying-Fish to run along the sea-reef as far as Round island before shaping her course for the Sandwich islands.

Passage of the Vincennes.

"The Vincennes and Peacock parted company on the evening of the 14th August. We stood to the northward on the meridian of 176° E., and kept the SE. trades until we reached the 8th parallel of south latitude, when we struck an ENE. wind and a long swell from the same direction. The weather was fine and the wind light. On the 19th we made an island in the neighborhood of the position assigned to Kemin or Gardner island, and remained a day to make observations. The next day another island was discovered from the mast-head. I called it McKean island, after the man who first sighted it. It is about a degree north of Gardner island. On the 21st we had showers of rain accompanied with a light wind from the westward. On the 22d we again had a light breeze from the northward and westward, and, what surprised me, a heavy rolling sea from SW., toward which quarter we experienced a current of some strength. On the three following days the wind was light and variable, and the weather squally.

"We remained 10 days among the islands of the Phœnix group.

"On the 4th September we crossed the line in longitude 167° 45′ 30″ W. with delightful weather, but met no westerly winds. I put the ship's head to the northward and ran up to 8° N. before meeting settled weather or the NE. trades. On the 12th the wind hauled to northeast, when I tacked to the southward and eastward, but soon went about again, deeming it advisable to run at once through the trades. On the 17th we were about 200 miles to westward of Oahu. I determined to beat up for it, and on the 20th made the island of Kauai. On the 23d September we anchored at Honolulu after a passage of 41 days, of which 10 were passed in exploring the Phœnix group."

"The Peacock took a different route. Captain Hudson continued to the northward and crossed the line on the 27th August at about 177° W., or 420 miles to westward of the Vincennes. Passage of the Peacock.

"The winds, until the latitude of 3° S., were from east, after which they became more variable between NE. and SE., accompanied with light squalls of rain and frequent lightning. The weather was at times hot and sultry. Between the latitudes of 5° and 8° N. the Peacock experienced a similar current with ourselves, setting northeast; wind and weather still variable. On the 8th September in latitude 14° N., the wind hauling to NE., she tacked to the southward until the 17th, when, having reached the longitude of 160° 27′ W., her head was again put to the north. They continued to have squalls and variable winds, with a current setting to the westward, and lost much time owing to the lightness of the winds. On the 30th September the Peacock reached Oahu; 50 days at sea.

"On the 10th September the Porpoise left the Samoas for the Hawaiian islands. In this passage they experienced similar weather and winds to those described in speaking of the passage of the Vincennes. They crossed the equator in 166° W. They had the ESE. and E. winds until 5° N. Between that and latitude 10° N., they experienced the same easterly current that we had done. In that latitude the NE. winds were fallen in with, accompanied with squalls of rain, and sometimes of wind. The whole passage from the Samoas was made in 27 days, and the time from the line to the Sandwich islands was 17 days, while the Vincennes and Peacock took respectively 26 and 33 days. Still I would not state positively that this result shows that the Porpoise crossing is the best, for it should be remembered that the Samoas are more than 500 miles to windward of the point where the other two ships crossed the same parallel. I think, however, that it is always well to cross the equator well to the eastward, as the SE. winds can be Passage of the Porpoise from the Samoas.

longer kept and the calms are fewer. The Porpoise also found the NE. trades steadier and more favorable than we did.

Passage of the Flying-Fish.

"The Tender crossed the equator in longitude 166° W., (same crossing as Porpoise.) She passed to eastward of all the small groups and kept the SE. trades longer than the other vessels. The weather they experienced seems to have been much of the same kind as heretofore described; there was little interruption of the easterly winds, and the Tender being a very fine sailer made the voyage in 33 days.

"The length of these passages can be considered as nearly the same when we take into consideration the different classes of the vessels of the squadron.

"The westerly current, on the same parallels, was found to be greater by the vessels which took the west route than they were by those which crossed the line well to the eastward. A northeasterly current was found to affect all of them in latitudes from 4° to 9° N., while a westerly current, running at a rate of 15 miles per day, was experienced in the NE. trade belt.

"It is therefore evident that vessels bound to the NE. portion of the North Pacific need not run up to the variable region north of the NE. trades, for this route takes them very much out of their way and leads them into bad weather. The direct route, however, which we followed, is only applicable to ships starting from points situated to eastward of the 180th meridian. Navigators deciding to stand across the NE. trades for the region of westerly winds, should always recollect that they are rarely found lower than 27° or 30° N., and that even when they are met with, it is very uncertain how long the wind will hold from the westward."

1st. The cape Horn route.

§ 143. Route from New Zealand to Europe.—Once to eastward of the group, steer as nearly SE. as the wind will allow, and gradually bear away for the parallel on which you wish to make the easting; keeping N. of 50° S. from October to April; and N. of 52° S. from April to October. For information concerning the end of the voyage, vide § 107.

2d. The Suez route.

Small auxiliary steamers will find it to their advantage to take this route, provided they leave New Zealand after the 15th April and before the 15th July. They should run to the north for the SE. trades, which they will find a little S. of the tropic, thence pass S. and W. of New Caledonia and head for Torres strait. For particulars concerning the remainder of the voyage, vide § 108 (2d part) and § 139.

The route south of Australia can only be followed by large ships. There will be no advantage in taking this route during the *southern summer* even if bound to one of the Mediterranean ports, as vast quantities of coal will be consumed, and the toll through the Suez canal is heavy.

§ 144. ROUTE FROM NEW ZEALAND TO THE WESTERN COAST OF AMERICA.—Once clear of the land steer to the southward and eastward, as stated in §§ 143 and 108. The reader should also refer to §§ 110, 136 and 137.

§ 145. ROUTE FROM NEW ZEALAND TO NEW CALEDONIA.—Starting from Otago, Port Nicholson, or Auckland, a vessel should make to the north along the eastern coast of New Zealand, as the weather is always bad off the western coast, (vide § 8.) She should run west of the Kermadecs, if the winds allow. From September to February, the SE. trades will generally be found near 30° S. During the remainder of the year they do not come much below 25° S. Havannah pass is the best entrance with the prevalent E. to SE. winds.

A sailing-vessel approaching Havannah pass and finding the wind from W. or even from NW. or NE. should give the isle of Pines and the great reef a wide berth; and run in through Dumbea pass, being first careful to get well to SW. of the light on Amédée islet, (vide § 111.)

Passage of the Bonite, Captain Jouan, from New Zealand to Noumea.—"During our first stay at Auckland, from the 11th December to the 17th January, the wind was generally fresh from W. with the sky clear; at times we had a very stiff wind, bringing rain. The regular sea-breezes did not set in until the day of our departure, and we had difficulty in leaving the gulf of Hauraki. Seven days gentle wind, varying from S. to E., brought us to Noumea.

"On my second voyage I remained at Auckland for two months; magnificent weather all the while; calms and light airs from the land during the night, and a gentle sea-breeze during the day. Only twice the wind hauled to N. with rain, after which fresh winds from W. Left Auckland on the 22d of January, 1863, with a light head wind from

NE.; and as the tide ran very strong we did not clear the gulf of Hauraki until the morning of the 25th, when we ran out, with a stiff breeze from E.; barometer 30^{in}.10. We kept this wind till we reached 24° S.; when we met calms and rain-storms. Sighted the New Caledonian reef on the afternoon of the 2d of February; land completely invisible in the fog. As the moon was up we ran through Dumbea pass that night. During our passage of 11 days the barometer stood between 29^{in}.88 and 29^{in}.63. The current was generally to S., at a rate of 20 miles per day."

Captain Richard Foy says, (*Ann. Hydr., vol.* 29:)

"The passage from New Zealand to New Caledonia can be made in 8 days, if you take advantage of every shift of wind, especially in the trades. The latter are often met at 26° or 25° S., that is long before a vessel enters the tropics.

1st. To Australia. § 146. ROUTE FROM NEW ZEALAND TO AUSTRALIA, SINGAPORE, AND CHINA.—From Otago, Port Nicholson or Auckland, and bound to Moreton bay or Sydney, vessels should keep E. of New Zealand as stated in § 144. Having reached the east wind near 25° S. they should run to westward; and finish the passage as stated in § 138.

If bound from Otago to Melbourne, we think that it will also be advantageous to keep N. of New Zealand and run to the northward as far as 30° or 25° S.; and thence to run down the E. coast of Australia and through Bass strait as indicated in § 178.

This route is about twice as long as the one by the S. of New Zealand, but quicker and easier. But captains deciding to run through Foveaux strait will have to wait for an east wind, (rare in this locality,) and even if they are lucky enough to carry it through the strait, they may meet a gale from NW. a few miles to the westward, (vide § 8,) in which case they will have to put back through the strait, to the eastward.

Navigators deciding to pass S. of Stewart island must look out for the Traps and Snares. Between Stewart island and Tasmania, the current and wind will both be contrary. The weather is, however, a little better during the southern summer.

2d. To China and Singapore. Vessels should at first follow the instructions given in § 145. Then those in § 139. The reader should also refer to the two easterly routes given in §§ 171 and 173; one of these passes to E. and the other to W. of New Caledonia, and a third through Torres strait.

§ 147. ROUTE FROM NEW ZEALAND TO TAHITI AND THE SANDWICH ISLANDS.—After a vessel is once well to the eastward of New Zealand, she should steer for 30° S. near 153° W., from April to November; and for 30° S. near 150° or 148° W., from November to April. Thence the track passes between Vavitao and Tubuai; it is best to sight one of these islands, whichever may be most convenient, and to make the northing so as to fetch to windward of Tahiti, (vide §§ 93, 94, and 140.) 1st. To Tahiti.

The parallel on which to make the easting will depend on the latitude of the port left. The attempt should be made to reach 30° S. between 135° and 133° W. Consequently a vessel should commence to make northing when this crossing bears ENE. or NE. The SE. trades once reached, pass east of the Paumotas and the Marquesas; and strike the NE. trades in the region between 133° and 138° W. and about the parallels 5° and 10° N.; thence to windward of Hawaii, (vide §§ 95 and 135.) 2d. To the Sandwich islands.

CHAPTER VII.

ROUTES FROM EUROPE TO CHINA AND RETURN ROUTES.

§ 148. SHOWING UNDER WHAT CIRCUMSTANCES THE SUEZ ROUTE IS PREFERABLE TO AND FROM CHINA.—The reader should refer to the second part of §§ 104 and 107. We will here begin a *résumé* of all the instructions that relate to the China voyage.

1st. The route to China.

It is understood of course that the Suez route is not practicable for sailing-vessels, for independently of the long Mediterranean passage and the cost of transit through the canal, they can only descend the Red sea from April to October. If they should clear the Red sea about the end of May, they will find the SW. monsoon, N. of the equator, in the Indian ocean, and can reach Singapore during the first days of July, and run up the China sea with the SW. monsoon. At no other season should this route be taken; and we can hardly advise it for sailing-vessels, even if they leave one of the Mediterranean ports of France in March or April.

The best route to China for mail-steamers and auxiliary-steam vessels, starting from Marseilles, is by the Suez canal. From March to June is the most favorable season, and a vessel will then find north winds in the Red sea; and the SW. monsoon, north of the equator, in the Indian ocean and the China sea.

From *October* to *February* ships are liable to be detained by strong head winds in the Red sea, and will use up all their coal if the attempt be made to steam against the NE. monsoon of the Indian ocean. They should therefore run south of the line for the winds known as the NW. monsoon; these are, however, no others than the SE. trades which are interrupted at this season, and vary to S., SW., W., and NW., accompanied with rain and squalls. In the China sea the NE. monsoon will be found.

2d. The route from China.

Sailing-vessels should not attempt to reach Europe via Suez, as the north wind prevails nearly the whole year round in the lower two-thirds of the Red sea.

Steamers and auxiliary-steam vessels wishing to take the Suez route should leave Yokohama about the end of October, and Hong-Kong in November. They will have the NE. monsoon in their favor, and after running through the strait of Malacca, should cross the Indian ocean north of the line. From October to April, and especially in January and February, the wind is generally from the southward in the southern part of the Red sea, but north of Jebbel Teer, steam will have to be used against the head wind. Voyages at this season are comparatively safe, rapid, and economical.

Vessels leaving China or Japan after the month of April will have to steam against the SW. monsoon of the China sea. This is also the season for typhoons, (vide § 20.)

After passing through the straits of Sunda they should run down their longitude in the Indian ocean, by keeping south of the line, with the SE. trades well abaft the beam. In the Red sea the wind will be ahead all the way to Suez. The voyage therefore at this season will be long and expensive.

§ 149. ROUTE FROM EUROPE TO CHINA, *(during the SW. monsoon, from April to October.)*—Instructions relative to this route will be found in the "*Navigation of the Atlantic Ocean.*" The reader should also refer to § 104 of the present volume.

The Indian ocean once reached, run down the easting between 43° and 46° S., keeping in a higher or lower latitude according to the weather, (vide § 104.) Thence steer NE. so as to cross 40° S. between 80° and 84° E. The SE. trades should be struck north of 30° or 28° S., and the tropic of Capricorn crossed near 102° E. According to Horsburgh vessels should then head for the strait of Sunda, being careful to make the meridian of Java head several degrees to the southward of the island and then run north. In March, April, and May, the prevailing winds haul to the eastward, making a westerly current along the south coast of Java; this fact will necessitate keeping well to windward, and making the landfall near Klapper point. From *April* to *October*, or the season of the SW. monsoon, the China sea should be entered by Banka or Gaspar strait. If going to Singapore, Banka strait will be preferable, *when the winds are light and unsettled*, as the approach to this passage is 1st. If bound to Singapore.

easy and a ship can anchor anywhere. Stanton channel is generally considered the best; it runs along the SW. coast of Banka; is 19 miles long, and 3 miles broad in its narrowest part. The depth varies from 7 to 20 fathoms in mid-channel. At the SE. extremity the depth is 7 fathoms, and 20 at the other. Once through, head for Rhio strait.

With a fresh, steady wind vessels should take Gaspar strait. The approach to this strait is dangerous in thick or bad weather, and the water very deep.*

Passage of the sailing-vessel Duperre, Captain Bourgois, (Ann. Hydr., vol. 23,) from the strait of Sunda to Singapore.— "As the Constantine and Didon had been detained for 15 days in the strait of Malacca, I decided to take the strait of Sunda. Passed safely through. As the water in Banka strait is shallow and requires frequent anchorages, and Carimata strait took me too much out of my route, I decided on Gaspar. This strait is itself divided into several channels; I ran through the one called Macclesfield. Left Anjer on the 16th May with a light breeze from SSW. to SSE., and during the night cleared the narrow channel which separates the Two-Brothers island from Schabunder bank. Sighted the entrance to Macclesfield strait on the evening of the 19th; ran through the next day. The breeze freshened as we went to the north, increasing our speed gradually to 6.5 knots. Barometer about 29^{in}.69. At daylight on the 23d the Duperre was among the low wooded islands of Rhio strait; picked up a native pilot acquainted with this locality and Singapore. Current running at the rate of ½ knot against us; did not clear Pan shoal, at the extremity of Rhio strait, till 9 p. m. Afterward passed between Affre reef and Johore bank. Made little or no headway during the night, the breeze being light and from SE., and the current setting to east. Anchored on the morning of the 24th at Singapore. During this 8 days' passage, though the moon was new, the sky was generally clear, and the land visible during the night; though we had to lie-to 2 or 3 times, we did not once anchor."

If bound to Saigon.

Vessels taking Gaspar strait should cross the line near 105° E. If they take Banka strait, they should pass be-

* Sailing-vessels bound to Singapore can also cross the line in the Indian ocean at 90° or 92° E., thence they can head to double Sumatra, with the SW. monsoon, and take the strait of Malacca.

tween Taya and Toejoe islands, and thence lay a course to cross the equator at 105° E. Once past the line the course is for Pulo-Aor, keeping well clear of Frederic bank, and to eastward of the reefs lying off Bintang. After leaving Pulo-Aor finish the voyage as stated in § 153.

When bound to Hong-Kong there is a choice of three routes: the *inner* route, the *outer* or *deep-sea* route, and the *Palawan* route. If bound to Hong Kong.

On leaving *Banka or Gaspar* straits in February, March, or April, vessels should take the *inner* route; to follow it they should first steer for Pulo-Aor; and thence keep along the coast of Cochin-China, passing W. of Pulo-Sapata and the Paracels. Although this is the best route at this season, the passage will probably be long and tedious under canvas. Complete instructions for the inner route from Pulo-Aor to Hong-Kong will be found in § 154. Inner route.

A ship running clear of *Banka or Gaspar*, between the end of April and the beginning of September, will find the outer or deep-sea route preferable. In this case she should steer to pass a few miles to the E. of Pulo-Aor, and finish the passage in accordance with the instructions given in § 154; passing E. of Pulo-Sapata, the Catwicks, and crossing Macclesfield bank. In order to follow this route Pulo-Sapata must be made before the beginning of October. Outer route.

A ship running clear of *Banka or Gaspar* during *September and October*, should take the Palawan route. Coming from *Banka strait*, cross the line near 105° E., as before stated. Thence steer to pass S. and E. of Barren island; thence E. of Low island and Hutton reef, and run between the Anambas and Soubi islands. Coming from *Gaspar strait*, during September and October, shape the course a little to eastward of St. Barbe island; thence head for the passage to the east of the Tambelan islands; rejoin the route S. of Low island and Hutton reef, and pass between the Anambas and Soubi. For information concerning the termination of the route, vide Palawan route in § 154. Palawan route.

§ 150. Route from Europe to China, *(during the NE. monsoon, from October to April.)*—As stated in § 149, instructions relative to this route will be found in the "*Navigation of the Atlantic.*" The reader should also refer to § 104 of the present volume. The Indian ocean should be crossed between 43° and 45° S. This is the season of the southern summer, and a ship can run as far south as 46° or 48° S., before meeting cold weather. The parallel of 40° S. should be crossed between 90° and 92° E.; 30° S. between 105° and 107° E.; thence the track is through the trades, for 20° S. near 112° E.; and for either Bali or Lombok, Allas, or Sapi straits, as may be thought best.

There is a choice between three routes at this season, viz: The first, by *Macassar strait*, can be followed in October, November, and March; the second, by *Pitt's passage*, is the surest, especially from the beginning of December to February; the third passes around the south of Australia, and is very little used. It is advantageous for vessels bound to Shanghai and Japan.

1st. The route to Hong-Kong by Macassar strait. This route should only be taken in October, November, and March. After crossing 20° S. near 112° E. any one of the following straits may be taken, viz: Bali, Lombok, Allas, or Sapi. The first two are generally preferred.

A ship *taking Bali strait in September and October* should bear north and pass between Sapoedie and Giliang, and afterward give a wide berth to the Kalkpen reefs and islands, situated north of Kangeang, passing them to the westward; Pulo-laut may be passed on either side, as circumstances allow.

The *Lombok-strait* route seems to be the best from the 15th January to the 1st March. Vessels should keep in the mid-channel, between Pandita island, on the W., and Lombok island, on the E., and afterward to the eastern side of the strait. Once out of this strait the course is NNE. for that of Macassar, the track passing between the most western island of the Pater-Noster group and Hastings island. After sighting Hastings, head to northward, for the Two-Brothers and Great Pulo-laut.

Vessels choosing *Allas strait*, should, when clear of it, head NNW., and afterward N. for Hastings island. After running east of it they should finish the voyage as if coming from Lombok strait.

On arriving from *Sapi* strait during the months of September and October, a ship would, according to the prevailing winds, pass to the east or west of the Postilions, and proceed to the north between Tanakeke and the Tongu islands; then pass at a good distance the isles and banks of Spermonde, and enter Macassar strait near the Celebes coast.

The route through Macassar strait, (*Horsburgh*, *vol.* 3.)—The Pater Noster islands divide this strait into two passages. The W. channel is from 30 to 33 miles broad, and the E. channel from 45 to 48 miles. The first is the most frequented, notwithstanding its dangers, as the depth of the water is less on the coast of Borneo, and vessels can anchor all the way to 1° N. In October and November the winds are exceedingly variable in this channel.

In the E. channel there is no anchorage on the Celebes coast, as the shore is steep-to all the way up the strait; however, this seems to be the best passage in October and November, when the wind is light from the southward.

A ship coming from Sapi strait should take the E. channel of Macassar strait; and from cape Mandhar to cape Rivers, follow the Celebes coast, keeping from 6 to 9 miles off the land, if the wind be light. But a vessel beating up against a head wind and current will have to approach the coast much nearer at times, especially in the bay S. of cape Temoel.

Vessels coming from Bali, Lombok, or Allas strait, and wishing to take the W. channel through Macassar strait, should double Great-Pulo-Laut to the SE., passing either E. or W. of the Alike islands, and thence steer for Shoal point. If compelled to beat they should stand on the off-shore tack until 12 or 15 miles from the coast, and in 14 or 16 fathoms of water, then go about and stand in until about 6 miles from shore, and in 7 or 8 fathoms.

There is generally a southerly current in Macassar strait. It is violent in January and February, and moderates in March. It sometimes sets to northward in October; this is, therefore, the most favorable month for going to China by this passage.

After having passed through Macassar strait, from September to the beginning of December, and especially in September and October, run into the Pacific, between Celebes and Mindanao, and then head for Hong-Kong; but in March it will be too late to take Pitt strait, or to enter the Pacific between Celebes and Mindanao, and the remainder of the voyage will have to be accomplished by running to westward of the Philippines. We therefore have the two following routes:

First route from Macassar strait.

A vessel clearing Macassar, from September to the end of November, should enter the Pacific by steering from cape Rivers to Sangui or Siao, and thence north of Morti. After passing between Siao and Tagolanda, or through one of the adjacent passages of Sangui, Horsburgh advises an E. course, so as to round the northern extremity of Morti. If the wind come out from NE., and the current be southerly, after passing S. of Siao, the ship will probably be drifted within sight of Gilolo or Morti, and perhaps Meyo and Tyfore islands. The southerly current generally ceases near these islands, and sets north along the W. coast of Gilolo. Still it is best to give Morti and Gilolo islands a wide berth when doubling them to the northward, as the southerly current, combined with the NE. swell, may sag the ship too close to the shore. After doubling Morti island (and always according to Horsburgh) do not go beyond 4° N. to make the easting, as the W. or variable winds appear to be dominant in low latitudes. On the contrary, if a ship run into a higher latitude she will strike the NE. monsoon. At the commencement of this monsoon—that is, at the season when this route should be followed—a vessel should keep south of 4° N. until the Pelew islands can be doubled to the eastward, and then bear N. In case of doubt it will be best to sight the most southern one of these islands, and then run up to westward of the group. In November, December, and January the wind is generally strong, and from NE., between the Pelew islands and Luzon; and, as the currents also set to W., it is necessary to make plenty of easting, in order to clear the N. point of Luzon and the Babuyans. As the wind and current are less violent in February and March, it will be safe to pass W. of the Pelew islands.

As an indorsement to this last observation we will quote, a few pages further on, the opinion of Captain Polack, who advises vessels leaving Gilolo passage not to lose time making to the eastward, but to bear north immediately, and keep along the coast of Luzon. According to this navigator it is useless to go toward the Pelew islands, unless for north winds. But we would draw the reader's attention to the fact that Captain Polack's observations refer to the passage from Gilolo to Bashees, which is generally made in

January, February, and March, when Horsburgh himself maintains that the W. currents and NE. winds are more moderate, and adds, "that vessels may pass W. of the Pelew islands."

On clearing the strait of Macassar in March and April a ship should pass W. of the Philippines; and, after leaving cape Donda, steer for the E. extremity of Basilan, being careful that the westerly current and east winds do not drift her upon the Sulu chain of islands.

Second route from Macassar strait.

In Basilan strait keep close to the coast of Mindanao, thus avoiding the shallow water off the Santa Cruz islands. By following this advice a ship can anchor, if necessary, off Samboangan, or on the coast of Mindanao, when she will have nothing to fear but the fishing stakes, as the dangers are well beaconed. The tides are here alternately strong and weak; the currents are generally very strong, and set sometimes to E., and sometimes to W., changing suddenly from one direction to the other.

Stand to the northward from Basilan, and keep near to the west coasts of Mindinao, Negros, Panay, Mindoro, and Luzon. While coasting these islands, from November to April, variable land and sea breezes will probably be experienced, although the direction of the dominant wind is N.

Between Mindinao and Negros, and between Payna and Mindoro, look out for the strong NE. winds and W. currents, which sweep through these passages, and may drift the ship near the Cagayanes. Being abreast of point Balagonan, (lat. 7° 46′ 30″ N., long. 122° E.,) with a steady SW. or S. wind, steer a direct course for point Naso, keeping a little to the eastward. If the winds are unsettled, light, and variable, keep along the coast of Mindinao to point Galera, or thereabout, prior to stretching off from that coast for point Naso; and, in crossing, endeavor to approach the west coast of Negros island. On leaving Panay head the ship for Ylin and Ambolon islands, and give a wide berth to the Buffaloes and the sand-bank off the Semirara islands.

When within 18 miles of Ambolon and Ylin islands stand to the westward, and keep 12 or 15 miles from that coast until the southern extremity of the islands bears SE. by E. ½ E. Double the reefs lying west of these islands to the northward, and approach Mindoro if the ship is to take the E. channel between Mindoro and Appo shoal. Keep about

6 miles from the islets off Pandan point, especially during the night; the E. extremity of Appo shoal being narrow and difficult to see if the wind is from the west. There are no breakers over the shoal.

The west channel of Mindoro strait is known as Northumberland passage, and is perhaps the best. It is from 15 to 18 miles broad. A vessel can here sight the Appo islands, and, if deemed advisable, run within a mile of the west coast of the largest of this group.

2d. Route to Hong-Kong by Pitt passage. This route should be taken from the beginning of December to February. Pitt passage is bounded on the W. by Bolton island; on the E. by Batanta and Salawati islands; on the N. by the Xallas and Obi Major; and on the S. by Bouro and Ceram. It was followed for the first time by Captain Wilson, commanding the *Pitt*, in 1758. It has three communications with the Pacific, viz: *Gilolo strait*, between Gilolo and Waygiou islands; *Dampier strait*, between Waygiou and Batanta; and *Pitt strait*, between Batanta and Salawati. Vessels coming from Europe should ordinarily take Ombay strait, leaving Sandalwood island to north, and thence, running between Timor and Ombay, pass the eastern extremity of Ombay, stand north close-hauled on the port tack, with the NW. monsoon, and thus endeavor to double Bouro to windward; that is, to westward. If they do not fetch to windward, the passage east of the island and west of Manipa can be used. Thence head ENE. so as to pass E. of Obi-Major, and through Gilolo or Dampier strait. This route can be followed after the 15th November, and is probably the best during December, January, and February. Pitt *strait*, between Batanta and Salawati, is rarely used.

Gilolo strait is to be preferred toward the end of November and during December, as the NE. wind is not then very fresh in that locality. It can also be taken in March when the NE. winds begin to moderate. It is broader than Dampier strait, and the shore near the islands is clean and steep-to, enabling a ship to beat through at night. The currents are rarely strong. As the NE. wind is often fresh in Gilolo strait during the months of January and February, and the Pacific swell heavy, it is best to take Dampier strait, where the wind is sometimes variable and favorable. But it should be remembered that there are in this locality several dangers, lying in deep water; that the tides are very strong, and that vessels should keep close to point Pigot, so as not to be set toward the coast of New Guinea.

After passing east of Obi-Major (in December and March) steer so as to double the southern extremity of Pulo Gasses very close aboard; while passing between Pulo Gasses and Kekik remember that the current often sets to east. After doubling the former island the highest of the Dammer isles will come in sight; (at a distance it has the form of a saddle;) thence run between Gebe island and point Tabo, (Gilolo.) Look out for Fairway ledge and the Weedah islands, if this passage be made during the night. As the currents often set to NE. and E., keep on the west side of the channel if the wind be light. If the wind be NNW., and the vessel cannot double Gebe to northward, pass between it and Gagy, and run into the Pacific, leaving the Syang islands to the eastward, if possible. If unable to double the Asia islands to northward, run between them and the Aiu isles; if absolutely necessary, the passage between the north coast of Waigiu and the Aiu islands can be taken. Thence steer east, keeping south of 3° N., and between 1° 30′ and 2° N., if possible, until the ship reaches 136° E., when let her run east of the Pelew islands, from October and December, and close to the westward of them from January to March. Horsburgh thinks that there are N. and NW. winds and S. and SE. currents between 1° 30′ and 2° N. Beyond 3° N. the winds are from NE. and the currents set to E., and even to the northward of E. When the NE. monsoon is well settled, they sometimes set to the westward. 1st. The passage through Gilolo strait.

The Dampier-strait route is taken in January and February. Once to the east of Gomono, shape the course for the passage between Pulo-Popa and the Bu isles. There is also a good passage south of Pulo-Popa, and between it and the Kanari isles. 2d. The passage through Dampier strait.

If the wind be from NW., the passage north of Pulo-Popa is the best. To follow it, round the low chain of islands,* situated NW. and WNW. of cape Mabo. When this cape bears S. keep 9 miles from Batanta, and do not let Pigeon island bear to east of ENE., as there are banks along the north coast of the channel. When 9 or 12 miles NE. of Fisher island the ship will be on soundings, but the bottom off the Batanta coast is foul, and if compelled to anchor till

* Probably the Tameay isles.—Translator.

the tide turns use a stream-anchor or a kedge. When standing along the west coast of Batanta, at a distance of 8 or 9 miles, steer NE., keeping Augusta bearing nearly NE. by E. If Mansfield island is visible keep it nearly in range with the southern point of Fowl island, and run south of all the shoal water.

When Augusta island bears N. by E., distant 4 or 5 miles, steer a little more to the northward, and pass 2 or 3 miles south of Pigeon island, and keep well to W. and N. of Vansittart shoal. To clear it the ship must be more than 4 miles from Fowl island, when she raises it between E. and SSE. Then lay an E. course and keep north of a line joining Fowl and Mansfield islands.

A ship clearing Dampier straits in December and January will be exposed to heavy northerly squalls and an accompanying swell. It is, therefore, best to keep within 2 or 3 miles of the small islands off Pigot point; then keep Pigot point, or the islet close to it, bearing west of W. 30° S. until they are out of sight. By following this precaution Buccleugh shoal will be avoided. Thence head NE., if the wind permit, and clear the coast of New Guinea well to windward.

According to Horsburgh, vessels wishing to pass east of the Pelews should run east, between 1° 30′ and 3° N., until near 130° E. A good weatherly ship can, however, even in November and December, pass close to westward of the group. Between the Bashees and the Pelews the current is generally westerly, with a speed of from 10 to 15 miles per day; the wind NE., and the sea heavy, especially in December, January, and the beginning of February. Near Luzon, the Bashees, and Formosa, typhoons are rare after the month of December, (vide § 20.) In February and March a vessel can easily run west of the Pelews, and weather the northern point of Luzon, as the trades often haul to ENE.

The end of the voyage. The northern extremity of Luzon once cleared, take any of the channels between the Babuyans and Bashees, except at the beginning of the monsoon, and when the winds are from NE., in which case run north of the Bashees and either north or south of Gadd rock. *In clear weather and in the day-time* vessels pass between the south point of Formosa and the Vele-Rete rocks. Between Formosa and China the prevailing wind is northerly. But during the night or in bad weather the Bashees must be doubled to the northward, and the ship kept well south of the Gadd and Vele-Rete rocks. In all cases try to sight Pedra-Branca or the China coast, and keep the lead going during the night. Enter Canton river by the Lema channel.

We give below the observations of Captain Polack, commanding the barque Esmeralda, (*Ann. Hydr.*, vol. 23.) It will be noticed that his advice differs materially from Horsburgh's instructions for this route.

"The passage from Hamburg to Hong-Kong was accomplished in 112 days. I was 10 days in going from Gilolo passage to the Bashees, and did not attempt to make any easting; while three other vessels, at the same time, took respectively 13, 15, and 16 days to make the run from Gilolo to the Pelew islands. Two Siamese captains told me that, although they often made this voyage, they never attempted to run to the eastward, but always bore right away north, during the NE. monsoon. They thus had no difficulty in reaching the Bashees, and always had a great advantage over those who sighted the Pelew islands. Consequently, it is advisable to make to the northward as soon as possible. In 10 days I encountered eight times a NW. current, and twice a NE. current, with rates of from 22 to 50 miles per day. It is, moreover, a known fact that vessels bound from Hong-Kong to Shanghai, during the NE. monsoon, pass east of Formosa, and beat up with the Kuro Siwo in their favor. Why, therefore, should you lose time by running for the Pelews, when you can bear north under the east coast of the Philippines, where the current is favorable? It should be well understood, however, that if you find a favorable wind for making easting on leaving Gilolo passage, the ship should be headed about NE., but not E.; because it is always prudent to clear the zone of calms as soon as possible."

The observations of Captain Polack.

Although this route is rarely used, it may be taken with advantage, under certain contingencies, in time of war, for instance.

3d. Route to China and Japan by the south of Australia.

Horsburgh quotes the passage of the Walpole, which left the cape of Good Hope on the 21st September, 1794, doubled Van Diemen's Land on the 31st October, passed east of New Caledonia, and arrived at Canton on the 5th January, 1795.

He also cites the Athenienne, which vessel doubled the island of St. Paul on the 11th October, 1804; entered Bass strait on the 28th of the same month; and after running east of New Caledonia, reached Pedra-Branca on the 28th December.

We think that this route can be advantageously followed by vessels bound to Shanghai, during the season of the NE. monsoon of the China sea; and we deem it the best route at this time for ships bound to Yokohama.

Vessels should first steer as stated in § 104, and afterward follow the instructions given in §§ 139 and 173. The reader should refer especially to the description of the two easterly routes, given in § 173. The second of these routes passes between New Caledonia and the Fijis, and is generally preferred by ships coming from Europe or Port Adelaide.

Route from Europe to Saigon and Singapore during the NE. and NW. monsoons.

Vessels bound to Singapore should, if possible, cross the tropic of Capricorn, in the Indian ocean, east of 82° E., and thence set the course a little to eastward of Engano island, so as not to fetch to leeward, that is to eastward, of Java head or Palembang. The easterly currents and west winds are strong in these localities. After running through the strait of Sunda, (especially with an auxiliary steamer,) either Banka or Durian strait can be taken. It often happens that sailing-vessels have to pass through Gaspar strait. In this case a sharp lookout must be kept during murky or overcast weather. The line should be crossed near St. Barbe island, and Singapore strait made with the NE. monsoon.

Auxiliary-steam vessels *bound to Saigon* should, if pressed for time, take the inner route given in §§ 149 and 153, and steam against the monsoon; they can touch at Singapore, in which case they should run through Banka or Durian strait.

Sailing-vessels bound to Saigon should take the route by Macassar, already described in this paragraph. They should ascend the Sulu sea and enter the China sea by Mindoro strait. They should cross the China sea, and head for Cape Padaran, making a good allowance for the current, and keeping well to windward of all dangers—the *Trident*, *Alexander*, *Minerva* shoals, etc.

For information concerning the termination of the voyage vide § 167. We would observe that the route by Macassar strait is a bad one in December, January, and February. Vessels bound to Saigon should not reach these waters at that season.

§ 151. ROUTE FROM CHINA TO EUROPE, (*during the NE. monsoon, from October to April.*)—From September to the end of February vessels descend the China sea by the *inner* route, described in § 167. This route seems preferable at this season, as it is shorter, and a ship can scud, in case she finds violent NE. winds on leaving the Great Ladrone. On the contrary, if the *outer* route be taken, a ship will, under like circumstances, have the wind and sea abeam, and be exposed to much danger, especially if ladened deeply.

The track for March and April runs by Macclesfield bank and east of Sapata, when it joins the *outer route*, described in § 167. Quicker passages are made by this route at this season, as it keeps well to the eastward in the China sea.

In all cases, the entrance to the Indian ocean is made by the strait of Sunda.

Vessels should keep a trifle to eastward of 105° E. while crossing the equator, and run through Banka strait, when they come from the *inner route;* still Gaspar can be used, in which case, after leaving Pulu Aor, the course is 15 or 18 miles east of Toty island, and thence for the strait. Coming from the *outer route* the rule is to take Gaspar strait, as it offers the most direct voyage; still it is a dangerous passage, and there are many instances of shipwreck in that locality. Macclesfield and Stolze channels are the best. On the homeward passage from Singapore either Rhio or Durian and then Banka strait are taken. Vessels leaving Saigon should reach Pulo Aor, as stated in § 170, and then take Banka strait.

Once clear of the strait of Sunda lay the course with the variable NW. winds, to pass east of the Keeling islands, and reach, as soon as possible, the parallel on which it is the intention to run down the longitude. The trades will be found steady between 18° and 19° S., except from February to May, when they are settled at 15° S. It is best to pass about 240 miles east of Rodriguez island, so as to avoid, as much as possible, the zone of cyclones, which are to be feared until April. Thence pass about 100 miles south of Madagascar. After reaching 26° or 26° 30′ S., head WSW. and sight the African coast near Port Natal or Algoa bay.

For further instructions vide the "*Navigation of the Atlantic.*"

General considerations.

§ 152. ROUTE FROM CHINA TO EUROPE, (*during the SW. monsoon from April to October.*)—Ships should return to Europe, at this season, by one of the *easterly routes* described below.

After *leaving Hong-Kong* about the end of April or the beginning of May they should take the *first easterly route*, follow the west coast of the Philippines, and take Mindoro and Basilan straits. Thence they can either run through Macassar and the strait of Sunda; Molucca passage, Salayer and Lombok straits; Molucca passage and Ombay strait, or finally passing the Sevangani islands, Gilolo passage, the strait between Buru and Manipa, and Ombay strait.

Vessels *leaving Hong-Kong, from the 15th May to the end of July*, can take the *second easterly route*, and enter the Pacific north of Luzon. Pass east of the Philippines and make Pitt passage by either Gilolo or Dampier straits.

It is not advisable to *leave Hong-Kong in August* unless absolutely necessary, or in a well-fitted-out and weatherly ship. However, if it be decided to sail during this month, take the *inner* route, following the coast of Cochin-China, as stated in § 167. The inner route will also be easy to follow *after the* 1*st of September*, (vide § 151.)

Starting from Saigon, conform to the instructions given in § 170.

Starting from Singapore, it is generally best to run through Durian and Banka straits.

In all cases, after the Indian ocean is once reached, run down the SE. trades, with the wind free on the port tack, and reach 19° or 20° S. near 82° E. Make the westing on this latitude, as the trades are here steady, and pass about 120 miles east of Rodriguez, then about 100 miles S. of Madagascar. When about 26° or 26° 30′ S. steer WSW. and sight the African coast near Port Natal or Algoa bay. Once around the cape of Good Hope follow the instructions in the "*Navigation of the Atlantic.*"

We will now give Horsburgh's instructions on the two easterly routes:

1st. Easterly route, when you leave Hong-Kong between the 15th April and the 15th May, or ater.

"Departing from the Grand Ladrone in May, steer southward to the Macclesfield bank. Then steer to the SE. by the wind; although unable to weather point Calavitte, variable winds may be expected near the coast of Luzon to carry you round the NW. end of Mindoro. If you adopt the eastern strait, or that formed between Mindoro and Appo shoal, keep within 10 miles of the coast in passing the latter in day-time. Keep about 15 miles from the islands Ambolon and Ylin in passing, when their southern extremity bears between SE. by E. ½ E. and ENE., to give a berth to coral shoals which lie to the westward of them.

"Northumberland strait, or that formed between Appo shoal and the Calamianes, should be chosen if the wind admit."

Whatever passage may have been taken when the wind is from W., and you have raised the southern extremity of the Calamianes, 16 or 18 miles to the westward, head S. by E. or SSE. and run toward Quinuluban island, to pass to the westward of Dry Sand bank. Thence bear S. along the west coast of Panay, with the prevailing S. to W. winds. The weather is frequently cloudy and rainy when these winds blow. The currents are moderate and sometimes set E. between Negros and the N. of Mindanao.

After leaving Naso point, if possible, head SSW., unless the wind is from E., in which case stand for point Balagonan. After reaching the SW. extremity of Mindanao run for Basilan strait.

Route by Macassar.—After leaving Basilan pass through Macassar strait, especially if bound to Batavia or through the strait of Sunda. European-bound vessels, taking this route, should shape their course as stated in § 167, and thus sight Tanjong Kaniongan and run down the west coast of Celebes east of the Little Pater Noster group. Thence they should attempt to fetch Allas strait, which can rarely be done without tacking, in order to double the Kalkoen islands and the shoals to eastward. Just here is the real difficulty of the Macassar route, as vessels generally have to work to windward at this season to reach Allas strait. Molucca passage is therefore perhaps preferable for a sailing-vessel, as it will enable her to strike the SE. monsoon well to windward.

Route by Molucca to Gilolo passage.—After leaving Basilan steer for the NE. extremity of Celebes, pass between Banka and Bejaren; thence bear south and run through the channel between Lisa-matula and Obi Major, or through Greyhound strait. The currents in the Molucca passage generally set to the N.; they are not, however, steady, and the head winds are light and variable.

Vessels arriving off the N. extremity of Gilolo or Morty, and finding the monsoon strong from SW., will do well to take Gilolo strait instead of the Molucca passage, (vide *second easterly route.*)

The Molucca passage may be pursued, keeping close to Obi Major and passing E. of Buru, and between it and Marripa.*

Thence, if the prevailing ESE. winds allow, run to windward of Ombay, between that island and Wetta. Follow the NW. coast of Timor, and enter the Pacific by the passage between Semao and Savu.

If unable to fetch to windward of Ombay, run through Alloo or Flores strait. But, these passages being narrow and subject to rapid currents, it is preferable, under such circumstances, to keep along the north coast of Flores, on the parallel of 8° S., thus avoiding the shoal water near the coast of this large island. The vessel can afterward enter the Pacific by Sapi strait.

2d. Easterly route, when you leave from the beginning of May to July.

After clearing Lema channel, head S., and thus place the vessel in good position for entering the Pacific, if possible, by the channel between the Babuyans and Bashees. SE. winds may be expected between Luzon and Formosa; typhoons are also common in this locality, (vide § 20.) Once within the limits of the Pacific, SW. winds and NE. or E. currents, with a rate of from 8 to 10 miles per day, may be looked for. After standing SE. and giving cape Engano and Luzon a wide berth, pass W. of the Pelew islands, if the wind permit, and a little to E. of Saint Andrew and Mariere islands and Helen reef. The currents are not very strong north of the latitude of the southern extremity of the Pelew group. But south of 6° N., and particularly between 5° and 2° N., during June, July, and August, an easterly current, varying from 30 to 60 miles per day, may be expected. Cross this zone as quickly as possible on a S. or SW. course, and, if drifted very far to the eastward, seek the westerly currents, which prevail between 2° N. and the equator. These have a speed of from 15 to 30 or even 40

* If you cannot run E. of Buru, double the W. extremity of the island, as stated under the head of *the second easterly route.*

miles per day near New Guinea and the N. coast of Waygiou. Near the entrance of Dampier strait the current sets to the east.

The best route is through Gilolo channel, and to follow it a vessel should, after reaching 2° N., steer for the Asia islands, rounding them to northward if possible; if not, to southward, between them and the Aiu islands. Afterward double Eye and Syang islands, and pass Gebe on either side, unless the weather is uncertain, when run west of this island. Thence a vessel will have to keep a little east of S., to allow for the westerly current, which prevails off the southern extremity of Gilolo. Finally enter Pitt passage by the large channel, between Pulo Pisang and the Bu islands, or by the passage between Kekik and Pulo Gasses, and after running between Buru and Manipa, make for Ombay strait. But, if unable to reach the passage east of Buru, or, as sometimes happens, the wind comes out strong from S., follow the north coast of Buru and run west of that island. Then, if the ship be kept close to the wind, she may still fetch Ombay strait. Look out for the St. Mathew and Welthoen islands, while standing between the west extremity of Buru and Ombay on the port tack. Make Ombay strait if possible; if impossible, run through one of the straits to westward—Sapi strait is the best. Instructions are also given on the last part of this voyage under the head of the "*First easterly route.*"

CHAPTER VIII.

Routes to the northward in the China sea.

§ 153. ROUTE FROM SINGAPORE TO SAIGON.—The following quotation is from the *Annales Hydrographiques*, vol. 23, and is a description of this route given by Captain Loftus, of the *Kensington:*

1st. During the NE. monsoon.

"On leaving Singapore in *December*, *January*, or *February*, it is advisable when the wind is strong from NE. to anchor under the lee of the Water islands in 9 or 10 fathoms. While these gales prevail it is generally rainy for 2 or 3 days. At the same time, the currents outside the straits set to SSE., with a speed of 2.5 or 3 knots. If once under way the port tack will take the vessel well to leeward of St. Barbe island; navigators are thus enabled to beat up under the west coast of Borneo. *After these NE. winds* the weather is generally fine; the wind shifts to N. and NW., and the current in the offing has a rate of only 1½ knots. The best time to heave up anchor at the Water islands is at the beginning of the ebb; the course thence is NE., rap-full, so as to pass between Subi or Low island and the Great Natunas. This is easy to accomplish at full and new moon, as the wind after several hours of calm often hauls to the west with rain-squalls, then shifts to SW. and S., from which point it blows moderately for *about 26 hours.* Vessels taking advantage of these circumstances, and passing to windward of Subi, will avoid the difficult channels which separate that island from the NW. coast of Borneo. If able to reach Low island, (107° 48′ E.,) and if the wind is steady from E., head about N. on the starboard tack, and keep at least 3 miles W. of the SW. coast of Low island. Thence continue to run N., and give Haycock island a berth of at least 3 or 4 miles, thus avoiding a coral bank which extends for more than 3 miles from its SW. shore. During the night it is dangerous to run east of this island. After passing Low and Haycock islands, no difficulty will be experienced in beating up for the SW. point of the Great

Natuna, as the islands break the force of the SW. current. If the weather be fine a land-breeze will be found at night off the S. coast of Great Natuna, but do not get closer than 2 or 3 miles if the wind be light, as the water is deep and affords no anchorage. Between Great Natuna and South Natuna, the passage is about 50 miles broad and clear of dangers. Koti passage, between Pulo-Pujang and Sirhassan, is equally good and about 10 miles broad. There is another good channel south of Sirhassan; it is not as broad, but by keeping close to the south coast of the island all dangers will be avoided. Api passage is the worst of all for a large ship, as the currents are irregular and sometimes set to SW. with great force; besides, she will have to beat through it close under the Borneo coast in 10 or 11 fathoms of water in order to avoid the current and take advantage of the land-breeze. Very strong squalls from SE. are, moreover, common in this locality; they are preceded by black rain-clouds and last about 2 hours, the wind hauling to the E. as they pass over.

"If the north wind holds as far as West island after leaving the straits, take Koti passage, keeping a good distance from Pulo-Pujang in order to avoid the reef situated south of Flat island. The coast of Sirhassan can be approached from the northward, as there are no dangers in that locality. After clearing the passage *do your best to reach* 112° *E.;* this will be easy to accomplish, as the wind is often N. and NNW. near these islands; but beyond cape Sirik (about 111° E.) the wind generally hauls to N. and NE. Thence, *stand on the starboard tack*, under all plain sail, for cape Tiwane, on the coast of Cochin-China. The current will be trifling until the west coast of the banks and 7° N. is passed; but farther north and until the mouth of the Cambodia bears W., distant about 70 miles, SW. currents exist, running at a speed of 3 knots during the strong winds which blow at the beginning of the monsoon.

"Toward the end of March and in April, east winds are often found to the eastward of the Anambas islands; these will be kept as far as the Brothers, about 24 miles W. by S. of Pulo Condore. Beat up inside of this island as far as cape St. James, keeping close to the coast of Cambodia, which is very low and hard to see during the night. After the mouths of the Cambodia are opened, the ebb tide sets to windward, and is consequently very favorable for vessels skirting the land on the starboard tack; on the other hand, they should not approach the mouths of the river during the flood, nor get inside of 11 fathoms during the night. The land can be seen for 10 miles from deck on a clear day. Keep the lead going when on the shore tack.

"In May, vessels make the quickest passage from Singapore to Saigon by keeping along the coast of the Malay peninsula and crossing the gulf of Siam. At this season squalls, calms, rain, and a light NE. current are often found off the gulf.

"*In December, January, February*, and sometimes in March, vessels are exposed to gales from NE. and NNE. between Pulo-Sapata (109° E.) and the coast of Cochin-China. They can be foretold by a gradual rise of the barometer and last 2 or 3 days, with a heavy sea and strong current; the barometer begins to fall before the end of the gale; the sky is overcast and heavy. If the land has been seen, and the meridian of cape St. James can be made during the gales, run for Pulo-Condore and find a sheltered anchorage in Great bay. If the wind be E., vessels will find a more snug anchorage at the harbor of Pulo-Condore, situated between the western extremity of the largest islands and Little Condore.

"Coming from the south during the NE. monsoon, make the landfall well to windward and thus avoid being carried too close to Bassok bank and to leeward of cape St. James by the flood tide and permanent SW. by W. coast current. The ebb tide is strong and to NE. by E.; at full and change it commences to run out of the river about midnight. In fine weather the wind commences about 90 miles from land to haul to ENE. and E. after 4 p. m., and is quite fresh and squally during the night. Should the ship be to leeward or on the meridian of cape St. James while running for the land, the ebb tide will set her to windward of cape Tiwan by morning. This cape can be seen for 40 miles. It bears E. 19° N. from cape St. James, distant 13 miles, and is generally the first land sighted coming from the south. Being abreast of cape Tiwan, cape St. James looks like two low islands. There is a fixed white light on this cape elevated 483 feet and visible 28 miles in clear weather, (10° 20′ N., 107° 05′ E.) Cape St. James is the first land sighted when coming from SW.

"During this season the wind blows from SE. to W. in the strait of Singapore; consequently, vessels can easily make to the eastward. Once clear of the strait they should lay a course that will take them between the Brothers and the western extremity of Pulo-Condore. Due allowance should be made for the current while crossing the gulf of Siam, as it sets to E. at about 26 miles per day in this locality. This current is strongest in June, July, and August, at which season there are strong W. squalls, heavy thunder, violent lightning, and plenty of rain. The wind remains steady at W. for 24 hours after these squalls, then hauls to SW. and moderates. When the western extremity of Pulo-Condore bears S., and the wind is W., and the current setting to the east, head N. for the bar at the mouth of the Cambodia. As the rapid currents from this and the Saigon rivers join with the regular easterly current, it is best to keep near the edge of the bar so as not to be set to leeward of the meridian of cape St. James. Keep the lead going, and when less than 10 fathoms are found edge away to the eastward, and thus raise cape St. James to NNE.; thence, head a little above it and run clear of the coral bank lying SW. by W. of the cape."

2d. During the SW. monsoon.

§ 154. ROUTE FROM SINGAPORE TO HONG-KONG.—We give below three different routes: 1st, the *inner route*, followed by vessels starting in March, April, and May; 2d, the *outer route*, used from the end of May to the beginning of October; 3d, *the route by Palawan passage*, to be followed after the 15th October.

1st. During the SW. monsoon.

The inner route runs along the coast of Cochin-China, and passes W. of the Paracels. The outer route passes E. of Pulo-Sapata, and crosses Macclesfield bank. The third route follows the W. coast of Palawan island, and the northern part of Luzon. Most of the following observations are from *Horsburgh*. He advises this route for March, April, and May, and adds that it will be preferable even in June and July if your ship is in bad condition. He remarks that the NW. and W. gales, which come from the gulf of Tonquin, are not very frequent, and that by taking this route, disabled vessels will be almost continually in sight of land.

1st. The inner route.

Vessels deciding on the *inner* route should, after leaving Pulo-Aor, run along the land as far as the Redang islands, then cross the entrance to the gulf of Siam. Thence double Pulo-Obi, and follow the coasts of Cambodia and Cochin-China as far as cape Touron. From this point make for the SW. point of Hainan, and skirt the eastern coast of that island to its NE. extremity, passing west of the Taya islands. Then, stand for the China coast near Haï-ling-shan. According to Horsburgh, all the islands lying between Tien Pak, Haï-ling-shan and Macao can be approached without danger; there are also several harbors of refuge in this locality. Typhoons are common after April, (vide § 20.) The reader should also refer to the instructions of M. Noel, given in § 157.

2d. The outer route. A vessel running out of Singapore, from the *end of May to the beginning of October*, should, when a trifle to eastward of Pulo-Aor, head for Pulo-Sapata, making due allowance for the easterly current which comes from the gulf of Siam. Horsburgh advises large vessels to steer NNE. after leaving Pulo-Aor, and until they reach the latitude of Charlotte bank. Thence they should head NE. by N., and sight Pulo-Sapata, and, after passing about 15 or 20 miles east of this island, stand about NE. ½ N., allowing for the easterly current; soundings may be obtained over the Macclesfield bank. The course from this point to the Great Ladrone is N. by W.; but when it blows fresh from S. or SW., keep this island to the north of N. by E. It will likewise be prudent not to sight the land west of Great Ladrone, and to keep clear of St. John island.

Vessels should not take this *outer* route unless certain of reaching Púlo-Sapata by the first of October. Toward the middle of October the currents about this island set strongly to the southward, and the northerly winds are often light or variable. A ship taking the *outer* route, and reaching Pulo-Sapáta after the 1st October, should not stand to the eastward until she reaches 13° or rather 14° N., when, the wind being from the northward and eastward, all the shoals will be passed to northward, provided long stretches to northward and short ones to southward be taken while beating over for the Luzon coast. Finish the voyage as in the route by Palawan passage, by running up the coast of Luzon to beyond cape Bolinao.

It is imprudent, especially at the end of October and in

November, to approach the China coast to westward of the Great Ladrone.

Passage of the Duperre, Captain Bourgois, (Ann. Hydr., vol. 23.)—Left Singapore on the 25th May with a variable breeze from S. to ESE. Experienced near Pedra-Branca light a strong squall from WNW., after which the wind shifted to SW. and SSE. Gentle and light breeze from SSE. till the 30th May, position 6° N. and about 108° E., when the wind hauled to WSW., and died away at S. on the following days. Position on the 3d June, 12° 40′ N. and 112° 10′ E.; light breeze from SE.; next day it shifted to NE., by the east, and after jumping around to SW. in a light squall, once more came out from SE. after a few hours' calm. From the 5th to the 6th moderate wind from S. to NE.; coppery clouds, and a typhoon sky. On the 7th, violent squalls and rain in torrents; then the wind hauled to S. and moderated. The weather grew fine and we cleared Macclesfield bank. On the 9th, light variable breeze from SSW.; 10th and 11th, calms and light airs from W. to NNW.; on the 12th, fine weather and light west wind; passed east of the Lema islands; beat up between them and Potoe island; crossed—with everything "touching"—the narrow channel between Lema and Hong-Kong islands; doubled Kellett bank to the north; and anchored off Hong-Kong in the evening.

3d. The route by Palawan passage.

On departing from Pulo Aor after the *beginning* of October pass south of the South Anambas, Low island, and Great Natuna, then head NE., so as to pass between Louisa and Royal Charlotte banks. Afterward edge away to ENE. so as to keep clear of both the Viper shoals, and to sight Balambangan. With a S. wind pass from 24 to 27 miles from this island; but *when the wind is from W.* keep about 45 miles off, and head for Balabac island, rounding it at a distance of 27 or 30 miles; finally if the wind be easterly keep closer to both Balambangan and Balabac, as a strong westerly current runs out of the strait when the wind is from that direction. A vessel standing NNE. after leaving Balabac will pass well clear of the banks around Palawan island, and to W. of Half-Moon, Investigator, Bombay, and Carnatic shoals. The channel is 27 or 30 miles broad; and it is well to sound during the night, especially between 9° and 10° N., as in many places there is a depth of 50 fathoms close to the edge of the shoals. It is best to pass the SW. extremity of Palawan at about 30 miles, if the wind will permit, and keep along the coast at the same distance. A sharp lookout should be kept for the shoals and rocks, which extend for 15 or 18 miles in a NW. and W. direction from the SW. extremity of Palawan.

If the wind should come out from the E., as it is apt to do after the end of October, head for the NE. point of Palawan and the Calamianes when to northward of the shoals; and after sighting the islands steer for Lubang. Thence, coast along the shore of Luzon, and pass near the Sister and Serpent islands. *If the wind be from SW. or west*, do not keep so close to the shore, especially near cape Bolinao, which is surrounded by reefs; the currents may also set the ship toward the gulf of Lingayen if care be not used to keep at a good distance from its mouth.

Once past cape Bolinao, vessels can ordinarily cross the China sea by running E. of Pratas reef; unless the wind is strong from NE., and the current setting rapidly to SW. In this case it will be better, in an ordinary sailer, to run up the coast of Luzon as far as cape Bojador; from which point the China coast can be made to eastward of the Lema islands. Never run to leeward of this group.

2d. During the NE. monsoon. Independently of the Palawan route, just described, which can be followed at the beginning of the NE. monsoon, there are two other principal routes from Singapore to Hong Kong.*

The first route by *Macassar strait* can be taken when Pulo-Laut is reached, before the 15th November, or toward the middle of February. The second route *by Pitt passage*, which also runs through Gilolo or Dampier strait, is advised for December and January.

1st. The route by Macassar. After leaving Singapore vessels should follow the instructions given in § 114, for the run through Macassar strait. After doubling Mankap shoals and reaching 3° 50′ S., on soundings of at least 19 or 20 fathoms, they should head about E. by S., keeping along the coast of Borneo in about 18 to 25 fathoms. They should not get in less water until over the bank off Salatan point, where the depth is from 12 to 15 fathoms. This point bears E. ½ S. from Mankap

* We think that auxiliary steamers can steam through Palawan passage during the NE. monsoon.

shoals, distant about 276 miles. The track keeps in about 14 fathoms while doubling the highlands of Salatan point; and within 24 or 30 miles of it. Thence, head E. for the Moresses islands; but do not follow the 18-fathom line, as Horsburgh states there is a rock near that locality; however, careful navigators can run into as shoal water as 8 or even 7 fathoms. Do not approach the Moresses nearer than 3 miles during the night, and run either north or south of this group. Beyond this, there are two little islands, united by a reef, and called the Brothers. Keep a lookout also for the three islets lying near the S. point of the large Pulo-Laut. Button rock is off the E. coast of the southern island. Do not attempt to pass inside of these islets.

The SE. part of Great Pulo-Laut once passed, head up for Macassar strait, and follow the coast of Borneo, as stated in the second part of § 150. But vessels desiring to ascend the W. coast of Celebes should, after leaving the SE. extremity of Great Pulo-Laut, steer ENE., if possible, and thus reach Celebes a little to northward of cape Mandhar. Soundings should be frequently taken during this part of the voyage, as there are several shoals between Borneo and Celebes. The Triangles and Union bank should be left to northward, and Laurel and Waller banks to southward. In both cases the voyage should be finished as indicated in § 150 under the head of the *route by Macassar*.

This route should be followed from December to February. The voyage from Singapore to Pitt passage should be made according to the instructions given in § 114, under the head of the proper route during the NE. monsoon. Pitt passage can be reached either from the north of Buru, or from to eastward of that island, between it and Manipa. For information relative to the latter part of the voyage vide § 150, second route. Gilolo strait is most frequented, especially during December and March. Dampier strait should only be taken in January and February. 2d. Route by Pitt passage.

§ 155. Route from Singapore to Manila.—Either the *inner* or *outer* route can be taken as far as Pulo-Sapata, according to circumstances, (vide §§ 153 and 154.) From Pulo-Sapata stand to the northward and eastward, and finish the passage as described in § 158. In October take Palawan passage, (vide § 154.) 1st. During the SW. monsoon.

2d. During the NE. monsoon. The route by *Macassar strait* (vide § 154) is used in October, November, and even in December; also during the month of March. After having followed the W. coast of Mindoro as far as point Calavite, keep west of Lubang and Cabra islands, and steer for Corregidor islet.

But, from December to February, sailing-vessels are obliged to take the easterly route, described in § 154 under the head of the route by Pitt passage. Thence, they should run through San Bernardino strait, and pass between Mindoro and Luzon.

In short, sailing-vessels should avoid leaving Singapore between the beginning of November and the beginning of February, so as not to have to make the long easterly route. If possible, they should only take the direct route during the SW. monsoon; the Palawan route at the end of the SW. monsoon; or the Macassar route in October, November, December, and March.

1st. During the SW. monsoon. § 156. ROUTE FROM SINGAPORE TO SHANGHAE AND YOKOHAMA.—Make the northing in the China sea, by following—according to circumstances—one of the three routes given in § 154. Whether bound to Shanghae, Nagasaki, or Yokohama, run through Formosa channel. The passage from Van Diemen's strait to Yokohama will be easy, as both wind and current will be favorable. Auxiliary steamers can sail through the Inland sea.

It may be shorter, but is certainly less prudent, to run east of Formosa. Vessels bound to Nagasaki, and taking this route, should sight the Hoa-pin-su islands after leaving the E. coast of Formosa; and afterward head so as to pass close to the eastward of Meico-sima chain.

Those bound to Yokohama should leave the China sea north of the Bashees, and keep in the Kuro-Siwo, (vide § 37;) they should also pass east of the Meico-sima, and Loo Choo chain of islands.

We cannot advise the route via the east coast of Formosa, as vessels are very likely to meet typhoons in that locality, a danger they will probably avoid if they run through Formosa channel.

The reader should also refer to § 159, where he will find a few instructions for the end of this voyage.

The following observations are by an old steamboat captain, of great experience in the China sea:

"July and August are the worst months in the north of the China sea; that is, beyond the head of Formosa channel. In the southern part—from the lower end of Formosa channel to cape Varela—September and October are the worst. Navigation is therefore dangerous from July to the beginning of November. It can be stated, in a general way, that there are no typhoons in the Formosa channel; as these stòrms, ordinarily coming from the east, are broken by the high mountains of Formosa.

"It is advantageous at all seasons to take Formosa channel when bound north. I once attempted the outside route, and had a hard time of it, encountering a gale and a contrary current setting to S. at a rate of 80 miles per day. Another steamer that left the same time I did, took the channel, and, though a slow vessel, arrived the day after. She had experienced no gale, and found very strong easterly currents.

"The Meico-sima islands seem to be especially attractive to all the Pacific storms; besides, the hydrography is uncertain.

"Ill-conditioned vessels leaving Saigon for Yokohama will do well to put into Hong Kong, and wait there for a good spell of weather, and until after the new moon. Keep an eye on the barometer after leaving Hong-Kong.

"The beautiful and safe bay of Pescadores is in Formosa channel; you never lose sight of the China coast, and its numerous harbors, and the typhoons of this locality never attain the violence that they do off the east coast of Formosa. A vessel can make Chusan, if necessary; and thence run for Nagasaki. After waiting here for good weather, she can take the Inland sea, and reach Yokohama with a smooth sea all the way.

"In this manner all the dangerous localities are avoided, while the voyage from Saigon to Yokohama is only lengthened by about 150 miles. The most important point is to be acquainted with the Japanese ports of refuge. Simoda, notwithstanding all that has been thought to the contrary, is a safe, deep anchorage.

"The Kuro-Siwo, or black current, is very changeable. It is hardly 40 miles wide; is sometimes found near the coast of Japan, and at others far from it. It does not appear to have much depth, and too much reliance must not be placed upon it.

"The regular S. monsoon, between Saigon and Hong-Kong, is variable, and uncertain between Hong-Kong and Yokohama. In July and August northerly gales are quite common near Formosa."

2d. During the NE. monsoon. Vessels leaving after the beginning of October will do well to take the easterly route, by Pitt passage, described in § 114. They can make Pitt passage, either from the north of Buru, or by passing between this island and Manipa; and, afterward, make their northing according to the instructions given in § 150.

It will be particularly advantageous to vessels bound to Shanghae and Yokohama to run to the east between 1° 30′ and 3° N., until they can pass to eastward of the Pelew islands. We think this recommendation important. Once beyond 3° N. and the Pelews, the wind will generally vary, (from December to April,) from W. to E., passing by the N. point. The prevailing direction, however, will be between NE. and NW., until north of 10° N. As a general rule, captains should choose that tack which will give them the most northing. From 10° to 20° N. the prevailing winds are NE., varying sometimes to ENE. and E. A good sailer can cross 20° N. about 132° or 134° E. Between 20° and 30° N. the winds are as often from NW. as they are from NE.; at rare intervals they blow from SE. It is, therefore, generally easy *if bound to Nagasaki* to cross 30° N. near 132° E. and thence to take the Van Diemen passage. Captains *bound to Yokohama* should aim to cross 30° N. between 132° and 134° E.; the current will be favorable; and north of 30° N. the variable winds will be favorable. If bound to Shanghae, run between the Loo-choo islands and Ou-Sima island; and thence stand for the Yang-tse-Kiang, with the wind abeam.

1st. During the SW. monsoon. § 157. ROUTE FROM SAIGON TO HONG-KONG.—The track follows the *inner route* described in § 154. Typhoons may be met at this season, on the voyage from Saigon to Hong-Kong. The reader should therefore pay great attention to the observations contained in § 20; and will find much interesting information in the following quotations:

Passage of the steamer Cambodge, Lieutenant Noel—"Messageries imperiales"—from Saigon to Hong-Kong, (*Ann. Hydr.*

vol. 28.)—"Left Saigon on the 27th September, 1865, and met a fearful storm between Hainan and the Paracels.

"From St. James to cape Varela, fresh SW. monsoon; barometer 29in.96. On the 28th, blowing hard off Padaran; ran in toward the land, where I found a calm and smooth sea. Threatening sky during the day, barometer falling slowly. Ship 50 miles from Batangan, at daylight of the 29th; very heavy swell from N., which shifted to NW. during the day. The barometer still falling at 7 a. m. The sky becoming of a grayish-lead color, with a black scud chasing in all directions; land birds and insects falling on the decks; blasts of wind from the N., and heavy, oppressive atmosphere. Barometer, at 4 p. m., 29in.69; at 5 p. m., 29in.65. About 8 p. m. on the 29th the wind burst upon us from NW. in a very violent squall, carrying away the fore-stay-sail and main-sail; raining in torrents; shipped a heavy sea. The barometer continued to fall during the night; squalls frequent; wind already very strong and hauling successively to WNW. and WSW. At 3 a. m. on the 30th barometer, 29in.53; soon after it reached its extreme fall, 29in.49; terrific wind from SW.; torrents of rain; torn-looking clouds; violent bursts of wind; enormous and broken sea. Luckily we were in the more moderate semicircle; and about 6 a. m., were about 40 miles from the center, bearing NW. From the evening of the 29th till 2 a. m. of the 30th ran NE., thus leaving the center and approaching Hong-Kong. Without a stitch of canvas, and with the engine only just turning over, we ran off from 10 to 11 knots, ship behaving admirably, steering well notwithstanding the heavy sea, and rolling very easily. At 3 p. m. fewer squalls and less rain; set the foresail, close-reefed. At 4 p. m. the barometer commenced to rise rapidly and in a little over six hours reached 29in.73. Though we were all the while leaving the center, we still experienced a frightful sea, coming from several directions; at 9 p. m. it reached its worst, (we were then at the point over which the center had passed the night before.) The wind moderated considerably about 9 p. m., after having hauled to S., SSE., and SE. At 10 o'clock the ship was out of the track of the typhoon. On arriving at Hong-

Description of the typhoon.

Kong we found that the storm had passed over the island,* and lasted two days."

Remarks of Lieutenant Noel.

"This typhoon was traveling from E. to W. when we struck it, and seemed stationary from the morning of the 30th at 4 a. m. until 4 p. m. of the same day. When the typhoon broke upon us at 6 a. m. on the 30th, I had not been able to obtain any observations for 36 hours, and was very uncertain of my position; for the currents had probably set us within dangerous proximity to Hainan, while on the other hand it was quite reasonable to suppose that the currents had been deflected from the neighboring coasts and were drifting me to SE. or toward the Paracels.

"In 1859 a French vessel, while on the dangerous side of a typhoon, went ashore on Hainan, at the very moment when the captain thought she was dangerously near the Paracels. By analogy, as I was on the moderate side, and feared Hainan, there was a chance of my bringing up on the Paracels. Moreover, if I had to go ashore, I preferred Hainan, as there we had some chance of saving our lives. As I ran N. from the afternoon of the 29th till the next morning, I must have been tolerably clear of the Paracels. My observations on the 1st October put me 60 miles to SSE. of my dead reckoning; and if I had commenced to bear toward the northward and eastward on the 29th it is highly probable that the ship would have been lost on the Paracels, as these 60 miles of SSE. current occasioned by the typhoon, (the westerly current of the storm was deflected by the coasts of Hainan and Cochin-China,) took place from the 29th to the 30th, at which time the swell was heaviest from NW., and the wind strongest from NW. and W.

"My opinion, therefore, is that, whenever you find yourself off *the mouth of the gulf of Tonquin*, and have *certain signs of an approaching typhoon*, the only thing to do is *to look for an anchorage on the coast of Cochin-China.* Nor should you get under way *until the barometer* and state of the sky clearly show that the storm has passed."

2d. During the NE. monsoon.

When the NE. monsoon has once fairly set in, the best route is that by Palawan passage. Cross the China sea with a clean full, making due allowance for the SW. current.

*As I write the above there comes, by telegraph, the account of the fearful typhoon at Hong-Kong on the *27th September*, 1874. A singular coincidence.—*Translator.*

Pass well to the northward of Charlotte bank, and afterward S. of Luconia reef. The aim should be to fetch Louisa reef, especially in an auxiliary steamer. Pass either north or south of this reef, according to circumstances, but keep well clear of all the banks lying north of Luconia shoal, viz, Sea Horse, George, and Friendship. They extend for a long distance, and generally compel vessels to run south of Luconia shoal.

In all cases, whether trying to reach the passage between Louisa and Royal Charlotte, or that S. of Luconia, invariably give a wide berth to the Viper shoal, and sight Balambangan island. Finish the voyage as described in § 154 under the head of "the Palawan route."

This is a long, tedious voyage, but surer and subject to less bad weather during the strong winds of the monsoon.

Vessels starting from Saigon in October, or at the beginning of November, (especially auxiliary steamers,) can greatly shorten the route by keeping along the coast of Cochin-China; taking advantage of the land and sea breezes, they can beat up, on short tacks, as far as cape Padaran; from which point the port tack will fetch them across the China sea to windward of all dangers, after making plenty of allowance for the current. The route then keeps along the coast of Luzon, and again crosses the China sea, as stated under "the route by Palawan passage" in § 154.

§ 158. ROUTE FROM SAIGON TO MANILA.—According to Horsburgh, vessels should sight Pulo-Sapata, and after passing the island either to the southward or eastward, head NE., *but not to eastward of that point*, until they have reached 12° 30′ N. Beyond this parallel the track is for Lubang or Cabra island, and keeps north of all the banks in this eastern part of the China sea. After passing about 9 miles west of Fortune island, which is about a mile in extent and high and rocky, head for Corregidor island, situated in the middle of the entrance to Manila bay. 1st. During the SW. monsoon.

We can only repeat the instructions given in § 157. The Palawan route is generally taken, especially by sailing-vessels; but the route across the China sea north of all the shoals is perhaps preferable for vessels leaving Saigon in October or at the beginning of November; it is especially adapted to auxiliary steamers. For further instructions the reader should refer to § 157. 2d. During the NE. monsoon.

§159. ROUTE FROM HONG-KONG TO SHANGHAE.—Maury's instructions contain a description of this route, which is due to Captain Potter, of the *Architect*, and runs as follows:

1st. During the NE. monsoon.

"Vessels departing from Hong-Kong, bound to Shanghae, in the northeast monsoon, should be in good condition to contend with rough weather, and to carry sail. Upon leaving, the Lyemun or Laruma channel can be taken, the latter being preferable in a large vessel. When clear of the islands, the wind will be found to be about ENE. generally, or as the line of coast trends, and when the monsoon is not heavy, periodical changes of wind occur. At such times vessels should be close in with the land, early in the morning, and tack off shore at about 8 o'clock, standing off till about 2 p. m., and on the inshore tack standing boldly in to the coast, making such arrangements during the night as will bring the vessel in a position inshore again in the morning. When the monsoon is moderate, vessels should not stand far into the bays, as they will, by so doing, experience light winds, and often calms, and, on the contrary, when the monsoon is strong, they should stand as far as possible in to the bays, and not stand farther off than is actually necessary, especially as the changes of wind above alluded to seldom occur at such times. It would be well to add here, that vessels almost always go faster inshore than they do off, as there is a ground swell heaving after them when in with the land.

"During the severe monsoon gales, which last about three days, vessels should seek shelter in one of the numerous good anchorages to the westward of Breaker point, when, upon the breaking up of the gale, they can make a fresh start, and perhaps get around Formosa before encountering another, especially after the month of November.

"Having reached Breaker point, vessels should then stretch over for the south end of Formosa, and upon getting to the eastward, the wind will be found to veer northerly, or more, as the coast of Formosa trends; and a good sailing-vessel will be almost sure to fetch the south cape or Lamay island to windward. Upon getting in with the land, light variable winds and calms are often met with, but the strong current to the SW. will very soon drift the vessel down, when she will find the breeze coming on fresh again. In passing South cape in the daytime, vessels should keep close in to the land, and the nearer the shore the stronger the favorable current, there being no hidden dangers. In passing round in the night, however, and when there is no moon, it will be advisable to pass to the southward of the Vele-Rete rocks, and to tack to the NW. when nearly in the longitude of Gadd reef, or sooner if it is daylight, as the South cape of Formosa is very low, and rather unsafe to approach in a dark night. When a gale comes on, and a vessel, being to the westward of the cape and near it, is obliged to heave to, a strict lookout should be kept during the night, as several vessels, under these circumstances, have found themselves to the eastward of the cape in the morning, having been drifted to windward during the night, and passed, probably, within a dangerous proximity of the Vele-Rete rocks. The current sets sometimes with incredible velocity round the cape, and then up northward, along the coast, and the stronger the northerly gale, the stronger the weather current, gradually diminishing in strength toward the north end of Formosa. After rounding the cape, vessels should work short tacks along the east coast of Formosa, keeping close inshore to get the benefit of the current.

"Having reached the northeast cape of Formosa, and the wind does not veer to the eastward, which is sometimes the case, vessels should keep between the meridians of the Barren islands and the islands off the north end of Formosa, and not stretch in for the coast of China until able to make a lead in for Video or Leuconna."

The following description of this passage is from the log of the British ship *Wanderer*, (Nautical Magazine:)

"We left Macao roads on the 28th December, 1842, but did not get under way from Harlem till the 3d January, the wind being light and ahead; beat up, close to the coast, as far as Breaker point, against a double-reefed topsail, NNE. and ENE. wind, and chop sea.

"Off the entrance to Formosa strait the northerly wind became more settled; sighted the South cape of Formosa on the 6th. On approaching the Bashees the wind came out again strong from ENE., with a heavy sea abreast the northern part of the group. The wind freshening as we made to the southward and eastward, went about, and on the 9th were off Botel-Tobago; on the 10th, doubled

Samasana; on the 11th sighted Kumi, bearing E.; and on the 12th left the high rocky islands of Hoa-pin-su and Ti-ao-yu-su to the westward; the wind was then E., but quickly shifted to S. and W., and came out violently from NNE. On the 13th, being in 27° 28′ N. and 120° 26′ E., went about to the northward and westward, and crowded on all sail possible to reach anchorage to leeward of the islands. On the 15th rounded the Kweshan islands and let go two anchors off the Buffalo's Nose, but the weather was not good enough for entering the inner harbor until the 19th. Passage 22 days.

"It will thus be seen that the wind was almost invariably from the northward and eastward. It may be stated as a general rule that if the wind does get to the southward of E., it invariably returns, with the sun, to the northward, and then blows with redoubled fury. NW. winds sometimes blow for several days end on, on the north coast of Formosa, and are felt for some distance at sea.

"Easterly currents are strong, as far as the strait of Formosa; here they generally set into the China sea, and to the southward. We found little current to leeward of the Pescadores. The current divides near Botel-Tobago, one of its branches setting strongly athwart the Bashees, while the other runs north along the coast of Formosa.

"In the open sea north of this island the movement of the waters is irregular and governed by the prevailing winds, setting rapidly to the southward during the northerly gales frequent in this locality.

General remarks.

"This is a dangerous voyage during the strength of the NE. monsoon, and ships generally experience a double-reefed topsail breeze and nasty sea until they pass the Bashees, when the weather becomes considerably better.

"After leaving the Lemas the best plan is to keep well in with the land and make to the eastward as much as possible. As the current is contrary it is well to beat up in the smooth waters of the bays during the day, and in case of heavy winds to anchor until the weather moderates. Do not be afraid to work to the eastward whenever the wind permits, and run in for the land when the wind comes out ahead. This is the worst part of the voyage, and requires much attention.

"Once clear of the South cape of Formosa, vessels can run either east or west of the Ty-pan-san group, (Meiaco Sima islands,) but vessels keeping close to the coast will find a northerly current as far as cape Formosa. When the wind varies from N. to E. off shore, it sometimes blows strong from NW. inshore, and it is then advisable to take that tack which will keep the ship out of the influence of the land-breeze, and to stand on until a good anchorage can be fetched to leeward of one of the Kweshan islands.

"With northerly winds the barometer is always very high, and is therefore of much use in this locality. After reaching the south point, the wind often jumps around suddenly to N. Look out for this shift of wind, and reduce sail in time. The squalls never last long."

Captain D. W. Stephens gives the following route as the best in his opinion for reaching the south point of Formosa in March and April, (*Ann. Hydr.*, 1870:)

"One of the usual routes at this season runs through Bashee channel, or through one of the passes between Formosa and Luzon, but vessels sometimes by so doing are delayed a week, beating along the coast for Breaker point against the NE. monsoon and a westerly current. A better route from Lema channel is to run SE., with everything full. The westerly current will thus be avoided, and easterly or southeasterly winds found near the coast of Luzon. A ship by following this track can then head NNE., and, the strong current helping her, will probably fetch to eastward of Formosa in less time than if the attempt had been made to beat up for Breaker point along the China coast."

If reference be had to § 156, it will be seen that Formosa channel is rarely visited by very violent typhoons. The following quotation also gives additional testimony on this subject: 2d. During the SW. monsoon.

Observations of Captain Potter.—"Regarding the passage to Shanghae in a fair monsoon, little can be said excepting that coasting vessels, when without observations, are in the habit of sighting the land to verify their reckoning.

"From the month of July to the latter part of September, and sometimes October, is considered the typhoon season, and at this season a barometer cannot be watched too closely. Typhoons have happened in May and June, but very seldom. These storms appear to originate to the east- Typhoons.

ward in the Pacific ocean, and passing the Bashee islands, traveling to the southward of west, their centers pass nearly over the parallels of Hong-Kong and Macao. A falling barometer, with a northerly wind, is almost a sure symptom of the approach of a cyclone in this vicinity. These storms, coming from the eastward, are sometimes probably turned off from their usual course by the highland of Formosa intervening between them and the China coast, and at such times they travel up north, curving again to the westward. This inference somewhat accounts for the fact that Amoy is seldom visited by typhoons, and they are never felt there with such a degree of severity as at the other ports to the northward and southward of Formosa. These storms are also generally preceded by a heavy swell from NE. to E."

Passage of the sailing-vessel Duperré, Captain Bourgois, from Hong-Kong to the mouth the Yang-tse-kiang, (Ann. Hydr., vol. 23.)—"Sailed from Hong-Kong on the 18th June; during the first few days fresh breeze from W. and SW., with heavy rain; afterward light wind from E.; squally. Ran through Tartani channel and cleared it during the night. We then tacked up along the China coast for Formosa channel. On the 19th and 20th, gentle breeze from E. to S.; on the 21st and 22d wind SW. and W., shifting to S. and E. on the 24th. On the 24th the wind hauled completely around the compass, from SE. to SW., NE., and again to SE., with lightning to NNW. On the 25th, wind ESE., very foggy weather. While steering for the Brothers, sighted, during a flash of lightning, Video island, toward which the vessel had been drifted 20 miles during the previous 24 hours; current WNW. Prior to this current, the flow had been to NE. with the same velocity. On the 26th, the wind was from S., the weather overcast and very rainy. Sighted the Saddle islands and rounded them at a distance of several miles. During the evening a calm compelled us to anchor NW. of these islands; position, 30° 54′ N. and 128° 26′ E.; bottom, muddy; depth, 8 fathoms."

1st. During the SW. monsoon.

§ 160. ROUTE FROM HONG-KONG TO JAPAN.—Vessels should take Formosa channel, and first run as if bound to Shanghae, (vide §§ 156 and 159.)

Sailing-ships should go through Van Diemen strait, or the two passages more to the southward; and thence can easily reach Yokohama. Steamers and auxiliary steamships can follow the same route, but it is probable that they would do better if they took the inner route, as described in the extract given in § 156.

Whether bound to Nagasaki or Yokohama, vessels should, on starting, follow the instructions given in § 159, for reaching the S. point of Formosa, and for ascending the E. coast of this island. Thence they should keep to westward of the Meiaco-Sima and Loo-Choo chain, and thus take advantage of the Kuro-Siwo or black current. The NE. winds are not here so persistent or strong as they are to eastward of the islands, and there is a chance of finding variable winds. If the wind allow, stand well to the northward and eastward, and pass north of Ou-Sima island; but if headed off, take the port tack and run west of the Linschoten group. As soon as the ship strikes favorable winds, which she will generally do about 30° N., head her to the eastward, and take one of the passages N. of the Linschoten group, or strictly Van Diemen strait. The run from here to Yokohama is easily accomplished. 2d. During the NE. monsoon.

Passage of the screw corvette Dupleix, Captain Bergasse-du-Petits Thouars.—"After leaving Hong-Kong on the 20th January, and running out of the China sea by the S. point of Formosa, I had the choice of standing well to the east and outside of the Loo-Choo islands, or of keeping in the Japan current and inside of the Loo-Choos, etc. I took the latter route, as I wished to take some observations of the current, and also expected to find a smoother sea. I ran from the S. point of Formosa to Van Diemen strait in 9 days, with a favorable current all the way, and my observations proved that the waters of the black current after running along the E. coast of Formosa, flow directly toward the S. point of Japan, without entering the channel between Meiaco-Sima and the Hoa-pin-su islands. I found a thick fog near Van Diemen strait, and was compelled to retrace my steps a little, in order to pass through the passage between Alcmene and Pacific islands, which I could have easily taken the night before. Once outside of the Linschoten archipelago I found stiff NW. winds. Anchored at Yokohama on the 10th February after a passage of 21 days from Hong-Kong."

§ 161. Route from Manila to Hong-Kong.—This is an easy passage at this season. After crossing the China sea, head directly for the Great Ladrone. If the wind is 1st During the SW. monsoon.

free and from SW., Horsburgh advises ships to make this island, bearing N. by E., or N. when approaching it from the China coast. If the wind incline toward NE., it is best to sight Great-Lema and enter the neighboring channel. Remember that typhoons are frequent from July to November, (vide § 20.)

During the NE. monsoon. Run up the W. coast of Luzon to cape Bolinao; a good vessel can cross the China sea from this point if the wind draw aft enough to let her head N. But when the wind is from NE. or N., and the current sets to SW., it is preferable, especially with an ordinary sailer, to beat up the coast of Luzon as far as cape Bojador, and then cross the China sea. For information concerning the latter part of the voyage, vide § 154, under the head of the route by Palawan passage.

1st. During the SW. monsoon. § 162. ROUTE FROM MANILA TO SHANGHAE AND YOKOHAMA.—This route does not present any difficulties. Vessels should steer straight for Formosa channel, and thus avoid the typhoons frequent among the Bashee and Meiaco-Sima islands. For further information, vide § 159 or 160.

2d. During the NE. monsoon. At this season keep along the coast of Luzon until near cape Bojador. On leaving this cape—or even point Banqui, with a settled NE. wind—make a long stretch, on the starboard tack, to NNW. However, as it is not well to get too far from the Bashees and the Babuyanes, the ship should, now and then, go about, and stand toward them on the port tack, heading about ESE. The S. point of Formosa should then be reached as soon as possible. Auxiliary steamers can, of course, keep their engines turning over slowly, and, lying a little nearer the wind when on the starboard tack, reach Formosa in less time than sailing-vessels. Pass east of this island, and stand to the northward with the current in your favor. For information concerning the latter part of the voyage vide § 159 or 160.

§ 163. ROUTE FROM SHANGHAE TO JAPAN.—As this route lies north of 30° N., the variable winds beyond that parallel will materially assist vessels on the voyage to Japan. When bound to Nagasaki they can steer a direct course. Sailing-vessels bound to Yokohama should, as a general thing, take Van Diemen strait, or one of the more southern passages, if forced by the wind. Steamers, and auxiliary steamers, bound to Yokohama, should generally pass through the Inland sea, especially during the summer, (vide extract in § 156.)

Ships bound to Hakodadi will find useful information in the following report by Captain Scott, (Merc. Mar. Magazine, 1863:)

"After leaving the light-ship off the mouth of the Yang-tse-Kiang, I usually ran close to the Amherst rocks, when the wind was favorable; passing them on either side, according to circumstances. I then headed for Fsu-Sima in the strait of Korea, and left it to port. Thence I sailed through the mid channel, between Hornet and Oki, where the current is strongest. Sometimes, when the wind was strong from the eastward, I had to keep north of this route, and run between Quelpart island and the Korean coast. I found this a safe passage, the current being rapid and favorable. After passing Hornet rocks it is best to stand for the coast of Nippon, and cape Yokoiso; behind which there is a very high mountain, visible 60 miles in clear weather. In order to counterbalance the effect of the current, head about half a point to eastward of the direct course, until beyond 39° 30′ N. As the current here commences to run to NNW., and beyond that parallel often attains a speed of from 40 to 50 miles per day, (when the wind is strong from S.,) a whole point should be allowed for drift.

"Due allowance should especially be made for the current when the weather is overcast or foggy, and no observations have been obtained.

"Notwithstanding the fact that we always allowed for drift, we invariably found that it set us well over toward Ou-Sima and Ko-Sima islands, when we were standing for the western coast of Japan. These islands form an excellent landmark to the entrance of the Tsugar strait. They can be seen for 25 or 30 miles on a clear day, and a ship can pass on either side of them, or between them, the shore being bold and clean. It is, however, always preferable to leave them to port; that is, to northward, as they bear due E. from Matsumai point, which forms the NW. limit of the strait.

"We do not think it advisable for navigators unacquainted with this locality to run through Tsugar strait at night, as the breeze often dies away about sunset. Mistakes may be made about the different points, and the ship be drifted into the Pacific; in which case it may take her a week to make up for lost time.

"On approaching the land during the day it is best to keep close to cape Tsiuka, situated about 18 miles north of Hakodadi, on the north coast of the strait. Thence keep along the north coast for Mussel point, thus taking advantage of the favorable eddies in the current, if the wind die away."

CHAPTER IX.

ROUTES TO THE SOUTHWARD IN THE CHINA SEA.

§ 164. ROUTE FROM JAPAN TO SHANGHAE.—Sailing-vessels, leaving Yokohama during the winter, (from October to March,) should stand out to sea, and make to the southward. They will generally find the NE. monsoon south of 30° N., at this season. Thence they should head for Ou-Sima; and passing to northward of this island, make due allowance for the current, which sets to N. and NE. In summer it is possible to keep near the land, and in the strength of the Kuro-Siwo current. Sailing-vessels should afterward run through Van Diemen strait, unless they wish to take the route through the Inland sea.

Auxiliary steamers will find it greatly to their advantage to take the route through the Inland sea, especially during the summer. From May to October it is never safe to approach Ou-Sima island, as typhoons are frequent in that locality.

The following observations, on the voyage from Hakodadi to Shanghae, by Captain Scott, may be of interest:

"The westerly winds render it sometimes difficult to cross Tsugar strait; however, I incline to the opinion that this route is better than the one to eastward of the Japan islands, as it is about 300 miles shorter; and the weather, in Van Diemen strait, is worse, and the contrary currents much stronger, than they are in Tsugar strait. It is a known fact that vessels taking the easterly route make long passages.

"When I run out of Hakodadi, with the wind between W. and SW., I always stand for Mussel point, which is situated on the opposite side of the bay. After rounding this point as close as possible without danger, I beat up along the coast of the bay, making short tacks, and thus keeping within the counter-current. In this manner I generally reach cape Tsiuka before dark, where there is an excellent anchorage in from 6 to 12 fathoms off the little village. Cape Tsiuka bears S. from this anchorage, distant a little less than a mile. I think this the safest spot in the bay, as a ship can always get under way without difficulty, if the wind should suddenly shift to E.; and the SE. counter-current is not very strong. It is certainly a far better anchorage than the one marked on the Admiralty chart, as it is difficult to beat up from the latter, in case the wind should come out from E.; the holding-ground, also, is not good, the bottom being of hard sand, while the bottom off cape Tsiuka is a kind of mud, formed by the *débris* from the adjacent mountains.

"If the west wind hold, get under way at daylight, and run along the north coast of the strait as far as point Matsumaï. Thence stand across the strait on the starboard tack for cape Greig; the wind will generally draw aft one or two points abreast the entrance of the strait. Between cape Tsiuka and point Matsumaï the coast is not clean, but there is no necessity of running close to it, as the current moderates considerably beyond cape Tsiuka. After clearing the strait, and reaching 37° N., it is well, when the wind is S., to stand in for the coast of Japan, and run inside of the Oki islands. By so doing a ship will be out of the strength of the current, and may even find a favorable drift as far as Hirado channel. I have often passed inside of the Mino-sima, Katsu-sima, and Oro-no-sima islands, and found the coast clear as far as Oro-no-sima. ESE. of this island lies a small rock, 8 or 9 feet above water, with a considerable outlying reef. During the night this rock is easily mistaken for a small vessel or a junk.

"After this it is best to make for Korea strait; but it is to be hoped that when the survey of these localities is complete, vessels will be able to keep inside of the Gotto islands, and thus avoid the strong currents of Korea strait, when the wind is strong from the west.

"Once clear of Korea strait, it is not usually difficult to reach the China coast, as the monsoons generally blow from N. or S. This part of the passage is ordinarily made in from 3 to 5 days, and sometimes in 36 hours when bound north. During either monsoon it is best to make the land near Shaweishan island or Amherst rock, where there is an anchorage in 6 or 7 fathoms, if the wind die away. Vessels are also likely to have one or two tides in their favor in this locality. Near the Saddles the southerly currents and tides are often strong during the NE. monsoon.

"The passage from Shanghae to Hakodadi has always seemed to me shorter and easier than the return voyage; the current being more favorable. We made our quickest passage to the northward in May, and our longest in October; the length of the respective voyages being 9 and 16 days. Our quickest passage to the southward was made in 11 days, during November, while our longest was 29 days, in June. Passages have been made in 6 or 7 days."

§ 165. Route from Shanghae to Hong-Kong.—In § 159 we quoted Captain Potter's instructions for the voyage from Hong-Kong to Shanghae. We will now reproduce his observations on the return route, first remarking, however, that vessels should always pass through Formosa strait:

"In the NE. monsoon there is a constant current down the coast, running with more or less velocity, according to the strength of the wind; and the wind generally blows along the line of coast, that is ENE. from Hong-Kong to Breaker point, NE. in the Formosa channel, and NNE. from Formosa north. The first part of the monsoon is very strong, and frequently in the month of October it is almost an incessant gale; in the latter stage, from January to May, SE. winds are not uncommon, and the more frequent as the season advances; there is considerable thick weather in the latter part of the monsoon, and a SE. wind to the northward of Formosa almost invariably brings a dense fog with it. During the NE. monsoon.

"The passage from Shanghae to Hong-Kong in the SW. monsoon is very tedious from the frequent calms and squalls and constant strong current up; and coasting-vessels generally use their kedge when there is not sufficient wind to make any progress. In working down it is well to keep in with the coast, stretching into bays and by headlands to get out of the current, if there is sufficient wind to preclude the probability of getting becalmed. During the SW. monsoon.

§ 166. Route from Hong-Kong to Manila.—According to Horsburgh, vessels should aim to leave Macao for Manila when the wind hauls to SE. or E., so as to make to SSW. or S. toward Macclesfield bank. This bank once reached, it is easy to fetch to windward of Manila unless 1st. During the SW. monsoon.

the wind comes out from SSE. and S. If the wind is from the southward, sight Goat island or the land south of the bay.

It will be perceived that this route is the same as that given in § 152 in the description of the first easterly route. We think it is not advisable to leave Hong-Kong from May to September, as it is exceedingly difficult to reach Macclesfield bank during these months. Of course the difficulty will be less with steam to fall back upon.

The reader should remember that typhoons are frequent in these parts, (vide § 20.)

2d. During the NE. monsoon. This is the favorable season for making this passage. According to Horsburgh vessels should run out by the Lema channel, and head up well to the east for cape Bolinao, if possible, as the currents set to leeward. When nearly as high as cape Bolinao they should keep some distance from the banks along the coast of Luzon. After doubling cape Bolinao and the Sisters, (*Las Hermanas*,) keep along the land at a distance of from 12 to 18 miles, until past the islands and rocks off point Capones, when run in closer and steer for Manila bay.

1st. During the NE. monsoon. § 167. ROUTE FROM HONG-KONG TO SAIGON AND SINGAPORE.—From September to the end of February it is best to take the *inner route*, along the coast of Cochin-China, (vide § 151.) This route is especially advantageous for deeply laden vessels, as they will have the wind aft all the way from the Great Ladrone. In March and April the *outer route* is preferable.

1st. The inner route from September to March. Captain Blake, of H. M. S. Larue, adding to his own experience that of several commanders of opium clippers, gives the following remarks:

"In running down the China sea with the NE. monsoon, the direct line mostly adopted is nearly mid-channel between Hainan and the Paracels, holding rather to the latter, when a southerly current of 30 to 50 miles a day is usual, and between 14° and 11° N. I have known it to reach 60 miles in twenty-four hours. Thence making the coast of Cochin-China about Varela, and shaping a course southward, so as to pass 30 or 40 miles outside of Pulo-Sapata, whence the course to Singapore is clear, giving the Anambas a berth of about 40 miles, and always, if possible, sighting Pulo-Aor, to insure the reckoning; more especially should the weather be thick, when the lead should be constantly attended to."

The following instructions are condensed from Horsburgh's work:

"The course from the Great Ladrone is SSW. ½ W., the route passing between the Taya islands and St. Espiritu shoal, and at a convenient distance from the Paracels. The currents are very strong along the China coast, and run at the rate of from 15 to 24 miles per day in the offing. When the wind is ENE., between the Paracels and Hainan, it causes a current in the opposite direction. Between the Great Ladrone and the coast of Cochin-China the westerly currents have a speed of 15 miles per day in moderate weather, and of 24 miles with strong winds; vessels should not, therefore, steer too southerly, but keep between SSW. and SW. by S. ½ S. prior to reaching 17° N. and the channel to westward of the banks. This course should fetch them 3° to the westward of the longitude of the Great Ladrone, when on the 17th parallel; the course thence to cape Varela being S. ½ W. or S. by W. If the weather be overcast or the wind be inclined to haul to the eastward, (as the current sets in toward Quin-hon,) it is well not to approach land before up to cape Varela. Beyond 15° N. the littoral currents are strong and southerly, reaching a rate of from 40 to 60 miles in 24 hours, from 14° 30′ to 11° 30′ N. If land was not sighted above cape Varela it is advisable to run in for it, and keep 12 or 15 miles from the coast.

"The route to Saigon is near land and to westward of Holland bank. Abreast the southern part of false cape Varela the course is SSW. or SW. by S. ½ S., across the gap of Padaran. The SSE. current should be attended to in this locality, and at a moderate distance from shore the lead will show 40 or 50 fathoms. Cape Padaran is easily recognized on a SSW. course, and cannot be mistaken, as the low land at the bottom of the gap gives the cape an isolated appearance. It should be passed at 5 or 6 miles on a SW. by W. course, which will carry the ship about 6 miles outside of Pulo-Cecir de Terre; cape Padaran can, however, be approached much closer, and Pulo-Cecir de Terre still be doubled. The island once passed at a distance of from 4 to 6 miles, in daytime, steer SW. by W. until it bears N. by E. ½ E.; it will probably be lost on this bearing. Then, after

Bound to Saigon.

running 18 or 21 miles SW. by S., the ship will be west of Holland bank, and the remainder of the passage is easy. As at night Saigon light is visible 28 miles, and during the day cape St. James is easily recognized, the only precaution is to round Britto shoal to the eastward.

Bound to Singapore. "After passing west of Holland bank, the course is SW. by S. for Pulo-Aor; but it is preferable to keep outside of Pulo-Sapata; in which case the course from false cape Varela is S. or S. ½ E., and to eastward of Pulo-Cecir de Mer and Pulo-Sapata. In overcast weather it is best to be on the safe side and steer SSE., as the currents set rapidly to SW. and WSW. toward Pulo-Sapata, and a ship may be forced to pass the night between the island and Catwick.

"In very clear weather vessels may pass close to eastward of Pulo-Cecir de Mer, and run west of Great Catwick on a SW. course; whence the course is straight for Pulo-Aor. When east of Pulo-Timoan in thick weather, keep in 32 or 33 fathoms.

"After passing from 6 to 12 miles from Pulo-Aor, the course for Banka strait is SE. by S. ½ S. or SSE., depending upon the strength of the wind and current, being careful to keep in at least 23 or 24 fathoms when between 0° 40′ and 0° 50′ N. After passing outside of Geldria bank a SSW. course will carry the ship 12 or 15 miles from the east point of Linga; this offing is necessary, as Ilchester bank bears south from the point (distant 8 miles) and the current sets toward the shoal.

"When abreast of the east point of Linga, distant 15 miles, set the course at SW. by S. ½ S., and run between Pulo-Taya and the Seven islands, and then head S. by W. for point Batakarang, passing it at a distance of 20 miles, as Frederick-Hendrick rocks are on the east side of the channel."

2d. The outer route in March and April. This route passes over Macclesfield bank and east of Pulo-Sapata. Horsburgh says: "Vessels should adopt the *outer* passage in March and April. They should run for Macclesfield bank on a S. by E. ½ E. course, with moderate winds, or on a SSE. course with strong winds and a heavy sea and current. From the bank to Pulo-Sapata the course is SW. But if soundings are obtained over the bank, it is best to head SW. ½ S. until the parallel of Pulo-Sapata is reached, and then if the island be not sighted, to steer SW.

by W. or WSW. until in 35 or 36 fathoms. This caution is necessary, especially in thick weather or at night; in fact, at such times it is advisable to keep well to the eastward of Pulo-Sapata until near 10° N., when head SW. by W. or WSW., and run for soundings. Some captains keep SW. by S. from Macclesfield bank, and in March, April, and May run a great distance to the eastward of Pulo-Sapata. If this route be taken, look out for the SE. currents, which are liable to sag vessels over toward the shoals to eastward of Pulo-Sapata.

"The parallel of 10° N. once reached, the course is SW. by W. and WSW. until bottom is found at 35 fathoms, then S. 28° W. or S. 30° W. for Pulo-Aor or Pulo-Timoan. When near 7° 6′ N. do not get in less than 30 or 32 fathoms, and look out for Charlotte bank. European-bound vessels should, in March and April, stand well to the eastward, and passing between the Natunas and Anambas, run for Gaspar strait."

According to *Horsburgh*, if a good sailer take advantage of the series of E. or SE. winds which often blow for several successive days at this season, she can make the voyage from the Great Ladrone to the Indies by the *inner route*. In this way the voyage from Hong-Kong to Singapore is sometimes made in 20 or 30 days; but in order to succeed they must, from the beginning, take advantage of every favorable circumstance, and while standing to the southward and westward keep as close as possible to Hainan and the coast of Cochin-China. However, the very best sailing-vessels should not attempt this route in June, July, and August unless absolutely necessary, but take the easterly route.

During the SW. monsoon.

Captain Blake advises the *outer route* at this season. He says:

"Vessels going either up or down the China sea when the SW. monsoon is blowing in full force, should pass to leeward of the Paracels, the Pratas, and Scarborough bank, as the current is strong with the wind.

"Ships descending the China sea against the SW. monsoon, should stand to the southward after they have passed Macclesfield bank, being careful to keep between 115° 50′ and 118° 50′ E. They should profit by every shift of wind."

Bound to Saigon.

In short, we think it important, if not indispensable, that sailing-vessels bound to Saigon should start before May when they can make quite a rapid passage by the *inner route.* But even the *outer route* will be difficult for a sailing-ship in June, July, and August; the head-winds and currents, the reefs to be avoided, and the danger of meeting a typhoon, all combining to make the voyage long, and even dangerous. The *inner route* is not so bad for a staunch, well-fitted out, auxiliary steamer, and can be followed without great difficulty. Still, she should not attempt to economize coal, but get up steam as soon as possible.

Bound to Singapore.

Vessels bound to Singapore should follow the instructions given in § 152. Thus, if they leave Hong-Kong from the 15th April to the 15th May, they should take the first easterly route; if they sail from the 15th May to the end of June, they should take the second easterly route. They should avoid starting in August, and if they leave in September, should take the inner route described at the beginning of the present paragraph.

First easterly route by Macassar strait or Molucca passage.

After leaving Hong Kong from the 15th April to the 15th May, vessels should run through Mindoro and Basilan straits, as described in § 152. The winds and currents from Basilan to cape Donda are very variable. If the wind be found steady from E., the aim should be to sight cape Donda to SSE. or S. Ordinarily, the winds haul to the westward on approaching Celebes; these, of course, making an easterly drift, it is prudent to keep well to the westward and sight Tanjong Kaniongan. A ship fetching to the eastward of cape Donda will find it slow work beating up for the strait.

Keep along the coast of Celebes while in Macassar strait, and thus find the vessel well to windward on striking the SE. monsoon. The only place where it is important to keep an especial lookout is near the latitude of Laurel bank and the dangers situated N. of the Noesa-Seras islands. For information concerning the remainder of the voyage the reader should refer to § 154, where will be found instructions relative to this route under the head of "the route from Singapore to Macassar strait during the NE. monsoon."

If the wind is foul for Macassar strait, head for Molucca passage, as stated in § 152, and pass successively between Banka and Bojaren, (NE. point of Celebes,) then between

Lisa-matula and Obi-Major, or through Greyhound strait, situated west of Xulla Taliabo. As the wind prevails from SE. and E. beyond Obi-Major, a ship will be in a particularly good position for making from that point to the NW. part of Buru, and for finishing the voyage, as indicated at the end of the second easterly route.

Starting from Hong-Kong, from the beginning of May to the end of June, a vessel should shape her course as stated in § 152, for entering the Pacific by one of the passages north of Luzon. After running down the E. coast of the Philippines as far as Gilolo or Dampier strait, head for the NW. part of Buru. Observations on this part of the route will also be found in § 152. After reaching the NW. extremity of Buru head about SW., and if the wind be fresh from the SE., allow half a point for northerly drift. Navigators should, if possible, manage to make the most northern of the Token-Bessi islands during the day, and double it a distance of 2 or 3 miles. Thence they can easily weather the south point of Buton, and head W. ½ N. for Middle island in Salayer strait, being careful to avoid Cambyna island if in that locality at night. It is not advisable to run between Middle and South islands after dark, unless at home in the neighborhood, as these islands are easily mistaken for each other. After leaving Salayer strait make to the westward, running on either side of Brill (*Taka-Ramata*) shoal, as most convenient. Pass close by Great Solombo; keep clear of the reefs off Pulo-Mankap, and make for Carimata passage. Thence shape the course for the northern extremity of Banka and run through Singapore strait. Vessels should reach this strait by giving a good berth to Geldria bank, and then running for Pulo-Panjang, afterward sailing close around the north coast of Bintang and passing between that island and Pedra-Branca. Second easterly route, by the Pacific and Gilolo or Dampier strait.

§ 168. ROUTE FROM MANILA TO SAIGON.—Once in the offing, stand across the China sea, making due allowance for the SW. current, so as to pass well to windward of all the reefs, particularly Trident, Alexander, and Minerva. Make the coast of Cochin-China near cape Padaran, and finish the voyage by following the inner route given at the beginning of § 167. During the NE. monsoon.

The passage will be rough, long, and very difficult at this season. Sailing-vessels should avoid it during June, July, During the SW. monsoon.

and August. There are, however, three routes from which to choose.

First route. In March, April, and the beginning of May, it is possible to cross the China sea, after leaving Manila bay, by keeping north of Macclesfield bank and the Paracels. After running close to the coast of Haïnan the aim should be to follow the inner route described in § 167. Good vessels can do this even in June and July, but should not attempt it, unless absolutely necessary, after the SW. monsoon has reached its maximum strength, as typhoons are then common.

Second route. This route runs through Mindoro strait; E. of Palawan; across Balabac strait, and thence along the W. coast of Borneo, until far enough south to cross the China sea. As the currents are strong to N. and NE., it is necessary to keep well to windward of all reefs, especially Luconia and Vanguard.

Vessels can also follow an analogous route through Palawan passage, but only in April, or at the beginning of May. Both these routes will, of course, be greatly facilitated by the assistance of steam.

Third route. This route passes through Macassar strait, and, though longer than the others, will be found easier for average sailing-vessels, (vide § 167, 1st easterly route.) By taking this passage they will avoid the chance of meeting a typhoon, and will always find favorable winds beyond Great-Pulo-Laut, from SE. to southward of the line, and from SW. to northward.

During the NE. monsoon. § 169. ROUTE FROM MANILA TO SINGAPORE, THE STRAIT OF SUNDA, AND EUROPE.—Horsburgh advises vessels to steer W. by S. from Manila bay to the parallel of 12° N., and recommends them to keep on this course if they are not certain of being at least 9° west of Goat island. They should afterward head SW. for Pulo Sapata, thus allowing for drift, as the westerly currents are at times very violent, and the dead-reckoning is not to be relied upon. If the island is not sighted after the parallel of Pulo-Sapata is reached, the course should be SW. by W., until soundings are found; thence SSW. ½ W. for Pulo-Aor, as stated in § 167, outer route. For information concerning the end of the route, vide § 151.

During the SW. monsoon. We think that the 1st easterly route, described in § 167, can be followed during the whole period of this monsoon. Vessels should run through Mindoro and Basilan straits, and then make for Macassar strait, when the winds will allow, that is to say, in April and May; otherwise, Molucca passage may be taken. In this case they should double Buru to NW.; Buton to SE. and S., and then take Salayer strait and Carimata passage.

It does not seem that there is any especial advantage in taking the Pacific or 2d easterly route. It can, however, be followed; in which case the track is through San Bernardino strait, and reaches Pitt passage, by Gilolo or Dampier strait, as described in the description of the 2d easterly route, § 152. Finish the voyage as indicated at the close of § 167. European-bound vessels should follow the same route and end the voyage as stated in § 152.

During the months of March, April, and May, vessels can still take the Palawan route; and, in relation to this subject, we give the instructions of Captain Stephens, commanding the Harkaway, (*Ann. Hydr.*, 1870:)

"Vessels leaving the coast of China, or Manila, and bound to the strait of Sunda, during March, April, and the beginning of May, should expect to make a long passage if they follow the old route, which passes near Pulo-Sapata. If, on the other hand, they keep along the coast of Luzon, cross Palawan passage, run along the Borneo coast, pass near Direction island, double Soruetou, take Carimata strait, running close around the North Watcher, and thence steer for Saint Nicholas point on the island of Java; they will probably meet E. winds, fine weather, and a smooth sea all the way. This is a direct route, and clear of calms. Prior to the end of May the currents are more favorable than by the other route.

"On nearing the strait of Sunda, vessels should stand over for the coast of Java, and in May keep close to the land, as the wind is then light; from SE. during the night, and from NE. during the day. By following these directions they will avoid all danger of being set toward Button islet by the westerly current. The current always sets to SW. in the middle of the strait; its force is diminished during the short flood tide, but increased during the long ebb."

§ 170. ROUTE FROM SAIGON TO SINGAPORE.—The following instructions are due to Captain Loftus, of the ship Kensington, (*Ann. Hydr.*, vol. 23:)

1st. During the NE. monsoon.

"After leaving cape St. James pass east of the coral shoals and Pulo-Condore; thence head for Pulo-Aor, allowing for the currents which set toward the gulf of Siam. Approaching Polo-Timon during the night, or in bad weather, it is necessary to look out for the SW. current, which tends to set vessels into the bay on the coast of Malay, to westward of Pulo-Timoan. When the weather is overcast and the wind strong it is best to bear up and double Pulo-Aor to leeward, so as to wait for a favorable opportunity for running through the straits. The current sets to SSE., between Pulo-Aor and the eastern point of Bintang; it is also well to steer S. 28° W. from Pulo-Aor to the Malay coast; and then to keep along the land at a convenient distance. Vessels should pass through the north channel, doubling the Romania islands at a distance of 3 miles. They will then be on the windward side of the straits, and can anchor under the lee of Water island if the weather be bad. Vessels keeping too far to the southward, after leaving Pulo-Aor, are often drifted to leeward of Bintang by the current and ebb tide, which runs out of the straits. If this happens they are obliged to take Rhio strait. In March the east winds are fresh, the current slow, and the weather fine. In April the east winds are interrupted by calms and squalls. The monsoon dies away between the end of April and the 15th May. In May the winds and weather are variable, with thunder-storms and squalls. Toward the end of this month the northerly current commences to flow.

2d. During the SW. monsoon.

"At this season a fair wind will be carried from the river to cape St. James. The wind along the coast is frequently from N. and NE. at night, and at times from these points for a few moments during the day. A ship taking advantage of these local winds, while running to the SW., can sometimes carry them for 40 or 50 miles to the southward and westward of Pulo-Condore. Quick passages have been made by keeping under the coast of Cambodia, as far as the Brothers and Pulo-Obi, and thence by crossing the gulf of Siam with a strong NW. wind, beating up the coast of Malasia with the tide, and passing inside of Pulo-Timoan, Siribuat, and Pulo-Sibu. Thence make for Singapore strait, taking advantage not only of the tides, but of the land and sea breezes, which are frequent at this season during fine weather. This route is used by vessels returning from Siam, and sometimes from Saigon; but the passage to *east-ward* of the Great Natuna is considered the *best*, especially for large ships.

"Vessels heading SE. on the long starboard tack meet the strongest current near Charlotte bank. It afterward diminishes and becomes slightly favorable SW. of Great Natuna. Here a SE. or E. wind is generally struck, and a bad sailer can often pass between Subi and Low islands, and then lay a course for Singapore strait. Strong west winds and rain are common at the beginning of the monsoon, though the wind is sometimes from S. Bad sailers should therefore be very careful how they approach the coast of Borneo near cape Sirik; and make the land between Tanjong-Datou and Boerong islands, by taking the Api passage. This they can easily do, as they will find land and sea breezes, and only a slight current near the shore. In Api passage, between Tanjong-Datou and Tanjong-Api, a ship should not go closer than 2 or 3 miles to the islands, nor in less than 14 fathoms of water; it is also advisable to have an anchor ready for letting go.

After leaving Boerong, the attempt should be made to fetch Pulo-Panjang on the port tack, passing either N. or S. of the Tambelans. Pulo-Panjang is off the east coast of Bintang. Take the south channel, passing 2½ or 3 miles from the NE. point of Bintang, in order to clear the Postillion and make allowance for the tide by keeping on the southern or windward side of the strait. When off Rhio strait remember that the flood sets toward it."

Passage of the sailing-vessel La Forte, Captain Bourgois, (*Ann. Hydr., vol.* 23.)—"The Forte was towed out of Saigon on the 11th May. Once clear of the river we beat up against the SE. wind, in order to double the banks off the mouth of the Cambodia. The currents in this locality are sometimes very violent.

"On the 12th, wind from NW. and then from SW., blowing fresh. I wished to cross the line as far to the eastward as possible, as in all probability I would find the SW. monsoon north of the equator, and the SE., south of it. I therefore steered for the passage east of Great Natuna, leaving the island to windward during a violent NW. squall.

"The peak of Great Natuna can be seen for 60 miles; the small island north of Pulo Laut is visible for 30 miles; the eastern coast of Great Natuna is clear of all dangers. Passed the island on the 16th; beyond this, light winds from SSE. to SSW., with moderate rain-squalls from S. to W. Beat through the passage situated between Great Natuna and Low island on the right, and Subi and West islands on the left; current about 1 knot to NE. On the 21st, 22d, and 23d frigate becalmed near West island; drifted from 15 to 18 miles per day, to the NE. On the 24th bore to the southward and lost sight of Great Natuna peak. On the 27th, while beating up with a light SSE. breeze, sighted Tambelan island, distant from 45 to 55 miles. Between Borneo, Tambelan, and the Direction and Pulo-Datou chain of islands, there is a large basin where a vessel can beat up against the monsoon, and even anchor in 20 or 25 fathoms of water, muddy bottom. On the 27th, Direction island bore S. 9° W. and Tambelan peak N. 62° W.; breeze moderate and from SW.; sky cloudy and overcast. We had scarcely hauled the frigate up on the starboard tack, and taken two reefs in the topsails, before the wind jumped around to W., blowing fresh. Squalls from west all the rest of the day and night; between them, however, the breeze was regular and moderate from SSE. to SW.

"In Bouillet's Dictionary these squalls are called the 'Borneos.' Commencing from the SW. they shift to W. and NW.; consequently, if well clear of the coast ot Borneo a captain can use them to make to the southward on the starboard tack. As the weather became better, the wind returning to SSE., we ran out of the basin on the evening of the 29th, passing 5 miles to leeward of Direction island on the port tack. The wind in this locality followed to a certain extent the motion of the sun, blowing from SE. in the morning, and from SW. in the evening.

"This point of the voyage reached, Captain Bourgois states that he decided to keep to windward, and if possible reach Carimata passage by the north of Billiton; and thence, if the wind should haul to the southward, to still keep on the starboard tack as far as Borneo, when he intended to go about and stand for Java; or if, on the other hand, the wind should come out from tlre eastward, to round close to the NW. point of Billiton *on the port tack*, and thus reach the Java sea through Stolze channel. This idea proved successful. The Forte beat to SE. from the 30th May until the 4th June, to reach the north point of Billiton. The weather, however, was not very good, squalls being frequent from ESE. to SE. during the day, and from SSE. to S. during the night. On the morning of the 4th, when we had nearly finished the starboard tack, we sighted Carimata island, distant 36 miles to the E.; but the wind had as usual shifted to ESE. Though we endeavored to reach Billiton, so as to be ready to go through Stolze channel the next morning, a NE. squall, followed by a calm, compelled us to anchor before we were up to the west coast of the island. Our anchorage was about 10 miles NE. from the outer one of the Eleven islands, and was well protected from the winds of this season; bottom mud; depth 16 fathoms. We got under way on the morning of the 5th and rounded the islands at a distance of about 2 miles; reeling off about ten knots during a heavy NE. rain-squall, which hid all the land.

"When the sky cleared we were abreast of Tjeroutjoup roads, where we found two Dutch men-of-war at anchor. This seems to be a very safe anchorage during the south monsoon. During the evening of the 5th the wind died away, and we anchored in 18 fathoms, bottom sand and gravel, to west of a beautiful bay with a sandy beach, situated between Pulo-Batu and Mendanao islands. From this anchorage the islet off the W. point of Mendanao bore ESE., and the one off the W. point of Pulo-Batu NE. Got under way on the 6th, after a heavy SE. squall, and stood for Stolze channel with a gentle ESE. breeze.

"Stolze channel may be divided *into two parts.* The *first* is the basin, bounded on the NE. by the Mendanao islands; on the SW. by the North, South, and Table islands; and on the SE. by the group of Six islands. In this basin, open to NW., and 7 or 8 miles wide, a ship can beat up against the SE. wind with some chance of success, the current being moderate and the sea quite smooth. It will not take over a day to reach the anchorage to leeward of the Six islands. Another 24 hours on the port tack will carry the ship through the second part of Stolze channel. *This second part* lies south of the first, and has a channel 5 miles wide; stretching to SW., between Table island and the

Six islands. Leave Vansittart bank to NW. It will be necessary to beat up *against a S. wind*, even after doubling the Six islands. Besides, it has already been stated that it is best to haul up on the starboard tack before entering Stolze channel with a S. wind; and to keep on this stretch through Carimata passage, as far as Borneo. There, the wind will probably shift to the eastward and enable you to fetch the strait of Sunda on the port tack. But with a SE. wind a ship can easily clear the strait on a SW. course. Soon after leaving the Six islands, Shoal-water island will be sighted at a distance of 15 miles in clear weather. The bearings of this and the other islands (especially Saddle island) will serve as ranges for doubling Shoal-water island to NW., for clearing Embleton rock, and for passing on either side of Fairlie rock according to the direction of the wind. Finally, if the wind come out from east after doubling the Six islands, or even if it tends a little more to the northward, head between S. and SSE.; and run to windward of the line of dangers formed by Shoal-water island, Sand island, and the adjacent shoals. But the want of an anchorage and the SE. swell will generally prevent vessels from following this route, except during steady leading winds.

"To return to the case of the *Forte*. We left the Six islands on the morning of the 6th June, on the port tack; wind ESE. and E. As the wind moderated and hauled ahead, we did not succeed in passing to eastward of Shoal-water island; the SE. swell also drifted the vessel to the westward; and the weather becoming overcast it was not deemed advisable to beat up among the reefs between Sand island and Pulo-Selio; we therefore bore up and ran to westward of Shoal-water island. The difficulty consisted in doubling Embleton rock during the squalls, when no bearings could be taken. The wind freshening, took 3 reefs in the topsails. Sighted a rock just awash, near the position of Embleton, also other reefs in different directions; the man at the masthead also reported breakers between Shoal-water and Sand islands. The islands were not in sight at the time. Wind steady from ESE.; weather better. After rounding Shoal-water island, steered S., full and by. As the wind died away at sunset, anchored 8 miles SW. of Shoal-water island in 10 fathoms, bottom sand and gravel.

At daylight on the 7th got under way with a SE. wind and doubled Fairlie rock on the port tack, taking our bearings from Shoal-water island. After sighting the little island called North Watcher, we reached the entrance to the strait of Sunda on the 8th; and anchored during the evening at Anjer, after a passage of 28 days from Saigon."

CHAPTER X.

ROUTES FROM THE AUSTRALIAN PORTS TO ASIA AND CHINA.

During the SE. monsoon, from May to August.

§ 171. NORTHERLY ROUTE FROM AUSTRALIA TO INDIA, BATAVIA, AND SINGAPORE.—1st. *From Sidney to Torres strait.* Steamers and auxiliary steamers should, after leaving Sidney, follow the inner route advised by King, and keep along the coast of Australia inside of the reefs.

Sailing-vessels, starting from May to the end of June, should follow the outer route, which passes E. of Gato bank and Wreck, (Naufrage,) Kenn, and Diana reefs. Thence, they can cross Torres strait, by the passage near Raine island or the one near Bligh island. The Bligh passage is preferable.

In support of these opinions we will give the following instructions from the *Australia Directory:*

Relative merits of the two routes.

"The outer route is more likely to interest commanders of merchant-vessels, whose chief object is generally to make a quick passage with the least amount of labor. This route, no doubt, possesses this advantage; but it must be borne in mind that the passage through the Great Barrier reefs, from the Coral sea into Torres strait, is frequently attended with danger, and sometimes the loss of the vessel, notwithstanding the recent surveys and the erection of the beacon on Raine island. These disasters, however, would doubtless be less frequent were the Great Northeast (Bligh) channel more used, as it may be generally navigated by night, so that the time and labor saved by not being compelled to anchor so frequently as in the route by Raine island, would more than compensate for the 90 miles which the former route exceeds the latter in distance.

"Notwithstanding all that has been said in favor of the *inner route*, supported by the weighty authority of Captain King, the *outer route* is undoubtedly preferred by nearly all the merchant-vessels bound from Sydney to Torres strait, more especially since the late survey of the Coral sea."

Captain Blackwood remarks as follows:

"Opinions are divided as to the respective merits of the various routes through Torres strait. During the season of SE. winds, that is, from May to September, either the channel by the Raine-island beacon or that near Bligh island, situated north of Darnley, furnishes a quicker route than the passage along the coast inside the reefs. Such being the case, merchant-vessels make use of this *outer route*, as much time is gained thereby.

"Steamers bound to India during the W. monsoon, that is, from November to March, will find it greatly to their advantage to take the Torres-strait route. And I think Captain King's inshore route is the best for them, as wood may be obtained along the east coast, nor do I see why the voyage from Sydney to Singapore need take more than five weeks, as the SE. winds will be carried up to 14° or 15° S. before the west wind sets in."

We will finish these considerations on the best route by giving the principal results obtained at the *Dutch Observatory* relative to passages from Australia to Java.

Below will be found the mean crossings of 14 ships *which took the Raine-island passage.* One of these vessels left in April, three in May, four in June, five in July, and one in August.

30° S.	crossed at	156° 20′ E., after	2.1 days at sea.
25° S.	"	157° 20′ E., after	4.6 "
20° S.	"	156° 20′ E., after	6.7 "
15° S.	"	150° 20′ E., after	9.6 "
140° E.	"	10° 30′ S., after	16.8 "
130° E.	"	10° 12′ S., after	20.5 "
120° E.	"	10° 12′ S., after	25.2 "
110° E.	"	9° 12′ S., after	27.3 "

The points of departure for the above table were Sydney, Newcastle, and Bass strait. The vessels which left Bass strait took a mean time of 2.5 days to reach 35° S. at 154° 20′ E., and 2.6 days to go from 35° to 30° S., crossing the latter parallel at 156° 20′ E.

We will now give the mean crossings of 21 vessels *which took the Bligh-island passage.* Three of these left in April, 4 in May, 7 in June, 4 in July, 2 in August, and 1 in September. The points of departure were the same as those in the preceding table, except four, which were from Melbourne.

30° S.	crossed at	156° 20′ E., after	3.3 days at sea.
25° S.	"	157° 20′ E., after	5.9 "
20° S.	"	155° 20′ E., after	9.1 "
15° S.	"	151° 50′ E., after	11.7 "
150° E.	"	13° 30′ S., after	12.7 "
140° E.	"	10° 36′ S., after	20.0 "
130° E.	"	10° 00′ S., after	23.5 "
120° E.	"	10° 54′ S., after	28.1 "
110° E.	"	9° 12′ S., after	30.2 "

The four vessels which left Melbourne took a mean time of 6 days to reach 35° S. at 153° 20′ E. Those which left by Bass strait were 3.2 days in reaching 35° S. at 152° 20′ E.; and afterward from 2 to 3 days in reaching 30° S. at about 157° E.

By comparing these two tables it will be seen that 27 days were consumed in reaching 110° E. when the Raine-island passage was used on the route from Australia to Java, and 30 days by the Bligh-island passage.

Notwithstanding this result, we still are in favor of the route by Bligh-island passage, as it is less dangerous and only gives an increase of two days' voyage for vessels *leaving the same Australian port.*

2d. *The Bligh-island-passage route.**—After doubling cape Rodney and the SE. part of New Guinea, vessels should reach 9° 10′ N. and make the westing on that parallel, passing north of the Eastern Fields and Portlock reefs. Thence run 4 or 5 miles south of Bramble key, a sand-bank about 9 or 12 feet above low water, and visible 7 or 8 miles from the mast-head.

During the night it is well to avoid the southern part, between Bramble key and the ledge of reefs to northward of Darnley island, and anchor north of Bramble key in about 22 fathoms of water. However, a ship can lie-to or stand off and on to the northward of the key, being careful not to get inside of 7 fathoms on the coast of New Guinea.

Near Bramble key the flood comes from NE. or E. at a speed of, sometimes, 2 knots; the ebb comes from the opposite direction, and is swifter. It is generally safe to count

* Indications concerning the beginning of this voyage, from Sydney to the end of the reefs, will be found under the head of "3d. The Raine-island-passage route."

on a knot of westerly current, unless the strait is approached after a squall, when a greater allowance should be made.

Water may be obtained on Darnley island, but the men should never go on shore in these localities without being well armed.

We will now cite several accounts of different trips made through Bligh passage:

Passage of the Pactole, Captain Allaire, through Torres strait, (Ann. Hydr., vol. 30.)—"I was off Portlock reef on the 16th August. Shaped my course for Bligh passage; at six shortened sail; squally; strong breeze and rain from SE. At 5 a. m. got bottom at 31 fathoms, sand. During the night ship drifted 15 miles WSW. in 12 hours, from which I inferred that the current sets toward the strait at this season. Headed for the strait at 7 o'clock; sighted Darnley island, bearing SSW., also the breakers NE. of that island; passed 1 mile from the reef. Darnley island is very high; it can be seen for 25 or 30 miles in clear weather, and is a good mark for entering the strait. At noon on the 17th Darnley island bore N.; position, 9° 37′ S. and 143° 22′ E. Passed between Darnley and Marsden; at 1 p. m. doubled Rennel island to the northward, at a distance of 1 mile; at 3.30 I was in mid-channel between Cocoa-nut and Dove islands. Headed for Bet-island pass. Strong SSE. and SE. breeze all day, with squalls. Ship making 10 knots with a double-reefed topsail breeze; pressed on more sail in order to clear the difficult passes before dark. At 5 p. m. anchored under the lee of Bet island. Vessels can anchor to leeward of any of the islands in this locality. Found a westerly current at our anchorage until 11 p. m.; about 2 hours afterward it changed. The ship swung to the ESE. current, notwithstanding the strength of the wind. After running at about 2.5 or 3 knots until 5 a. m. the current again changed and set to the westward. Got under way and at 9 a. m. passed between Ninepin rock and Saddle island; at 10.30 I was abreast of Double island; at 11.45 on the meridian of Ince point, at the entrance to Prince of Wales channel; at 1 p. m. on the meridian of the west point of Goode island, and consequently out of all danger. I was, therefore, 29 hours in Torres strait, 13 of which I lay at anchor. I consider Bligh passage the best at all seasons, on account of its numerous anchorages. I never found more than 11.5 fathoms of water, nor less than 8 fathoms."

Observations by Captain Croudace, (*Mer. Mar. Mag.*, 1864.)—"Left Sydney July 1st, 1860, becalmed 40 miles from the Heads for five days, then a nice westerly breeze took us down to the trades; pursued the new route through the Coral sea, as far as lat. 18° S., long. 155° E., from which point, instead of steering for Raine island, we shaped a direct course NW. ¾ W. by compass, to pass midway between the Eastern Fields shoal and the coast of Guinea, and on the morning of July 17th, at 7 o'clock, we sighted the breakers on East cay, having seen nothing from the above given point of departure until sighting the cay itself; the weather was very hazy on this run of nearly 700 miles, and the nights very cloudy; we felt little or no current, and the ship made the course good very accurately; we ran down fearlessly until we sighted the Cays, never having reduced sail; at 8 a. m., the 17th, we were abreast of Anchor cay, the opening to Bligh passage, and having passed it, we hauled up, and by 6 p. m. (same day) we had reached Dove islet, and anchored under the lee of an island, (without a name,) bearing E. by N. from Dove islet; the next day the weather was very unsettled and squally, with showers; remained at anchor, having been joined by the ship Storm Cloud; on the 19th, at 5 a. m., we both weighed and proceeded for Prince of Wales channel, passing close round Bet rock or islet, and thence close past Ninepin rock. The course from this rock to the entrance of the Prince of Wales channel is SW. ½ S. by compass, and there appear to be no dangers in the way, with depths of 8 and 9 fathoms.

As we neared Double island and Mount Earnest, the look-out from the fore-topsail yard reported breakers on the starboard bow. As we approached them, the ship's position was taken as accurately as the opportunity afforded, and the shoal at the same instant being judged nearly one mile distant, was placed as follows: A line drawn from the center of Double island to center of Mount Earnest will pass through the center of the new reef, (now called Campbell reef;) it is 6¾ miles from the former,* and 5¼ miles from the latter; it appeared to trend SW. and NE., about 1 mile

* On the charts of 1873, the center of Campbell reef is 1 or 2 miles east of the line given by Captain Croudace.

long, and very narrow; should say it may have 4 to 6 feet over it when smooth water. It lies much in the track of vessels coming up from Bligh passage.

Captain Campbell, of the Storm Cloud, who passed it at the same time, assigns the same position to it. Both ships anchored at Booby island at 1.30 p. m., having come from our last anchorage—80 miles—in 7½ hours. For large, or indeed for any ships, I consider the Bligh passage preferable to the Raine island passage, there being no sunken reefs or dangers, and smooth-water anchorages the entire lngth, with a bank of soundings as you approach the entrance of the channel; no embarrassment from the glare of the sun, as in the passage from Raine island. As a general rule also, a ship will only require to anchor once, and even that may not always be necessary, for, if taking the channel early, as we did for example, had the wind been more easterly, we should have passed the narrows before dark, and could then have hove-to until daylight, keeping mount Adolphus in sight, or any of the islands in that locality, but being the syzygy we found weather unsettled, and the winds far from the south."

Observations by Captain Ankers, (*Mer. Mar. Mag.*, 1865.)—"The ship *Queen of the East*, under my command, left Sydney on the 29th June, 1864, with horses, bound to Madras, and, although I had never been through Torres strait, I determined to take that route rather than attempt a winter passage round cape Leeuwin in a large ship with the certainty of losing some of the horses. I accordingly took the ordinary route to the northward, but owing to a succession of very light winds, and four days' calm, I did not sight the Eastern Fields until the 14th July, and from the same cause did not pass Bramble cay until daylight on the 16th, having passed the north end of Portlock reef on the previous evening. The sea breaks high on both Portlock reef and the Eastern Fields, and they are plainly visible from the mast-head at the distance of 6 or 7 miles in clear weather.

"July 16th, at 7 a. m., Bramble cay bearing N., ½ a mile distant, Darnley island peak being clearly visible bearing SSW., steered a direct course for Stephens island, making due allowance for the tide, (the flood setting pretty strong on the weather beam,) the shoal patches northward of Darnley island, showing quite plainly, being covered with a white sand or ground coral. Carried all sail during the day with studding-sails on the port side, the wind being light from ESE.; at 3 p. m. the weather thick and squally, the rain at times completely obscuring the land, I anchored in 9 fathoms water, the NE. end of Stephens island on with the peak of Darnley island, about half a mile off shore, the water smooth and the holding-ground good.

"Sunday, 17th July.—Got under way at daylight with a fine breeze from ESE. and clear weather. Set all sail and proceeded, giving her all studding-sails as we went along, passing south of Dalrymple island, and north of Campbell, Marsden, Rennel, and Arden isles, between Dove and Cocoanut islands, and north of the Three Sisters, Bet, Sue, and Poll. The extensive reef running out to the eastward from the northernmost of the Sisters (Bet) is steep and free from outlying dangers; I sailed along within a cable's length of it, in order to give a berth to some shoal-patches to leeward, which are, however, plainly in sight from the mast-head. After rounding Bet, the Ninepin rock and Saddle island can be seen if the weather be at all clear—the Ninepin rock being not unlike a vessel under sail. I then passed between the Ninepin and Harvey rocks; it is better to keep close to the Ninepin, in order to avoid some shoal-patches which lie to the north of Harvey rocks, some portions of which are only awash at low water. I then steered down for Double island, where I anchored at 11 p. m., in 7 fathoms of water, half a mile from shore.

"Monday, 18th July.—Got under way at daylight, made all sail and steered down for the Prince of Wales channel, rounding Wednesday spit at the distance of 80 yards, found the tide running like a mill-stream; rounded Hammond rock at 30 yards distance, the eddies so strong that the helm was almost useless. From the top-gallant yard I coursed the ship through between Ipili and Sunk reef, keeping within 120 feet of the edge of Ipili reef, as Sunk reef being covered, renders it the more dangerous of the two.

"The Prince of Wales channel, which I thought the best out to the westward, requires very great caution; in passing Hammond rock and Ipili reef, I feel certain the ship was going over the ground at the rate of 10 knots per hour, the current having such power on the ship that I was compelled to assist the helm by working the sails, although I had a good breeze at the time. I should therefore consider it advisable, especially if the wind be light and baffling, to take the Prince of Wales channel as near slack-water as possible, and am of opinion that a commander would be justified in sacrificing a few hours in order to accomplish that object, especially in a dull-sailing or badly-steering ship, rather than run the risk of touching on the reefs when the passage of the straits may be said to be accomplished. I reached Booby island at noon, hove to, and deposited a few necessaries in the cave for the use of those less fortunate than myself."

Captain Ankers then goes on to state that, after reading some logs which he found on Booby island, and thoroughly considering the relative merits of Bligh and Raine passages, he is still in favor of the former. He does not think the objection, that this route is too far to leeward, well supported by facts. As to the inhabitants of these islands, they are a miserable race of savages, without good huts or the least civilization. No provisions are to be found on the islands, the natives not even cultivating bananas or corn.

3d. *The Raine island passage.—Instructions of Captain Blackwood:* "Captains wishing to take the outer route from Sydney are recommended to sight Cato shoal or Wreck reef. They should on no account pass west of 153° east longitude before reaching 17° south latitude. Lihou reef will be avoided on this track, and the current will be found variable, though generally setting to the westward, at a little over one knot per hour.

"Osprey reef should be avoided by passing vessels. It is situated in lat. 13° 51′ 10″ S., long. 146° 36′ E. After leaving Osprey reef the course should be for the Great Barrier, at lat. 11° 50′ S. and 144° 11′ E.; this will keep the ship from being drifted to northward of the Raine island passages.

"Raine island lies nearly in the middle of the large opening in the reef, and has a clean, safe channel on either side. The islet is low, of coral-formation, and has no fresh water upon it; the soil being covered with a thick underbrush. A reef extends 1¼ miles to the ESE.

"A substantial round beacon of stone was erected in 1844, on the SE. point of the island. It is 60 feet high,*

* Without the dome, which has decayed and fallen in, vide Australia Directory.—*Translator.*

and painted with alternate red and black vertical stripes. It should be visible 8 miles in clear weather.

"It is high water at Raine island, full and change, at $8^h 10^m$. Spring tides rise 10 feet, the flood runs to WNW.; the ebb to the eastward, the strength of the stream sometimes exceeding 2 knots at full and change.

"Provisions and water have from time to time been deposited in the chambers of the tower on Raine island."

Captain Charles Harrold gives the following full and complete instructions on the Raine island route, (Mer. Mar. Mag., June, 1860:)

Captain Harrold's instructions.

"Having made the passage from Australia to India through Torres strait upward of a dozen times, and hence having had considerable experience of this route, the following remarks may prove acceptable to those commanders who have not been through before.

"A ship leaving Sidney early in the season, from the beginning of April to the middle of May, should at once get an offing; later in the season it is better to keep the land on board (and with strong westerly winds) as far as Solitary isle or Mount Warning; but should the wind veer to the southward, with a rising glass, shape a course at once to pass 40 or 50 miles to the eastward of Cato bank; steering on a northerly course, keeping at least a degree to the eastward of Keen reef, after passing which, a NW. course may be pursued, taking care to give the Cayes and Lihou reefs a good berth of 50 or 60 miles. After leaving these dangers there is nothing on the track to the Great Barrier; the Osprey reef and some others to the southeastward are quite out of the way, and a ship has no occasion to go near them.

"In approaching the Great Barrier considerable care is necessary, if no observations for latitude have been taken; and an allowance must be made for the northwesterly set, so as not to get to the northward of Raine island beacon. The plan I always adopted was not to run down on the Barrier and work to windward all night; but should the ship be upward of 200 miles off at noon, and not be able to enter the next day, I would reduce sail at once and steer slowly on a course, from 2 to 5 knots, according to distance, so as to reach within 30 miles of the Great Detached reef by the second morning at daylight. Should no observations for latitude have been taken, (or whether or no,) when breakers are sighted from the mast-head, ahead of the ship or a little on the port bow, as you draw toward the reefs, or should you see them on the starboard bow as far as can be seen from the mast-head, you may be sure you are to the northward, and may haul up immediately on the port tack; and if the wind is far enough to the eastward to lay along the reef, you will soon see clear water, and sight the beacon from the mast-head; or should it be the Great Detached reef that you sight, clear water will be seen on the starboard bow, and by edging to the WNW. you will shortly see the beacon. I have always found this a good guide for Raine island, for if you get too far down toward the reefs before you find out your mistake, you will have great difficulty in working to windward. Should you be to the southward, which will rarely happen when steering for the Great Detached reef, you will be sure to see Yule reef, and could then take Stead passage, which, however, is not very easy to make.

"After making the beacon and steering down with it, a little on the starboard bow, the edge of the Great Detached reef will be distinctly seen, also the reefs off Raine island; when the beacon bears N., haul up SW. by W. After a few miles a very small sand-bank will be visible from the mast-head, on the starboard bow, and if you see it distinctly from the deck of a small vessel you are falling to leeward, and should haul up more to the southward. This is one place, I think, that should have a mast with a sort of basket-top placed on it as a beacon; it would be a good guide for ships entering late in the afternoon; for, after passing this, you are near the edge of soundings, and with night coming on, after getting on soundings, you should haul well to the southward and bring up at dark, taking care to give at least 60 or 70 fathoms of cable, as the holding-ground is here very bad, (hard coral.) From this anchorage keep well to windward of the course until you sight either Ashmore banks or the Middle banks. Ashmore banks show much higher out of the water and whiter than the Middle banks, which latter are at high water nearly covered. This also would be an excellent place for a mast-beacon. If there is sufficient daylight to reach Ashmore banks, or under the lee of Cockburn reef, a better anchorage may be obtained; or if it should be blowing a gale, with squally weather, good anchorage in smooth water may be obtained under the lee of Sir Charles Hardy islands. In going in, keep close to the weather island, and anchor well to windward in a small, sandy bay; by so doing you will be in a better position for weighing in the morning. In anchoring here one voyage the chain fouled on the windlass, and before the other anchor was let go we were in mid-channel. Next morning the tide would not let the ship cant any other way than head to the northward. The top-sails were single-reefed and yards hoisted to the mast-head before commencing heaving. The ship broke ground with the fifteen-fathom shackle in the hawse, and before we got the anchor off the ground and sail made we were close down on the leeward island, and only cleared it by scarcely a stone's throw. In coming out from here you will have to keep NE. by N. for a few miles, and might have to tack should the wind be to the eastward; but you can see the Middle and Ashmore banks quite plainly. If there is sufficient daylight to reach Cockburn reef, it will be found a still better anchorage than under Ashmore bank. It is quite smooth, especially at low water, although you require a good scope of cable, as the ground appears to be hard coral. In weighing from here you will soon see the edge of the reef that stretches to the NW., and you may run down toward it until you can see it distinctly from the deck; then edge away along the reef, and as you draw toward the end you will see two sand-banks; the nearest one is small, and probably may be covered at high water, although I have never seen it quite covered, but you will be sure to see it from the fore-yard before you get too close, and very likely before they see it from the mast-head. If bound to Bird island, haul close round the small sand-bank, leaving the large bank on the starboard hand. This bank is high out of water, with an extensive reef running off the weather side of it, which is also seen distinctly from the deck. This track I prefer to going close round the end of Cockburn reef and having to haul to the southward. If bound on to Cairncross island, leave the large bank also on the port hand, and, steering for the Hannibal island, reefs V and W will be seen, with a small sand-bank on reef V, also the Boydong cays.

"After passing V and W reefs, the great danger of Torres strait is past, and all sail may be carried down to cape York; or anchor at Cairncross, and the next day, with a fresh breeze you will get out in good time.

"If the wind should be light, and a ship not able to reach cape York or mount Adolphus before dark, anchor close under the lee of a small sand-bank at the end of reef X, which is far preferable to Turtle island; but should the night be fine and moonlight, having reached thus far, a ship could run on, and passing Albany isles, anchor round cape York. If rounding this cape in daylight, anchor abreast of two rocky islets, a short distance past the cape in 7 or 8 fathoms water, cape York bearing E. ½ N., Peaked hill SSW.; this is a very good anchorage, and you will be in a position to choose either Prince of Wales passage or Endeavor strait. If intending to pass through the Prince of Wales channel, (which is as good as buoyed and beaconed by Hammond rock and the Ipili and North-West reefs, with no danger excepting a strong tide,) should the wind be light and contrary, after passing the Albany isles steer to pass point Ince half a mile off, taking care to avoid rock A; the North-West reef can be seen from the mast-head a long while before you reach the point; at any rate should it be hazy or squally, by passing point Ince close, the North-West reef will be avoided, and Hammond rock will be seen; steer to pass close to, and before reaching it the Ipili reef will be distinctly visible, the rocks on it sticking up above water. Borrow toward Ipili reef to clear Sunk reef; after passing this danger, the last in the straits, Booby island will be seen. Should you intend to anchor, leave the island on your port hand, as the reef extends to some distance off the other side. The landing is very bad at low water, but you can pull close to the entrance of the cave at high water. I was once detained from noon till 9 p. m. to land a few casks of water sent by the Government.

"If intending to go by Endeavor strait, which is quite safe, after passing Possession isle, you are soon out of the strong tide. I have always passed close to a high rugged island (Entrance island) west of Great Woody island, as far preferable, and it leads you clear of McKenzie and Gibson rocks. My last voyage I towed through with the boat ahead, and shortly after a breeze came from the SW., and squally, and we worked down within 3 or 4 miles of Red Wallis isle, and brought up when the tide turned. Should the weather be clear and fine as you draw abreast of Red Wallis island, a good mark for mid-channel is cape Cornwall on with Peaked hill, from which steer to pass out to the south of Rothsay banks.

"In conclusion, I again repeat, that in running for the Great Detached reef, 25 or 30 miles is quite close enough to come to it until daylight; for if a ship get to either Ashmore bank or Cockburn reef the first night, or even only just on sounding, she will reach Cairncross island the next day, and out clear the day following, and she would not do any better by getting in earlier. She might reach Bird island the first night, but she would have to anchor under cape York the next, excepting she were a very fast ship, with wind and tide in her favor, and she then *might* get through; but it rarely happens that a ship passes through with once anchoring; and it is certainly not worth the risk of running down close on the barrier for the sake of twelve hours. I have often gone through it in 48 hours, viz, 26 at anchor and 22 under way. I am satisfied that nearly all the wrecks have taken place through the anxiety of masters to get in early, and to running down too close on the Barrier. The wrecks inside I attribute to going off the tracks laid down, and running in squally and thick weather. Great improvement might be made by placing a few mast-beacons (with basket tops to distinguish them by) and buoys at various places, and the passage by Raine island rendered much easier for a stranger going through for the first time."

4th. *Route from Torres strait to Singapore.*—After leaving Booby island or clearing Endeavor strait, vessels make to the westward, with a fair wind. A good lookout should be kept for the banks beyond Booby island, as they are not all accurately located; especially Proudfoot, Lucius, and Aurora shoals. Wessel islands should always be passed about 20 miles to the northward; thence the track lies between Croker island and Money reef. Afterward the course should be shaped so as to run about 20 miles south of Damo or Dana, (to southward of Rotti;) this will take the ship close to Echo bank, and well clear of Hibernia and Ashmore shoals; all of them being passed to northward.* Thence leave Hockie island (south of Savu) far to the northward, and give a wide berth to the south coast of Sandalwood.

Either Lombok or Allas strait can be taken, and the reef south of Baars island passed several miles to SW.; thence the route lies through Sapoedie strait, between the island of the same name and Giliang. Vessels sometimes head straight for Urk island after leaving Allas or Lombok strait. The passage between Urk and Kangerang is better than the one to westward of the former island.

The passage across the Java sea to Carimata strait is quite easy. After leaving Sapoedie strait or Kangerang passage it is safest to pass south of Bawean, a bright lookout being kept for Hastings rock; thence the route runs 10 or 12 miles north of the Crimon-Java islands. The departure may be taken from Parang island—the most northwesterly of the group—for Carimata, and that island will soon after be raised. Run east of Discovery, Lavender, and Cirencester banks; and when Carimata peak is sighted N. 20° W., head for it, until able to take the position of the vessel by cross-bearings on Carimata and Soruetou islands. Thence bear to the westward, and pass between Soruetou and Ontario banks. Auxiliary steamers can head for Rhio strait from this point; but sailing-vessels should make for Singapore strait, passing east of Bintang, and looking out for Pratt reef, Frederick rock, Geldria bank, and all the dangers off Panjang. This island should be doubled to NE., when the course should be for Horsburgh light. The South channel, 3 or 4 miles to southward of the light, can be used.

It is hardly advisable to take the route through Torres strait from September to May. For though it is possible to run through Bligh passage at this season, the danger of losing your vessel among the reefs is great, and much time will be lost while at anchor during the squally weather frequent in that locality. Vessels with steam-power can take this route if they have enough coal on hand to reach Kupang, (Timor,) as it is doubtful if they can fill their bunkers on the NE. coast of Australia. During the NW. monsoon.

The best route for sailing-vessels leaving Sydney during the NW. monsoon consists in following at first the route for Bligh

* Vessels can coal at Kupang, (Timor.)

passage. Keep well to west of Bampton reef, pass east of Mellish shoal, and round San Christoval island (the most eastern of the Solomon group) to the eastward. Some vessels prefer passing west of Mellish shoal, coasting the southern shores of the Solomon group, and running north of New Ireland, where they sight St. John and the Green islands. Whether passing east or west of the Solomons, always keep clear of the New Guinea coast, particularly in November, December, and January. A ship should reach the neighborhood of 6° N. as soon as possible; here the prevalent NE. wind will probably enable her to head for the Pelew islands. After passing these islands to the southward enter the Celebes sea, by the Serangani islands; the Sulu sea, by Basilan strait; and the China sea, by Balabac strait.

For further information the reader should refer to the examples given in § 139. Information concerning the first part of the passage will also be found in § 173; and the voyage should end by following the return route from Singapore to Palawan passage, (vide § 154.)

§ 172. Southerly route from Australia to India, Batavia, and Singapore.—We do not advise this route, and the reader may perceive, by referring to the preceding paragraph, that Torres strait may always be taken from May to August, even if the port of departure be Melbourne.

From September to May the route passing north of New Guinea, though longer, is, we think, preferable to that by the south of Australia, especially for poorly-fitted-out vessels or slow sailers. Still, this route can be taken, from October to March, by vessels leaving Port Adelaide, Melbourne, or even Sydney. Fine passages are often made at this season, but the voyage is generally boisterous, and the weather bad. From cape Otway to cape Leeuwin both wind and current are frequently contrary; nor is it rare to strike a NW. or SW. gale and heavy sea in this part of the voyage. Vessels leaving Sydney can run through Bass strait, especially in January, February, and March. Of course, if vessels have steam to rely on in case of necessity, and are well equipped in other respects, they can take the southerly route.

We, however, would never advise this route for vessels leaving Sydney, or even Melbourne, during the southern winter; and captains bound to Port Adelaide should endeavor to make their arrangements so as not to leave that port for Batavia or Singapore during the winter months.

We will now quote the instructions of the *British Admiralty*, *Horsburgh*, and the *Dutch Observatory* on this southerly route:

"*British Admiralty route to the westward, south of Australia.*—Ships bound from Sydney to Europe or Hindostan may, from the 1st September to the 1st April, proceed by the southern route through Bass strait, or round Tasmania, easterly winds being found to prevail along the south coast of Australia at that season, particularly in January, February, and March, when ships have made good passages to the westward, by keeping to the northward of 40° S., and have passed round cape Leeuwin into the southeast tradewind, which is then found to extend farther south than during the winter months. In adopting this route advantage must be taken of every favorable change of the wind, in order to make westing; and it is advisable not to approach too near the land, on account of the southwest gales which are often experienced, even in the summer, and the contrary currents, which run strongest in with the land. The prevalence of strong westerly gales renders the southerly route very difficult; indeed, generally impracticable for sailing-vessels in the winter, although the passage has been performed at that season by ships in good condition that sailed well; but the northern route through Torres strait is preferred in the winter months."

Horsburgh's instructions: "Ships bound from Sydney to Europe, or Hindostan, may adopt the southern passage, through Bass strait, or round Tasmania, if they depart between the beginning of September and the end of March. In the months of January, February, and March, SE. winds frequently prevail about Tasmania, and near the coast of Australia, enabling ships to make considerable progress to the westward; they ought, however, to preserve a *considerable distance* from the south coast, in order to benefit by every change of wind in their favor, and to avoid being driven too near the land by southerly or SW. gales, which are likely to happen at times. The strong westerly gales which prevail here in winter render the southern passage difficult; yet it has sometimes been performed, even in that season, by ships which were in good condition and sailed well.

"Captain Middleton, however, is of opinion that the westerly winds are not so strong or so constant near the south coast of Australia, as they have been experienced in the winter months at a great distance from the land. While he lay in King George sound, a colonial brig arrived in June from Hobart-Town in nineteen days; in which month, also, an open whale-boat, employed sealing along the coast, arrived from the eastward; and, in July, a small vessel, about twenty or thirty tons burden, arrived in thirty-nine days from Launceston, which was thought to have touched at Kangaroo island, and thereby prolonged her passage. Captain Middleton sailed from King George sound August 12th in the ship James Pattison, rounded cape Leeuwin, and reached Swan river a week after his departure from the former place."

A recent publication by the Dutch Observatory contains the data from which the following tables have been compiled. The reader will perceive that each of the tables contains two lines for every month. The 1st line shows the crossings made by the shortest passage under canvas; the 2d line, the crossings by the longest route.

Table of routes from cape Otway to cape Leeuwin, showing the shortest and longest passage for each month.

Number of voyages examined.	Time of leaving the meridian of cape Otway.	Crossings and time taken to reach them.									
		135° E.		130° E.		125° E.		120° E.		115° E.	
		Lat. S.	Days.	Lat. S.	Days.	Lat. S.	Days.	Lat. S.	Days.	Lat. S.	Days.
		°		°		°		°		°	
5	January...	37.8	3.5	37.8	5.0	38.0	7.0	38.0	8.5	36.0	10.5
		37.5	7.0	36.8	9.0	35.8	13.8	36.0	16.0	36.2	20.0
7	February..	39.2	4.5	39.0	6.0	39.2	7.5	38.2	9.0	37.0	10.5
		38.2	8.0	37.2	12.5	37.5	15.8	39.0	20.5	36.0	23.5
8	March.....	38.8	1.5	37.0	3.5	36.2	4.8	36.2	6.0	35.0	9.0
		40.2	4.0	40.5	7.0	42.0	10.8	41.0	18.0	38.0	21.8
5	April......	36.8	2.8	36.8	5.0	36.2	7.5	36.2	9.5	35.0	12.5
		37.8	6.5	35.2	11.0	36.0	14.8	37.5	16.8	35.5	20.0
1	May.........	38.0	2.5	38.2	4.5	38.5	5.8	39.8	7.2	40.2	11.2
2	June	36.0	2.0	36.8	4.0	37.0	5.8	36.0	9.5	37.0	11.0
		39.2	4.0	39.8	6.0	39.0	7.8	39.8	11.0	39.5	12.5
3	July	36.2	2.5	36.0	4.5	36.0	6.0	37.0	9.0	35.8	11.0
		41.0	5.2	42.2	7.2	40.8	11.2	41.0	12.8	37.0	17.2
3	August....	39.8	4.5	37.8	6.8	40.5	8.5	41.0	11.2	36.2	15.5
		36.5	4.0	36.2	6.8	35.8	10.0	36.8	13.8	35.5	17.0
11	September.	37.2	3.8	37.5	5.8	37.2	6.8	37.2	8.0	37.5	9.8
		39.5	6.8	38.5	13.0	39.0	14.8	40.5	19.2	36.0	31.8
10	October ...	36.2	4.2	36.8	6.8	38.0	7.2	38.0	11.0	35.8	12.8
		44.5	9.2	46.2	13.0	45.5	22.2	42.8	26.8	43.5	31.8
6	November.	36.5	6.8	36.5	8.5	36.5	9.8	36.5	11.8	35.8	14.5
		39.8	3.0	37.5	6.5	37.8	8.8	37.2	17.2	36.5	19.5
6	December .	38.2	2.8	37.0	6.5	36.5	8.8	36.8	10.5	35.5	14.0
		36.5	5.8	37.8	9.5	40.0	16.2	38.2	18.8	36.5	21.8
67	Means ..	38.3	4.4	37.5	7.1	37.5	9.8	37.7	12.9	36.4	17.0

Table of routes from cape Leeuwin to Java, (strait of Sunda,) showing the shortest and longest passage for each month.

Number of voyages examined.	Time of leaving the meridian of cape Otway.	Crossings and time taken to reach them from cape Otway.												Days at sea from cape Otway to Java.
		Lat. 35° S.		Lat. 30° S.		Lat. 25° S.		Lat. 20° S.		Lat. 15° S.		Lat. 10° S.		
		Long. E.	Days.	Long. E.	Days.	Long. E.	Days.	Long. E.	Days.	Long. E.	Days.	Long. E.	Days.	
		°		°		°		°		°		°		
5	January	114.0	13.2	108.6	15.5	106.8	17.2	104.3	19.2	104.0	21.0	102.5	23.5	27.2
		113.0	13.0	109.2	15.0	107.5	17.0	105.1	19.0	103.0	21.2	101.8	23.5	36.0
7	February	110.5	12.0	108.2	14.2	106.8	15.8	106.0	17.2	104.8	20.0	104.2	23.0	25.8
		113.5	24.5	110.5	26.8	107.0	20.8	105.0	33.0	104.8	35.2	104.5	37.8	42.2
8	March	114.0	18.0	110.0	21.0	108.8	23.0	107.8	25.0	106.0	26.8	105.5	20.0	30.2
		112.5	23.8	107.5	25.8	104.5	27.2	102.8	28.8	103.0	32.2	104.8	36.0	37.5
5	April	113.5	15.0	109.2	19.0	108.2	23.0	111.8	26.2	107.5	31.0	106.0	32.5	33.2
		113.0	20.5	110.0	24.5	111.5	27.5	109.5	29.5	106.5	31.5	105.5	33.0	34.5
1	May	112.2	17.8	108.5	23.8	108.5	27.2	107.2	29.8	106.0	31.5	105.2	33.5	34.5
2	June	113.0	12.8	109.5	15.2	110.5	17.5	109.8	19.8	107.2	21.0	106.0	23.2	24.5
3	July	113.2	11.5	110.0	14.8	109.0	18.2	107.2	20.0	106.0	21.2	105.5	23.0	24.5
		112.5	16.8	108.2	20.0	108.0	23.8	107.5	25.5	106.8	27.0	106.0	28.5	29.8
3	August	115.0	16.8	111.8	21.8	110.0	23.8	108.8	25.5	107.8	26.8	104.5	28.5	29.5
		113.2	19.5	112.2	21.5	110.0	23.5	108.0	25.8	106.8	27.5	105.5	29.2	30.2
11	September	113.2	12.2	111.0	15.5	109.0	18.2	107.8	20.0	106.5	22.5	105.2	25.2	26.5
		112.8	32.8	109.5	35.8	107.8	37.5	106.6	39.5	105.2	41.2	105.5	43.2	44.5
10	October	114.8	13.0	112.0	15.0	109.5	16.5	108.2	18.2	106.2	20.5	105.5	22.5	24.8
		109.8	40.5	107.0	42.5	105.5	44.0	103.5	46.2	104.0	48.5	104.2	50.2	52.8
6	November	113.0	17.8	106.5	20.5	101.0	22.8	98.5	24.8	99.0	26.5	101.0	28.8	30.8
		117.3	24.0	112.2	25.5	110.5	27.2	107.8	31.2	105.0	33.8	105.0	35.5	40.0
6	December	114.2	14.5	110.2	16.2	110.5	18.0	105.0	19.8	104.0	21.2	103.8	22.8	26.0
		112.8	14.0	107.8	16.0	105.5	18.2	102.2	19.2	101.5	21.5	101.0	25.2	40.0
67	Means	113.1	18.3	109.7	20.7	107.6	22.3	105.5	25.0	104.0	27.0	103.5	29.3	31.9

1st. During the SE. and SW. monsoons.

§ 173. ROUTE FROM AUSTRALIA TO COCHIN-CHINA, CHINA, AND JAPAN.—Vessels bound to Saigon, and even those bound to Hong-Kong, from *May* to *August*, should follow the route indicated in § 171 and pass through Torres strait. They should enter the China sea by Carimata strait and finish the voyage as stated in §§ 153 and 154.

Vessels bound to Hong-Kong can also take the route north of New Guinea. After passing the Serangani islands and Basilan strait, they should enter the China sea south of Mindoro. If it be decided to follow this route, the course from Sydney should be ENE. so as to pass to southward and eastward of Ball pyramid. The northing should be made between 159° and 161° E. A good watch should be here kept for the reefs marked on the charts, as well as for those which are supposed to exist between 23° 30′ and 18° S. After reaching 14° or 13° S., the course should be about NW. for St. George's channel; this will also carry the ship clear of Pocklington bank and Laughlan islands, as well as the west coast of Bougainville. Vessels should be careful not to let the current set them to the westward, as they may then be unable to make St. George sound. If this does happen, Dampier strait, between New Britain and Rook island, can be taken. This strait is little frequented and can only be crossed by watching the reefs carefully from the mast-head.

After clearing St. George's channel the course should be W. or WNW. to pass to northward and eastward of the Admiralty isles, and thence between the Hermit and Anchorite islands, or a little to northward of the latter. Thence the route keeps south of the line as far as the Providence islands, which should be doubled to the northward; thence passes east of the Asia islands, and half way between Morty island to the west and Lord North island to the east. During this last part of the route it is probable that the currents will first set to NE., then to E., and finally to W.

After leaving Lord North island the course is for Meangis and Sevangani islands and Basilan strait. A bright lookout should be kept for Iphigenia reef. The passage ends as described in § 150 in the description of the route to China (during March and April) for vessels which have taken Macassar strait, and afterward passed through Basilan strait and along the west coast of the Philippines.

Vessels bound to Shanghae and Yokohama can follow the same route until they are north of New Guinea. They should cross the equator near 142° E., and pass east of the Pelew islands, when they will begin to find the variable NE., NW., and SW. winds, generally known as the *SW. monsoon.* If bound to *Shanghae*, they should run south of the Loo-choo islands, and thence easily finish the voyage with the frequent SW. winds. Typhoons are frequent in this locality from June to November, (vide § 20.) If bound to Yokohama they should, after leaving the Pelews, head for the passage between the Borodino and Loo-choo islands; thence both the winds and Kuro-Siwo current will be favorable, (vide § 37.)

* During this season (from September to March) you will have a choice between two routes called the "Easterly Route." *The first* passes west of New Caledonia and the Santa Cruz islands, and east of the Solomon group. This is the shortest, but has the inconvenience of running through localities where reefs are numerous and not all located. The *second route* runs east of New Caledonia between the Fijis and the New Hebrides. It is longer, but safer and more clear of dangers. It also passes through a zone of steadier winds, and is probably the best for vessels leaving Melbourne or coming from Europe by the south of Australia. Passages by this route are quite as rapid as by the other. 2d. During the NW. and NE. monsoons.

The course out of Sydney should be ENE. and to the southward and eastward of Ball Pyramid. Route passing west of New Caledonia.

The northing should be made between 159° and 161° E., between which meridians a careful lookout should be kept for reefs. Keep well clear of the reefs lying to NW. of New Caledonia, especially as the wind is liable to be from SSW. to NNW. in that locality. Afterward head NNE. as far as 164° E., when follow this meridian to the north and thus pass between San Christoval and Santa Cruz islands.

Vessels to eastward of the Solomon islands, and the islands and reefs situated to the north, at *the beginning of the NE. monsoon*, should steer N. by W. or NNW. if possible and cross the Carolines between 155° and 149° E. A

* The instructions given in this second part are mainly extracts from *Horsburgh.*

bright lookout should here be kept for new reefs, as well as for those on the charts, as the positions of many of them are not accurately determined.

Once north of the Carolines, the course should be laid for the south point of Guam island, (the most southern of the Marianas,) which can be passed on either side. A vessel north of the island can take either the channel south of Tinian, or that north of Saipan. Enter the China seas by the Bashees and finish the voyage as indicated in § 150, under the description of the route from Hong-Kong to Pitt passage.

If the Solomon islands are doubled after January, when the SE. monsoon is no longer settled, the course should be NW., so as to run between Eap (or Yap) island *and the* Ngoli (or Matelotas) islands. Vessels can also go between the Goulou and Pelew groups;* thence round the NE. extremity of Luzon at a convenient distance and take one of the passages between Luzon and Formosa. Finish the voyage as stated in § 150, (vide route by Pitt passage.)

Route passing east of New Caledonia. This route passes east of Norfolk island and close to Mathew rock; the latter is visible for 25 miles. After running to eastward of Mathew rock the course should be N. by E. or N., a good lookout being kept and due allowance made for the westerly current. Keep between the 171st and 172d meridians without attempting to sight Erronan island, and thus double the New Hebrides. Vessels can pass between Erronan and Tamra if drifted to westward. It is, however, generally best to pass well east of all the islands; if the chronometers be faulty, Fataka (or Mitre) can be sighted at a distance of 22 miles. Thence the course is N. or NNW. for the line between 160° and 168° E. If the westerly current, which is strong in this locality, will allow, the attempt should be made to run through the Caroline group between 162° 50′ and 162° 20′. But if the equator be reached between 160° and 162° E., it is best to cross the Carolines between the meridians of 156° and 155°, as there are fewer reefs and islands in that locality.

Once north of the Carolines, the course should be about west in order to pass south of Guam, or through one of the passages situated in the northern part of the Mariana group, the one south of Tinian, or that north of Saipan,

*According to the N. Pacific Directory the "Goulou" island is another name for the Ngoli islands.—*Translator.*

for instance. Thence the vessel should be headed for one of the channels between Luzon and Formosa. For information concerning the end of the voyage vide § 150, (route by Pitt passage.)

When bound to Shanghae or Yokohama.

Vessels bound to Shanghae or Yokohama can follow either of the above-mentioned routes; the latter is, however, the preferable. Passing east of the Solomon islands they should cross the equator in the neighborhood of 166° or 168° E.; and, after striking the NE. trades, run on either side of Ualan and Providence islands, and thence north of the Marianas.

Vessels bound to Yokohama should keep west of the Volcano islands, and reach 30° N. to westward of their port. Between 28° and 31° N. the current will be found favorable, and the wind variable or westerly. Ships bound to Shanghae should run north of the Marianas, or between Grigan and Assumption; thence north of the Borodino and Loochoo islands, whence the voyage will be easy.

Observations of Captain Wm. Hall, (Ann. Hydr., 1870:) "Left Newcastle (New South Wales) on the 6th August, with a moderate SE. wind; SW. wind on the 10th; sighted Middleton atoll; it seems to be placed correctly on the charts; the sea breaks all around the reef, the water inside being smooth. From the 11th to the 17th violent rain-squalls and wind from WSW. to WNW. As we were too far to the eastward, did not sight Hunter or Geru island. As we rounded it the wind gradually hauled to SE., and we sighted Mitre island on the 23d. When first raised it looks like two islands; off the north coast there is a rock which resembles a ship under canvas. At noon we sighted Anouda or Cherry island; its position 11° 36′ S., and 169° 43′ 15″ E.

"Experienced westerly currents until we sighted Pleasant island at daylight on the 29th. Between the equator and 8° N. the wind was variable, and the current easterly, with a speed of from 25 to 30 miles per day. Though we headed for Ovalou, or Armstrong island, (?) we were drifted in sight of Baring island. Here a light SE. breeze came to our assistance; also a feeble westerly current, which we kept till we reached 14° N. at 155° E., the wind shifting from SE. to SW., with a heavy SW. sea running.

"We passed in sight of Providence, or Arecifos islands,

with the intention of making Alamagan or Grigan, and running thence north of the Loo-choos, but did not strike the trades until we had passed the Marianas, and even then they only blew feebly for 2 or 3 days. The wind then came out from the N., and obliged us to run to leeward of the Loo-choo group, and beat up for 3 or 4 days, notwithstanding the fact that we had the Kuro-Siwo in our favor. The wind then shifted to E., and two days afterward we sighted the Saddle islands after a passage of 62 days from Newcastle.

"We started in company with several other vessels; two of these entered the China sea through Van Diemen strait, and made the passage in 51 days. The worst sailer of all reached the light-ship on the same day as we did; although we were 14 days ahead of her when we passed Providence island. As she struck fresh winds beyond this point she made up all the time lost. This is just the contrary to what happened to Captain Brown, making the same voyage in October and November, 1865. It is probable, if we had struck the trades a little later, and not run quite so far north just before reaching the Loo-choos, that we would have made the shortest passage."

§ 174. ROUTE FROM PORT ADELAIDE, OR MELBOURNE, TO SYDNEY.—We will only give a *résumé* of this route, as information concerning it will be found in several other paragraphs. After leaving Melbourne, or Port Adelaide, the course is through Bass strait, unless there are indications of an east wind. January, February, and March are the months when a wind from this direction is to be feared. If Bass strait is taken, Horsburgh's instructions, given in § 104, should be followed. A NE. by E. course from the Kent group clears cape Howe by about 60 miles; it is, however, advisable to head much farther to the eastward, if the wind is strong from the south. By disregarding this precaution—as Horsburgh and the British Admiralty both remark—a vessel may bring up on a lee shore in the center of the bight, which is 150 miles long and lies between Wilson promontory and cape Howe, on the Australian coast. There is a fixed light on Gabo island, (south of cape Howe,) visible 22 miles. Cape Howe is a low, sandy, and rocky point, making out from the foot of the mountains.

Vessels passing west of Tasmania should keep well clear of the coast until they are to eastward of the island, when they should run up along the land for cape Howe, as stated in § 104, in the extract from the "Australian Directory."

The latitude is the surest guide in approaching the eastern coast of Australia. The soundings extend for 12 or 15 miles from the land.

CHAPTER XI.

ROUTES FROM CHINA AND ASIA TO AUSTRALIA.

1st. The southerly route.

§ 175. ROUTE FROM SINGAPORE TO AUSTRALIA.—*This route may be taken at all seasons.* The following observations, on the voyage from Singapore to Australia and New Caledonia, are due to *Captain Hunter*, (*Ann. Hydr.*, vol. 16:)

"The most frequented route to the SE. coast of Australia is that which passes south of that continent. The mean passage to Melbourne and Sydney is from 9 to 10 weeks. Voyages, under canvas, have been as quick as 7 weeks to Sydney, and a few days shorter to Melbourne. The voyage to Port Curtis and New Caledonia is 10 or 15 days longer. Only well-equipped and staunch vessels should attempt to make this route.

"From Singapore to Banka and Billiton the wind is N. and NW., between *November* and *March;* there is, however, a slight chance of striking a W. or WSW. wind. If the wind be fresh and steady, run for Bali strait; otherwise, time may be lost in attempting to beat against the SW. wind in the strait of Sunda. Latterly, Bali strait is used by the Dutch vessels going from Batavia to Europe, as they prefer to run to the eastward, along the southern coast of the island rather than to beat through the strait of Sunda. Bali strait is clean, and does not need a pilot; it is easy of exit from November to March, even when the wind is from SW., as it is open to SE.* Good water and provisions can be obtained at Banjoewangie. Vessels usually find calms and light, baffling airs, or southerly squalls, in the Indian ocean, between the monsoon region and the steady SE. trade-belt; sometimes the transition period only lasts a few hours. The zone of SE. trades lies between 10° and 32° S. The trades are occasionally found as far south as the 40th parallel. Ships bound to Port Curtis, or New Caledonia, should run south of Tasmania, and thus get clear of the

*Allas strait is perhaps better, (vide § 115.)

east winds, which, during the summer, are quite frequent in Bass strait. In the high southern latitudes the wind is cold and variable, but a well-clothed crew will not suffer much. After doubling Tasmania the course is straight for New Caledonia. Vessels bound to Port Curtis should not go within 300 miles of the east coast of Australia until they reach the parallel of Sandy cape, (24° 40′ S.,) as NE. winds are very frequent near the coast. If they need stores, they can stop at Hobart Town.

"The southerly route is certainly the quickest from May to September inclusive. The winds will be ahead and from SE. as far as the strait of Sunda; it is, however, easy to clear the strait and enter the Indian ocean, when the SE. trades will be found strong and settled as far as 28° S. Below this parallel are the 'brave west winds' to carry you toward your destination. These winds weather cape Howe and blow from S. and SW. along the eastern coast of Australia as far as the tropic, thus giving a fair wind to port Curtis or New Caledonia at this season. The faults of this route are: the SE. winds from Singapore to the strait of Sunda, and the cold winter weather in the high southern latitudes; a heavy cross-sea is also here frequent, caused by the sudden shifts of wind from NW. to SW."

The reader should refer to § 104, where will be found instructions on the advantages and disadvantages of Bass strait; also to Horsburgh's instructions, (vide § 104.) We would especially caution the reader concerning the danger of running too close to the west coast of Tasmania, (vide §§ 104 and 107.)

We are again indebted to the *Dutch Observatory* for the following table, which contains the crossings, etc., of 9 routes, from the strait of Sunda or Lombok strait to cape Otway:

Routes from Java to Australia.

Date of departure from the strait of Sunda and Lombok strait.	Crossings and days at sea to reach them.																						Days at sea to cape Otway.
	Lat. 10° S.		Lat. 15° S.		Lat. 20° S.		Lat. 25° S.		Lat. 30° S.		Lat. 35° S.		Long. 115° E.		Long. 120° E.		Long. 125° E.		Long. 130° E.		Long. 135° E.		
	Long. E.	Days.	Long. E.	Days.	Long. E.	Days.	Long. E.	Days.	Long. E.	Days.	Long. E.	Days.	Lat. S.	Days.	Lat. S.	Days.	Lat. S.	Days.	Lat. S.	Days.	Lat. S.	Days.	
	°		°		°		°		°		°		°		°		°		°		°		
March 4	103.2	2.5	98.8	4.7	95.0	6.8	93.2	8.7	92.0	11.0	95.8	14.0	36.8	20.5	37.5	22.5	39.0	23.8	38.7	25.2	39.0	26.2	32.0
May 12	103.5	3.2	99.8	5.5	95.0	6.5	92.5	9.2	96.2	15.5	114.0	28.0	35.2	25.2	36.5	24.0	28.2	26.2	38.5	27.2	38.5	29.0	33.5
May 17	103.5	3.2	99.8	5.5	95.0	6.5	103.5	14.2	108.0	17.2	112.5	24.0	36.2	25.0	37.2	26.2	39.5	28.8	41.0	33.0	40.2	35.0	39.0
May 18	103.5	2.5	101.7	4.5	100.5	9.2	98.0	12.0	100.5	15.0	101.2	18.5	41.8	29.0	39.8	30.0	40.2	31.8	39.8	33.2	39.8	34.5	39.0
July 22	115.0	1.0	110.5	2.8	107.5	5.8	101.5	9.8	99.8	11.8	99.8	16.5	40.2	22.0	40.8	24.0	40.5	25.2	39.8	26.8	39.2	29.0	31.8
August 18	103.2	1.8	99.8	3.8	98.8	6.0	99.8	10.0	96.0	12.8	96.8	17.8	38.5	25.2	39.5	26.8	40.0	28.0	40.0	29.2	40.0	30.5	33.5
September 9	114.5	0.8	108.0	5.0	104.8	8.2	102.5	14.0	101.5	20.8	105.5	24.8	37.5	28.2	38.5	31.2	39.0	33.2	39.8	34.8	40.8	38.0	42.0
October 26	101.8	2.8	96.8	5.2	97.5	8.0	97.2	11.0	104.5	21.0	111.2	24.0	37.0	25.2	38.2	28.0	39.5	29.5	39.5	31.0	40.5	34.5	38.5
November 5	104.8	5.8	104.0	11.2	99.2	14.8	96.8	17.5	98.0	20.0	109.5	26.5	37.0	28.5	38.0	30.5	39.0	31.8	39.2	33.0	39.2	34.2	36.5

This route should only be undertaken from May to September. Captain Hunter remarks that the track from Singapore to Australia, by way of the North Pacific, has never been tried, nor is it practicable for vessels bound to ports south of Moreton bay on account of the frequency of the southerly winds between the east coast of Australia and New Zealand. Though the delay caused by these winds will be great, the voyage may be accomplished by passenger vessels wishing to avoid the cold weather of high southern latitudes. 2d. The northerly route.

Vessels leaving Singapore, Saigon, or Hong-Kong during the season of the SW. monsoon, can pass through Formosa channel, if about to follow the northerly route; or, what is perhaps better, they can enter the Pacific by the Bashee island channel, south of Formosa, as they will then have the Japan current in their favor. According to Horsburgh, it is well to run for 167° or 172° E. before bearing south of 30° N.; a good weatherly sailer, however, may enter the trades even 10° more to the westward. It would also seem that the easting might be made between 25° and 30° N., with the SW. monsoon of the China sea. Captain Hunter states that both water and stores may be obtained at Peel island, one of the Bonin group.

The harbor on the western coast of the island is much frequented by whalers; vegetables and turtles being there found. Vessels can also put in at the Kingsmill islands or Pleasant island, (0° 35′ S. and 167° 10′ E.;) they are both abundantly provided with fresh provisions.

Further information concerning this route will be found in § 177.

§ 176. THE EASTERLY ROUTES FROM SINGAPORE OR BATAVIA TO AUSTRALIA, NEW CALEDONIA, AND NEW ZEALAND, (*when starting from the 15th November to the 15th February.*)—Captain Hunter has given interesting and detailed instructions concerning these easterly routes, between Singapore and Australia (*vide Ann. Hydr., vol.* 16.) We give his statements below in as condensed a form as possible.

Before deciding on a route the reader should, however, first refer to the preceding paragraph, where will be found instructions on the southerly route.

Though the *easterly route* is the most direct from Singapore to Sydney, port Curtis, and New Caledonia, the wind is not always favorable during the whole voyage. Still vessels leaving Singapore from the middle of November to the middle of February may count on a fair wind.

There are two easterly routes: 1st, that passing north of New Guinea, which is best fitted for sailing-vessels of over 150 tons; 2d, the one through Torres strait, which is best for all vessels with steam-power and sailers of less than 150 tons, especially if they are packets and have to stop for provisions or water.

1st. The route for sailing-vessels to the northward of New Guinea.

*The first part of this route was given in § 114. From Singapore the course is for Gilolo or Pitt strait. Provisions may be obtained at Bonthein, (southern extremity of Celebes,) or at Gebi, (Gilolo passage.) The anchorage at Bonthein is large, and safe during the westerly monsoon. Excellent water, wood, vegetables, etc., may be here obtained at moderate prices. At Gebi the anchorage is equally good; though fresh water is abundant, the provisions are inferior. Vessels bound through Dampier or Pitt strait lengthen their passage somewhat by stopping at Gebi. It is never advisable to anchor at port Dorei, (?) as the harbor is small and the natives are dangerous.

This passage is used by whalers; if set to leeward by the easterly winds, they are in the habit of running down toward the line, in which locality westerly winds are common from November to March.

Dampier ran through the strait, which bears his name, in 1699, and, coasting the northern shore of New Guinea, rounded New Britain. He found the westerly monsoon quite strong, sometimes even amounting to a gale.

Captain Hunter quotes from the logs of the Meandre, Captain Keppel; the Rattlesnake; and a vessel under command of Forrest Thomas, to prove the possibility of making this passage to the eastward at any season of the year. He then gives his own experience, as follows: In October, 1835, he was off the Asia islands, (situated north of Waygiou island,) and aiming to make easting with a light, variable wind and strong westerly current against him; headed N., and on the 19th was in lat. 2° 06′ N.,

---

* The author again repeats that the route south of Australia is the best, for sailing-vessels. The reader is also referred to the Appendix at the end of § 177.

long. 134° 11′ E. He here lost the contrary currents, and steered east, keeping between 2° 15′ and 2° 34′ N. On the 27th the ship was in long. 146° 59′ E. Steering SE., St. Matthias island was sighted. His orders being to cruise in this locality, he ran into St. George's channel and with a favorable current stood for the Solomon group. After cruising among the islands until the 19th December, he carried a westerly wind as far as 169° 36′ E. After passing south of Banks island on the 26th December, he was becalmed for two or three days, then with variable ESE. by-winds he made to the southward, and anchored in the bay of Islands on the 15th January. Here is an example of a passage being made by an average vessel, unprovided with studding-sails, in the most unfavorable season, and over a route generally considered impracticable. It is therefore possible to make this easterly voyage by keeping between the trades and the monsoons and north of the equatorial current.

On another occasion, being at 1° S. and 149° 20′ E., on the 23d June, and after having been drifted near New Hanover and St. Matthias islands, by a westerly current, running at a rate of from 2.5 to 3 knots per hour, Captain Hunter stood north and found a west wind near the equator. He then made to the eastward and reached 155° E. at 0° 45′ S. on the 27th. He sighted point Bourka (Bougainville) soon afterward, and found the westerly current as strong as at Matthias island. At this season the westerly current does not extend quite to the equator.

In September, 1840, the same captain could not hold his own against the westerly current, near the Admiralty isles, and had to run to 2° N. before losing the westerly set. He then went about to the eastward and made his southing under the eastern coast of the Monteverde islands, (154° 05′ E., 4° 45′ S.) Captain Hunter also made the passage from Morty to Bouka (Bougainville) in the month of August. He could, of course, have made the remainder of the voyage, but his orders took him no further than the Solomon group. The line of the westerly current rarely runs north of the parallel of 2° N. Finally, he quotes five vessels which made this easterly passage between January and April; one of them following the equator till east of the Kingsmill group, another keeping near the Solomon islands and New Zealand, and the remaining three running near New Ireland and the adjacent islands. From all these examples, he concludes that a vessel reaching Gilolo passage or Dampier strait, between the 15th December and the 15th March, can make a rapid passage to any port situated in east longitude. Moreover, he thinks the passage practicable at all seasons, provided the ship is kept north of the equatorial current and between the monsoons.

The following directions for the season of the west monsoon are also given by the same authority:

Northerly and northwesterly winds are common at this time near the cape of Good Hope, (New Guinea.) To eastward of this point the west wind is generally strong and steady, the current setting to the east, with a speed of from 2 to 2.5 knots, between the coast of New Guinea and 1° N. This wind and current will hold while passing the St. David isles, and running north of the Providence islands, (0° 20′ S.) Hence there is a choice between several passages. The strait to eastward of Dampier (between Rock island and New Britain) is perhaps the most direct for Sydney, but can hardly be advised until a more complete survey is made of the locality. St. George's channel can be taken. The best way to make for this passage is to keep along the equator as far as the longitude of the Admiralty isles, and then edge away to the southward and eastward, passing to westward of Matthias island; the reefs and low islands farther to the southward will thus be avoided. Sail may be carried boldly while cruising north of the Admiralty group, as there are no dangers near the line in that locality. The best route, however, especially when bound to any of the islands to the eastward, is that running north of the Solomon.

The following observations were published in the Indian Archipelago Journal, 1851:

2d. Route for steamers and auxiliary steam-vessels by Torres strait.

"The only vessels which have used Torres strait, up to the present time, are those making passage to the westward. The three or four ships which attempted the voyage from west to east experienced great difficulty, from the irregularity of the westerly monsoon, (from November to March.) One vessel, however, bound from Sydney to Port Essington, in April, via the Middle passage, met a NW. wind at 19° S. She made six knots per hour for five days, and afterward held the same wind to the Arru islands and Macassar. The fine season in Torres strait, as far as cape York, is from November to March; this is also the season of good weather in the Molucca islands, (vide § 13.) The winds are never strong enough in Torres strait to interfere with a steamer making passage in any direction."

Captain Hunter says: "The Torres strait route is the shortest, the distance from Singapore to Port Curtis being about 3,400 miles. It is advised for small vessels, as water may be obtained frequently.

"Vessels bound from Singapore to Torres strait follow the same route as those bound north of New Guinea until they reach Salayer strait; here the tracks separate; the former, bearing to the SE., keeps along the north coast of Timor. Provisions may be obtained at Manatuti,* (126° 55′ E.) This roadstead is sheltered from westerly winds, but it is dangerous to run in close when the wind is to the northward of west. Kisa bay (about 127° 05′) is a better anchorage, as small vessels can run well in with the land, the coast being bold and the harbor safe at this season, (November to March.) Vessels can anchor with equal safety on the south coast of Moa, (127° 55′ E.) Water and provisions are obtainable at both anchorages. The best roadstead at Moa is off the four villages, four miles from the SW. point of the island.

"Endeavor strait is the best, when coming from the westward. Vessels carry 4.5 fathoms, at low water, over the shoalest parts.†

"The western entrance to Endeavor strait is easily recognized. The depths across the head of the gulf of Carpentaria are about 36 fathoms, while about 120 miles from the entrance the soundings diminish to 30, 19, and 9 fathoms. Captains should be careful to keep south of the parallel of Booby island, to avoid the banks to WNW. In clear weather Prince of Wales island will be sighted at a distance of 30 miles, and before Booby or Wallis island, as the latter can only be seen for fifteen miles. There are several channels between the sand banks of Endeavor strait. The largest and best lies north of Red Wallis island. Steer

* Probably Mantotte.—*Translator.*

† As will be seen hereafter, the author prefers the Prince of Wales channel, when running for Bligh passage.

for Red Wallis, when it bears E. 20° S., and Booby island N. by E. (by compass) distant 10 miles. The strait can then be safely entered between the points extending from cape Cornwall and Wallis islands; and the two 3-fathom ledges will also thus be cleared. The depths in the passage vary from 4 to 8 fathoms, and, with the exception of Heroine and Eagle rocks, the strait is clear of dangers.

"MacKenzie's three voyages in the Heroine, in 1844–'5–'6, together with a dozen others on record, do not speak very favorably for this route; for after clearing the Barrier, the vessels were detained in the open sea by variable winds.

"In 1847–'8 a vessel taking the inner route found that the easterly winds, frequent outside the Barrier, rarely blew in with the land; NW. winds were often found, and lasted for several successive days. Generally the monsoon began as a land-breeze toward midnight, and, blowing all day, died away calm again in the evening. The inner route is clear of coral reefs, and the weather being fine at this season, the landmarks can generally be made out. Light squalls are, however, common, especially at night.

"On dark nights vessels should always anchor until they are 500 miles distant from cape York.

"Twelve days is a good passage from cape York to cape Curtis."

To complete the instructions already given in § 115, and which refer to the route via Endeavor strait, we give below Horsburgh's observations:

"Coming from the westward, for Endeavor strait, a vessel standing along the parallel of 10° 50′ S. will first sight at 20 or 25 miles' distance the high lands of Prince of Wales islands, extending from NE. to ENE. At 11 or 12 miles' distance the northern Wallis island will be raised from the mast-head, bearing S. 75° E.; also Booby island, bearing N. 5° E. North Wallis island looks at first like two islands, about a ship's length apart, the southern appearing the larger of the two. South Wallis island is low, flat, and wooded, the largest trees being on its northern extremity. North and South Wallis islands are separated by a dangerous channel a mile and a half wide; the channel south of the Wallis islands, between them and the mainland, should not be used, as it is full of shoals."

Steamers follow the inner route from Endeavor strait to

Sydney. *Auxiliary steamers* and *sailing-vessels* should head for the Bligh passage.

Once clear of Entrance island, they should head NE. for 30 miles; this will bring them near Harvey rock; they should then pass between the North Sister and Long island. But instead of Endeavor strait, the shortest and simplest route for auxiliary steamers and sailing-vessels is through Prince of Wales channel; this track lies north of Booby island. All experience goes to prove that the Bligh passage is the best; this opinion being substantiated by the following remarks by Captain Blackwood:

"Leaving Sourabaya on the 12th January, I took three weeks to reach Endeavor strait. Squalls and light westerly airs caused some delay. A good steamer ought to accomplish the 2,000 miles, to cape York, in ten days.

"From February to the end of March, when the monsoon changed to the SE., ending in a heavy NW. squall, the weather was never too bad to interfere seriously with our explorations; and if this one was a fair sample of the usual westerly monsoon, I regard this season as much better than that of the SE. monsoon.

"I think that Captain King's *inner route* should always be followed by a steamer, as the short delay occasioned by anchoring, during the first 5 or 6 nights, will be amply compensated for by the speed, during the day, in the smooth water of this sheltered route. Even with due allowance for the winter storms, common near Sydney, the 2,000 miles' distance, from cape York to Sydney, ought to be run in 15 days.

"I would recommend the Bligh passage for sailing-vessels, as the track is clear of coral reefs, and a ship can anchor anywhere." *

From Singapore to New Caledonia and New Zealand.

Instructions relative to the greater part of this route will be found on the preceding pages; we will, however, complete them with a few considerations bearing especially on the latter part of the voyage. Captain Hunter remarks as follows:

"When the route north of New Guinea has been chosen, the easting should be made along the line as far as the 152d

* Detailed instructions for the Bligh passage will be found in § 171. One of the advantages there enumerated is the fact that vessels bound to the westward have the sun behind them; of course this is an objection when going in an opposite direction.

meridian—if the wind allow. Once arrived at this longitude, and if bound for New Zealand, the attempt should be made to cross 10° S. near 171° or 172° E.; thence the course should be to the southward, close along the west coast of the Fijis, as the east winds draw to the southward in January, February, and March. Once beyond the reefs off the SE. extremity of New Caledonia, the track is clear. Vessels *bound to New Caledonia* should also do their best to work well east near the equator, as far as 152° E. It is especially necessary to cross 10° S. well to eastward, as the trades during the southern summer are from E. to ESE. as far as 20° S. In this way a ship can pass to windward of the New Hebrides, which, with New Caledonia, seem to form a barrier between the trades of the South Pacific and the variable winds of the Coral sea.

"Vessels taking the Torres-strait route can make the strait by Prince of Wales channel or Bligh passage. Thence the route lies along the south coasts of New Guinea and Louisiade; where the wind will be found from W. and NW., and quite steady during the months of December, January, and February, if not later in the year."

Auxiliary steamers can also take the *inner route* along the east coast of Australia. As an example of this latter route we quote the following:

Passage of the Guichen, (*2d rate*,) *Captain Perrier.*—"Left Batavia on the 16th December, 1869. Arrived at Sourabaya on the 18th and filled up with coal. Left on the 25th; steamed through Madura and Bali straits; banked fires; after passing south of Timor, reached Torres strait, wind very light and from SW. to NW. Ran through Torres strait on the 10th January. Took the inner route, anchoring sometimes at night. Although under the lee of the Great Barrier reef, we had to put into Cleveland bay and wait for a favorable slant to the strong SE. winds. Anchored off Townsville, a place of 500 or 600 inhabitants. Obtained 12 tons of coal and a little wood. Left on the 19th. A violent gale from ESE. on the 21st compelled us to anchor under the lee of Percy islands. Remained there 2 days; crew cutting wood on the islands. Got under way on the 24th. Anchored in Fitz-Roy river on the 25th, 17 miles from Rockampton; vessel drew too much water to go any farther up the river. Obtained all necessary supplies. Sailed on the 30th, ran out by Curtis channel, and headed for Noumea, where we arrived on the 4th February, after a 50 days' passage from Batavia."

§ 177. ROUTE FROM CHINA AND JAPAN TO AUSTRALIA.— The NE. monsoon prevails from *October to April*. Sailing-vessels from Saigon generally run south of Australia, touching at Pulo-Aor, (vide § 170.) They reach Carimata passage as stated in §§ 114 and 115, and follow the southerly route through the Java sea as far as Allas strait. Thence they shape their course for the south of Australia, according to the instructions given in § 175 and the latter part of § 104. 1st. During the NE. monsoon.

Auxiliary steamers, bound from Saigon to Australia, can take the Torres-strait route from the middle of November to the middle of February. The passage from Saigon to Pulo-Aor should be made as stated in § 170, and thence to Torres strait, as described in § 115. For information concerning the termination of the voyage by the inner route, vide § 176. It may be useful to remark, that it is not probable there will be any opportunity of coaling between Kupang and Sydney. From Saigon.

Auxiliary steamers can also run from Pulo-Aor to the Moluccas, (vide § 114,) and then, after passing north of New Guinea, finish the voyage as given for sailing-vessels in § 176.

It is hardly advisable for ordinary-sized auxiliary steamers to take the route through Torres strait except from the 15th of November to the 15th of February, as they will at any other season be compelled to use a great deal of coal. They should, unless absolutely impossible, take the route south of Australia at the beginning and end of the monsoon.

Sailing-vessels leaving Hong-Kong should run down the China sea by the outer route described in § 151. From Pulo-Sapata they should pass east of the Anambas, and then make for Carimata passage. They can also steer straight from Pulo-Aor to Carimata passage, but this will make the voyage a little longer. Thence the route runs through the southern part of the Java sea to Allas strait, (vide § 115.) For information concerning the route south of Australia, vide § 175 and the end of § 104. From Hong-Kong.

Auxiliary steamers leaving Hong-Kong should always follow the same track as sailing-vessels, and run south of Australia. If need be, however, they can pass north of

New Guinea, or through Torres strait, from November to March.

If it be decided to take the route north of New Guinea, it will first be necessary to run through Mindoro and Basilan straits; and, after crossing the Celebes sea, to reach Molucca passage as stated in § 152 under the head of the first easterly route. This route has already been advised in § 152, for the return trip from China against the SW. monsoon; it is also easy to follow it during the NE. monsoon. The Moluccas once reached, either Dampier or Gilolo strait may be used, as advised in the second route described in § 150. The voyage will end as stated in § 176.

There is still another route for an auxiliary steamer, namely, that which, after leaving Basilan strait, runs near the Serangani and Meangis islands. For further information concerning this route, vide the first easterly route in § 152. Steam will probably have to be used against the NE. monsoon during a portion of the passage. This route passes far to the eastward of Morty and the Asia islands, and north of New Guinea, (vide § 176.)

If it be decided to pass through Torres strait, Molucca passage should also be used. Take the channel between Xulla and Obi-Major, and then the one between Buru and Manipa. Thence, with the NW. monsoon, pass south of Banda, Ki, and Arru islands. Cross the Arafura sea according to the instructions given in § 115, and finish the voyage by following the route given in § 176.

From Shanghae. The southerly route is probably the best for *sailing-vessels* coming from Shanghae to Australia. They should descend the China sea as stated in §§ 165 and 167, and finish the voyage in the same manner as if they had started from Hong-Kong.

Auxiliary steamers starting from Shanghae, and not wishing to run south of Australia, can first go to Yokohama, (vide § 163;) and, after coaling there, proceed on their voyage as follows:

From Yokohama. *Both sailing and auxiliary-steam vessels*, after leaving Yokohama, should take the northerly route, and commence the voyage as if bound to California, (vide §§ 119 and 120.) The easting should be made north of the 30th parallel, or farther to northward, if necessary, to find the west winds. They should not bear south until the meridian of 165° or even 172° E. is reached. Thence, the voyage is easily accomplished. Pass west of the Ralick islands; and, if possible, east of Ualan island. The NE. trades will be lost between 5° N. and the equator, which should generally be crossed between 162° and 167° E. The variable winds near the equator predominate from the northward and westward at this season, and usually allow vessels to pass, first, between St. Christoval and the Santa Cruz islands; then, between Mellish and Bampton reefs; and, finally, east of Kenn reef and Cato bank. Thence follow the route from New Caledonia to Australia given in § 138.*

The SW. monsoon prevails from *April to October*. *Sailing-vessels from Saigon* run through Mindoro and Basilan straits, (vide § 158.) Thence, they have a choice between three passages: first, Macassar strait; second, Molucca passage; third, the passage near the Serangani islands and Gilolo strait, (vide the easterly routes from Hong-Kong to Europe in § 152.) After striking the SE. winds of the Indian ocean the course should be SW., clean full, for the west winds near 30° S. For information concerning the end of the passage, south of Australia, vide § 175 and the end of § 104. 2d. During the SW. monsoon. From Saigon.

Auxiliary steamers starting from Saigon can make a more direct course, and steam to Singapore and the strait of Sunda. Thence the voyage will be the same as that for sailing-vessels. Auxiliary steamers can also run through Mindoro and Basilan straits, and pass near the Serangani islands. Thence they can make their easting between 4° N. and the equator. They should run far enough to the eastward to make the Solomon islands with the SE. monsoon, and finish the voyage as described in § 176, (for the season from November to March.) Though several authorities state that easting can be made under canvas near the equator at all seasons, we can hardly advise this route for auxiliary steamers, (vide appendix to this paragraph.)

Sailing and auxiliary-steam vessels leaving Hong-Kong should—according to the time of starting—follow one of the two easterly routes described in § 152. After crossing the SE. trades of the Indian ocean they will strike the west winds, and finish the voyage as described in § 175, and the end of § 104. The northerly route can also be followed dur- From Hong-Kong.

*The reader should refer to the *Appendix* at the end of § 177.

ing the SW. monsoon; in which case the course is the same as if bound for Japan, (vide § 160;) and thence, as if the point of departure were Yokohama.

Auxiliary steamers leaving Hong-Kong can take the same route as that indicated for those leaving Saigon, and run north of New Guinea; this, however, is a route we should not advise.

From Shanghae or Japan. *Sailing and auxiliary-steam vessels*, starting from Shanghae or Japan, should take the northerly route, beginning the voyage as if bound to California, (vide §§ 119 and 120.) The west winds will not usually be found below 35° N. It is not advisable to bear south until beyond 167° E., nor is it well—especially in a sailing-ship—to enter the trades to westward of 172° E. This detour is more marked than that given for the route during the NE. monsoon; the doldrums and counter-currents of wind in the western part of the north Pacific are thus avoided. The voyage will end in the same manner as that described for vessels leaving during the NE. monsoon.

APPENDIX TO § 177.—We will complete the general instructions, given in the present paragraph, by quoting the following considerations on the voyage from China to Australia, (*Ann. Hydr.*, vol. 31:)

"In December, 1866, nine vessels left Fu-chu for Sydney; six took the China sea route, some ran through the strait of Sunda, and others through Allas strait for the Indian ocean. Two others and myself took the easterly route, and arrived at Sydney in 52, 54, and 56 days respectively, beating the other vessels by from 12 to 25 days; and this was the worst season I ever had for making easting near the line.

"On the 7th December, 1855, I left the Serangani islands with a NE. wind, and, steering to the eastward, ran through Saint George's channel, and south of the Solomon isles. After crossing 176° E. at 12° S. I experienced violent north winds until I reached the North cape of New Zealand on the 10th February.

"Rosser quotes some passages made by the easterly route, from May to September; that is, during the SW. monsoon.

"Several vessels, attempting this route, made very long passages; one took 101 days from Manila to Sydney, and another 120 days from China to the same destination; none of them, I believe, made a passage of less than 90 days.

"All the captains who made this voyage state that, after losing the SW. winds near the Marrianas, they had light east winds and calms, with strong westerly currents. From *May* to *October*, according to my experience, the weather between 2° N. and 3° S. is rainy and calm all along the northern coast of New Guinea after leaving the cape of Good Hope. During most years the currents between these parallels are westerly after attaining a speed of from 30 to 60 miles per day. Between 2° 30′ and 5° 30′ N. there is generally an easterly current at this season, but a merchant-vessel will find it hard work to make to the eastward, north of the trades, if she wishes to reach 170° E. It is doubtful if she can always clear Pleasant island. As the natives of this island are very savage and well armed, a vessel in these localities, and in need of stores, will do better to put in at Arongs (?) island, where there is an American mission at the harbor on the NE. coast. Ascension island (Ponapi) is also comparatively safe, though it is best not to trust too much to the inhabitants.

"From *November* to *February* I think that the best route from China to Australia consists in making to eastward, as far as possible, with the NE. trades, and then crossing the equator. There is a clear passage between the Pelews and the Matelotas, thence keep along the line to 141° E. There is a low, rocky islet or reef at 1° N. and 141° E. The route passes north of the Anchorite islands, and along the equator as far as 163° E., passing E. of Saint Christoval, W. of Bampton reefs, and approaching the Australian coast in the neighborhood of Moreton bay. It is best to keep near the coast, if bound to Sydney or Melbourne, as NE. and E. winds are frequent during the summer."

Passage of the Esmeralda, (vide Nautical Magazine.)—"'Left Fu-chu on the 24th September; passed north of Formosa, and ran to 30° N. and 150° E., in order to cross the line at 162° E.; passed east of the Solomon islands, west of New Caledonia, and arrived at Sydney on the 21st November, after a passage of 58 days.'—(*Observations of Captain Polack, master of the Esmeralda.*) After taking the above route, Captain Polack advises ships leaving China, from the end of October to the end of January, to pass north of Formosa, if they can, without losing time, (referring evidently to vessels from Fu-chu.) If unable to double Formosa they can run between that island and the Pescadores; and, after clearing the Bashees, stand to the southward and east-ward until able to cross the NE. monsoon with a topmast studding-sail set. The line will thus be made between 140° and 145° E. From this point there is a choice between two routes: the first—which appears to be the less desirable of the two—lies close to the equator, and passes between New Ireland and Bougainville; the second, and better one, runs along the north coast of New Guinea, and between that isl-and and New Britain. Once arrived at 10° S. and 157° E. the course should be made to pass some distance east or west of Fairway reef, (?) (about 161° 42′ E.) North winds will generally be found near the NE. coast of Australia. This route, from Fu-chu to Sydney, is some 600 miles shorter than the easterly route along 30° N. to 155° E. It is also 2,300 miles shorter than the route south of Australia."

Such is Captain Polack's advice. Nothing proves, how-ever, that he ever followed the route himself. Still, we will give his reasons for preferring the New Guinea route to that passing between New Ireland and Bougainville. He does *not* think that the SW. monsoon reaches as far to the east as Captain Hunter states. If the west winds reach 160° E. it is only in exceptional cases. An experienced captain from Sydney states that, during a three years' cruise in these localities, he never found he could count on a west wind. This opinion is also substantiated by a whaler at home in these latitudes. Captain Polack also states that, between the equator and 10° S., he found only calms and light northerly breezes, with no indications whatever of a west wind. He is firmly convinced that off New Guinea west winds are fresh and constant *from November to Febru-ary,* and he believes that the passage from China to Syd-ney may be made in from 35 to 45 days, especially if below the line in January.

§ 178. FROM SYDNEY TO MELBOURNE.—Abundant infor-mation has been furnished by Captain Flinders, on this voyage, and especially on the passage through Bass strait from east to west. Following are his remarks:

"The three months, (January, February, and March,) during which the voyage from Sydney to Melbourne may best be made, correspond to the season when the passage through Torres strait is uncertain, if not impracticable. Nor would it be advisable to enter Bass strait before the middle of December or even the middle of January.

"In coming from Sydney, or from any other port situated to the NE., the departure may be taken from cape Howe. Thence the course, by compass, should not be to the westward of SSW. until 39° 30′ S. is reached, as there is danger of the wind coming out from SE., and setting the vessel into the long bight between cape Howe and Wilson promontory. After reaching 39° 30′ S. steer about W. by S., leaving the Sisters, Craggy isle, and Wright rock (210 feet high) to port; that is, to the southward. On Deal island—the most westerly of the Kent group—there is a revolving light, 884* feet above the level of the sea, and visible 37 miles, unless hidden by the mist, which often happens on account of the great elevation of the light. It is situated in 39° 29′ S. and 147° 22′ E., and is a good landmark. After passing 3 or 4 miles south of the light-house the other islands will be successively raised to the south; pass these at about the same distance. The first island is a small one lying to SW., and S. of Judgment reef; next are the Sugar-Loaf rocks and Curtis island. After leaving the latter island the course for King island is about west, and the distance 126 miles. Here there is a fixed light, visible 24 miles, but it is best to pass 15 or 18 miles north of the island, if the wind allow.

"If the weather be thick or rainy, and the wind come out strong ahead, that is, from SW., there are many points where a vessel can anchor. The following are the best anchorages: 1st. West cove, in Erith island. This island is one of the Kent group, and is separated from Deal island by Murray pass. 2d. Hamilton roads, near the east point of Preservation isle. This small island is situated in the northern part of Banks strait between Barren island and Clarke island. It lies several miles east of a line joining the light on Swan island to that on Goose island. 3d. On the south coast of Swan island, suitable for small vessels; or under the lee of Waterhouse isle, situated in about 148° E. near the north coast of Tasmania. 4th. At Port Dalrym-

* 950 feet above the sea, according to *Aus. Directory*.—Translator.

ple, near the mouth of Tamar river, (north coast of Tasmania.) 5th. At Port Sorrel, (12 miles west of Port Dalrymple,) suitable for small ships. 6th. Several places among the islands of the Hunter group, off the NW. point of Tasmania. 7th. At Sea-Elephant bay, on the east coast of King island, where wood and water may be obtained. There is also an anchorage, sheltered from SW. winds, off the NE. point of this island. 8th. At Port Western, between Grant or Phillip island and the Australian coast, near the meridian of 145° 10′ E. Let go here as soon as you are under the lee. At this anchorage the wind will be fair for getting under way and clearing Bass strait. 9th. At Port Phillip, Melbourne roads.

"As the weather in Bass strait is variable and sudden shifts of wind frequent, it is advisable to take all precautions before coming to anchor in an open roadstead, and even when partially under the lee of the land it is well to be ready to get under way at a moment's notice, in case the wind should suddenly change. There is no other advice especially necessary, as navigation through Bass strait does not require more than the usual care and vigilance always requisite at sea."

www.ingramcontent.com/pod-product-compliance
Lightning Source LLC
Chambersburg PA
CBHW020307240426
43673CB00039B/731

* 9 7 8 3 3 3 7 3 1 9 7 3 1 *